Editorial Board Members' Collection Series: Nonlinear Photonics

Editorial Board Members' Collection Series: Nonlinear Photonics

Guest Editors

Luigi Sirleto
François Sanchez

MDPI

Basel • Beijing • Wuhan • Barcelona • Belgrade • Novi Sad • Cluj • Manchester

Guest Editors

Luigi Sirleto
National Research Council (CNR)
Institute of Applied Sciences and Intelligent Systems (ISASI)
Napoli
Italy

François Sanchez
Laboratoire de Photonique d'Angers
Université d'Angers
Angers
France

Editorial Office
MDPI AG
Grosspeteranlage 5
4052 Basel, Switzerland

This is a reprint of the Special Issue, published open access by the journal *Photonics* (ISSN 2304-6732), freely accessible at: https://www.mdpi.com/journal/photonics/special_issues/34894I3517.

For citation purposes, cite each article independently as indicated on the article page online and as indicated below:

Lastname, A.A.; Lastname, B.B. Article Title. *Journal Name* **Year**, *Volume Number*, Page Range.

ISBN 978-3-7258-5551-3 (Hbk)
ISBN 978-3-7258-5552-0 (PDF)
https://doi.org/10.3390/books978-3-7258-5552-0

Contents

About the Editors

Luigi Sirleto

Luigi Sirleto has been a researcher scientist at the National Research Council (CNR) of Italy since 2001. As a visiting scientist, in 2001, he spent 2 months at the University of Twente (NL) and, in 2003-2004, he spent 9 months at UCLA (University of California, Los Angeles), USA. In 2006, he founded the *Nonlinear and Ultrafast Optics Lab* at CNR and since then he has led the research activities at the lab. He was a member elect of the presidential board of the Italian Society of Optics and Photonics (SIOF-EOS) from 2009 to 2014 (six years). In 2018, he received his qualification as a full professor of Experimental Matter Physics. He has served as a reviewer of many international journals and as a committee member of a number of national and international conferences. He has been Co-Chair of the Workshops *Progress in Nonlinear Photonics and Applications- PNPA 2020; PNPA 2023; PNPA 2024; PNPA 2025*; under the umbrella of ICTON 2020; ICTON 2023; ICTON 2024; and ICTON 2025; respectively. He is Co-Editor of the books 'Nonlinear Photonics Devices' (2021, MDPI); 'Micro/Nanophotonic Devices in Europe' (2023, MDPI); and 'Advances in nonlinear photonics' (2023, Elsevier). Since 2021, he has been a Section Editor of *Optik* (Elsevier) and an Associate Editor of the *Journal of Nonlinear Optical Physics & Materials* (World Scientific). He is the co-author of over 220 papers, mostly on nonlinear/ultrafast optics spectroscopy/microscopy and photonics devices.

François Sanchez

François Sanchez was born in Paris in 1962 and obtained his PhD in 1987 from the University of Paris Sud (Thales Research & Technology). The subject of his thesis was the investigation of wave mixing in liquid crystals and semiconductors in the middle infrared. In 1988, he obtained an Assistant Professor position at ENSSAT, where he was involved in several projects on fiber lasers and microchip lasers. He became a full Professor in 1997 at the University of Rouen, where he initiated activities on double-clad fiber lasers. In 1999, he received the Arnulf Françon prize from the Société Française d'Optique (French Optical Society) for his book *Optique Non-Linéaire.* In 2000, he joined the laboratory POMA (now the Laboratory of Photonics of Angers) at the University of Angers, where his research includes nonlinear optics, fiber lasers, short pulse generation, high energetic pulses and square-wave generation. From 2010 to 2021, he was the head of the Laboratory of Photonics. He is the author or co-author of more than 400 publications and communications.

photonics MDPI

Editorial

Editorial Board Members' Collection Series: Nonlinear Photonics

Luigi Sirleto

National Research Council (CNR), Institute of Applied Sciences and Intelligent Systems, 00185 Napoli, Italy; luigi.sirleto@cnr.it

1. Introduction

Photonics has often been defined as the key technology of the 21st century. In recent decades, the overall trend of research and development (R&D) towards the miniaturization of devices and systems, pushed by the tremendous development of microelectronics, has led to the continuous development of micro/nanophotonic devices, which have huge potential for low-cost, scalable production, as well as the integration of high-density components [1]. On the other hand, nonlinear optical phenomena play a key role both in the knowledge of the light–matter interaction and in many fields of photonics applications [2]. The combination of integrated photonics technologies with nonlinear optics, which has led to the growth of nonlinear photonics, has also opened the way for groundbreaking new devices. Their evolution would enable advanced applications to emerge not only in optical processing and computing, but also in metrology, single-molecule sensing, imaging, microscopy, mid-infrared photonics, terahertz generation, microwave photonics, biomedicine, and quantum photonics [3].

The present Special Issue (SI) entitled "Editorial Board Members' Collection Series: Nonlinear Photonics", aims to highlight the current state of the art, some recent advances, and perspectives for further development. As a result, this SI contains two reviews and nine research articles, which can be divided into three groups.

In the first group of papers, three papers [4–6] highlight new effects. Reference [4] focuses on nonclassical states of light, which are an essential resource for high-precision optical techniques, relying on photon correlations and noise reshaping. Efforts toward engineering the spatiotemporal profile of quantum states of light have already found great success ([4] and references therein). This article focuses, instead, on shaping the noise statistics of quantum states of light. Deviations from Brownian diffusion, known as anomalous diffusion (AnDi), occur when the mean square displacement (MSD) growth with time has an exponent other than one. AnDi is investigated in reference [5]. In disordered photonic systems, we can find examples in which light propagation exhibits subdiffusive behavior ([5] and reference therein). There are limited works reporting the modulation of chirped Gaussian soliton molecules in extra-cavity optical fiber modulation systems. Reference [6] concerns this issue and reports the theoretically simulated modulation of a soliton molecule that has an initial chirped Gaussian pulse shape in a 1 μm extra-cavity optical fiber modulation system.

The four papers [7–10] of the second group are devoted to the hot issue of identifying efficient device integration platforms, which justifies the continuation of fundamental studies and the search for new or advanced materials with higher nonlinear coefficients and/or better overall properties. In reference [7], attention is given to graphene, which has properties that make it very promising for various applications in photonics and optoelectronics: combining graphene with other materials, one can create a variety of waveguiding devices with better characteristics compared to waveguides made of traditional materials. In reference [8], the development of optical materials based on a nonlinear optical single crystal of lithium niobate is described. $LiNbO_3$ is a crystal state that has high nonlinear optical coefficients, which is crucial for the development of functional materials. Composite magnetic nanostructures are a subject of high research interest, as they provide several exciting effects absent in live nature. Among them, much attention has been paid to the studies of exchange coupling in antiferromagnetic/ferromagnetic (AFM/FM) films, which leads to the pinning effect. In reference [9], comparative studies of linear and nonlinear magneto-optical effects, under the laser-induced switching of the pinning effect in IrMn/CoFe films of various thickness of the ferromagnetic CoFe layer, is reported. The study of localized surface plasmons (LSPs) in nanoscale structures is described in reference [10]. Nanostructures are an essential step towards identifying optimal plasmonic modes that can facilitate robust optomechanical coupling and a deeper understanding of light–matter interactions at the nanoscale.

The last group of papers [11–14], two studies and two reviews, are dedicated to some of the most important application fields of nonlinear photonics. They begin with a review [11] focused on plasmonic core–shell nanoparticles in photovoltaic (PV) applications. Core–shell nanoparticles have emerged as a promising new technology with unique structural attributes and widely tunable properties. Among label-free nonlinear optical techniques, stimulated Raman scattering (SRS) microscopy plays an important role in enabling imaging systems and performing label-free imaging with high sensitivity, high spatial and spectral resolution, 3D sectioning, and fast image acquisition. In the second review [12], applications and perspectives of SRS in bio photonics are described and discussed. In reference [13], a diode-pumped solid-state (DPSS) laser, used for intracavity pump-enhanced difference frequency generation (DFG) to create a 3.5-micron laser, is presented and potential future applications in free-space optical communication are also discussed. Finally, in reference [14], microwave photonic (MWP) signal processors are presented and their intrinsic advantages such as low loss, large processing bandwidth, and strong immunity to electromagnetic interference are discussed.

2. An Overview of the Articles in This Special Issue

Systems that utilize quantum information resources generally take advantage of coherent macroscopic superposition and/or entanglement, both of which are generically found in states supported by nonlinear systems. However, optical nonlinearities are typically weak for closed systems. With the advent of epsilon-near-zero materials and polariton-enhanced scattering, commonly facilitated by electronically induced transparency, large effective nonlinearities become increasingly realistic as well. When realized in a high-Q optical resonator, strong optical nonlinearities with relatively long evolution time and low loss facilitate deterministic quantum optical state engineering.

We note that in a material with second- and third-order nonlinearities comparable in strength at optical frequencies, the interaction between field states is strongly photon-number-dependent. With this in mind, in reference [4], a theoretical method for the deterministic shaping of quantum light via photon number state selective interactions is proposed. A class of nonlinear optical resonators is individuated, which can transform many-photon wavefunctions to produce structured states of light with nonclassical noise statistics. The devices, based on parametric down conversion, utilize the Kerr effect to tune photon-number-dependent frequency matching, inducing photon-number-selective interactions. With a high-amplitude coherent pump, the number-selective interaction shapes the noise of a two-mode squeezed cavity state with minimal dephasing. The requisite material properties to build the device and highlight the remaining material degrees of freedom, which offer flexible material design, are also reported.

Recurrent neural networks (RNNs) are promising methods for the classification of anomalous diffusion trajectories, identifying temporal dependencies and sequential patterns, which are essential for characterizing anomalous diffusion processes. The integration of statistical methods with RNNs has enabled the development of robust frameworks that can detect and predict the occurrence of anomalous diffusion, even in the presence of significant noise. By leveraging these techniques, reference [5] aims to improve accuracy for identifying diffusion types without the need for extensive preprocessing or noise filtering. Therefore, it contributes to the broader application of a combination of higher-order statistics and artificial intelligence in analyzing and predicting phenomena within stochastic and noisy environments. In reference [5], feature extraction through parametric and non-parametric spectral analysis methods is explored to decode anomalously diffusing trajectories, achieving reduced computational costs compared with other approaches that require additional data or prior training. The proposed methods deliver accurate results, even with short trajectories and in the presence of noise.

A soliton molecule consists of two or more solitons, and these solitons have fixed phase differences and temporal pulse intervals between adjacent solitons. Soliton molecules can be efficiently generated in fiber laser oscillators through increasing pump power or decreasing saturable power/energy of intra-cavity saturable absorbers. The possible experimental realization techniques of soliton molecules with chirped Gaussian pulses are the constructions of passively mode-locked fiber lasers, accompanied by intra-cavity management of pump power, group-velocity dispersion, saturable absorbers, and so on.

In reference [6], the extra-cavity modulation of a soliton molecule with a chirped Gaussian pulse shape is developed in a 1 μm optical fiber system. The optical fiber modulation system can efficiently modulate temporal and frequency properties of optical soliton molecules and their orthogonal components. Different soliton parameters in orthogonal polarizations are applied to achieve controllable optical solitons' output with specific properties in the time/frequency domain. These simulation results provide a beneficial reference value for the extra-cavity shaping of different solitons that come from nonlinear optical systems. Optimally, the reported results could pave the groundwork for industrial growth in ultrafast laser design.

It is well known that graphene has an extremely strong third-order nonlinearity compared to the commonly used dielectrics and metals; this nonlinearity results from the interaction of the charge carriers in graphene with strong electromagnetic radiation. Reference [7] focuses on the problem of a monochromatic terahertz TE-polarized wave propagation in a plane dielectric layer filled with a homogeneous isotropic medium; one of the boundaries of the waveguide is covered with a layer of graphene. On the one hand, in the study, energy losses both in the dielectric layer and in the graphene layer are neglected; the latter assumption is reasonable in the terahertz range of electromagnetic radiation (on which the paper focuses), where graphene has a strong plasmonic response and much less loss. On the other hand, this study considers the significant third-order nonlinearity resulting from the interaction of the electromagnetic wave with the charge carriers in the graphene layer. Besides studying the problem analytically, the paper presents some numerical results as well. In particular, the obtained results demonstrate how the nonlinearity in graphene affects the propagation constants and eigenwaves, providing the dispersion curves and eigenwaves for nonlinear graphene as well as for the linear one.

$LiNbO_3$:Mg (5.0 mol % or 0.85 wt %) is a $LiNbO_3$ single-crystal-based, optically highly perfect and compositionally uniform material. It has a low photorefraction effect and it is currently used for optical frequency conversion in the phase quasi-phase-matching mode on regular domain structures (PPLN—periodically-poled $LiNbO_3$). However, the strong single doping decreases the compositional and optical uniformity of a single-doped crystal compared with a nominally pure congruent crystal. Doping (including double and multiple doping) provides finer control than single doping, controls many properties of an LN crystal, and also influences the type and concentration of defects in a crystal, which significantly affects the magnitude of the photorefraction effect and the optical damage resistance of the crystal.

In reference [8], the features of the defect structure, compositional uniformity, and photorefractive properties in a double-doped $LiNbO_3$:Gd:Mg (Gd concentration of 0.003, Mg—0.65 wt % in the crystal) single crystal are investigated using several optical methods: laser conoscopy, photoinduced light scattering (PILS), optical spectroscopy, and Raman scattering. The crystal has been shown to have no photorefraction effect and high optical uniformity. Nonlinear optical coefficients are higher in perfect double-doped $LiNbO_3$ crystals, including those co-doped with magnesium, compared with single-doped crystals.

A promising method for the optical control of the pinning effect in antiferromagnetic (AFM) and ferromagnetic (FM) interface is laser-induced heating; namely, as the structure is heated above the blocking temperature, the pinning is destroyed. A complete reversal of the exchange bias at the AFM/FM interface can also be induced by a single femtosecond pulse of sufficiently high energy. Spin precession in magnetic structures subjected by ultrashort excitation provides extensive information on magnetic interactions. Laser pulse perturbation is found to modify the exchange bias, as evidenced by the shift of the hysteresis loop with respect to the applied DC magnetic field as observed by the magneto-optical Kerr effect [9]. In optical second harmonic generation (SHG), when applied in studies of magnetic structures, magneto-optical effects at the SHG wavelength appear with typical values of one to two orders of magnitude higher than their linear analogs such as the magneto-optical Kerr effect (MOKE). This nonlinear optical method allows one to distinguish various types of nontrivial magnetic states, such as magnetic vortices, magnetic toroidal moment, and gradients of magnetization [9].

Reference [9] is a comparative study of linear and nonlinear magneto-optical effects in IrMn/CoFe films of various thickness of the ferromagnetic CoFe layer. It demonstrates that the magneto-optical response of the pinned AFM/FM nanofilms appears with different hysteresis loop parameters in the transverse magneto-optical Kerr effect (MOKE) and interface-sensitive magnetization-induced second harmonic generation (SHG), indicating the diversity of the magnetic effects at interfaces compared to the bulk of the films.

In reference [10], localized surface plasmon (LSP) modes in a design comprising two coupled nano-ridges deposited on a gold layer with an interposing polymer spacer layer are numerically investigated. Such a structure is usually referred to as a particle-on-mirror structure. First, the LSP modes of a single nano-ridge and the analysis of its scattering cross-section in the visible and infrared ranges are examined. To enhance the plasmonic response, a thin polymer layer is placed at the middle of the ridge, which introduces additional LSP modes confined within the former. Then, the analysis is extended to the dimer configuration, which yields LSP resonances with a quality factor enhancement of approximately threefold relative to a single nano-ridge. Furthermore, the presence of the polymer layer within the ridges significantly improves plasmon field localization and the quality factor. These findings underscore the potential of nano-ridge-based structures in advancing optomechanical coupling and offering valuable insights for the development of high-performance acousto-plasmonic devices. In particular, the proposed device could help significantly improve the design of nano-acousto-optic modulators, operating in the visible or in the near-infrared ranges, that require an enhanced light–phonon coupling rate.

The field of photovoltaics (PV) continually seeks innovative material solutions to enhance the efficiency and the stability of their standard devices. In reference [11], the use of plasmonic core–shell nanoparticles in PV applications through various experimental validations is reviewed. Advancements in the design and in the control over the properties of core–shell nanoparticles are described and their integration into various solar cells, based on their ability to finely tune optical, electronic, and chemical properties, are highlighted. Experimental results for organic, perovskite, and dye-sensitized solar cells are discussed, where core–shell nanoparticles have been successfully deployed. Additionally, gaps in the current research are identified, such as the need for scalable synthesis methods and long-term stability assessments and promising new developments at the frontier of the field.

Today, fluorescence microscopy is used worldwide, despite suffering from drawbacks such as photobleaching, phototoxicity, and challenging quantification. Most small biomolecules, such as nucleosides, amino acids, fatty acids, choline, glucose, and cholesterol, are intrinsically nonfluorescent, so fluorescent tags are mandatory. Unfortunately, these tags, such as organic dyes, fluorescent proteins, or quantum dots, are all relatively larger than small biomolecules; consequently, they can severely falsify their native biochemical or biophysical properties. Therefore, label-free imaging, with high sensitivity and high chemical selectivity of unlabeled living cells, is preferable. In this framework, vibrational microscopy techniques have emerged as a powerful approach based on a chemically label-free selective contrast due to intrinsic molecule vibrations. Among them, stimulated Raman scattering (SRS) microscopy, which was adapted to microscopy only in the last two decades, is one of the most promising techniques.

Stimulated Raman scattering (SRS) microscopy is a high-speed imaging modality based on intrinsic molecular vibrations, producing chemical maps in living systems. Such capability, allowing for direct visualization without the perturbation of biological processes, has enabled a plethora of biological and medical applications. In reference [12], after introducing the basic theory and competitive effects of SRS, some crucial features for SRS microscopy implementations, such as noise, spectral bandwidth, speed, chemical sensitivity, spatial resolution, and quantum enhancement, are discussed. Finally, some SRS applications in biological and medical imaging are described. Even if certainly not exhaustive, the review aims to offer a broad overview, providing guidance for newcomers and hinting at a more detailed investigation to interested researchers in this rapidly growing field.

Free-space optical communication has seen a surge in interest, both for terrestrial and last-mile solutions, as well as optical satellite communication. Current efforts have focused on utilizing the short-wave infrared (SWIR) atmospheric transmission window, containing the 1530–1565 nm C-band used in telecommunications. However, SWIR communication suffers significantly from scintillation, scattering, and weather effects. The 3–5 μm mid-infrared (MWIR) atmospheric transmission window should offer improved signal-to-noise ratio and resistance to these adverse effects and could lead to higher link availability and reliability.

In reference [13], a diode-pumped solid-state (DPSS) laser, used for intracavity pump-enhanced difference frequency generation (DFG) to create a 3.5-micron laser, is reported. Using a 50 mm long periodically poled lithium niobate (PPLN) crystal inside the cavity of an Nd:YVO_4 solid-state laser at 1064 nm with 4.5 W pump power at 808 nm, and a

310 mW C-band signal at 1529 nm, up to 31 mW of mid-infrared output power at 3499 nm is obtained. The cavity requires no active stabilization and/or locking, and the entire cavity is <8 cm in length. The obtained output power corresponds to a black-box efficiency of 2.20% W^{-1}, which is the highest value reported to date for continuous-wave DFG based on a bulk nonlinear optical crystal with no active stabilization. Potential future applications in free-space optical communication are also discussed.

Traditional microwave signal processors relying on electronic devices exhibit significant loss and strong crosstalk, which make them suffer from limited operation bandwidths. To overcome this restriction, microwave photonic (MWP) signal processors that perform signal processing functions based on MWP technologies have found wide applications in telecommunication and radar systems. For MWP signal processors implemented by the transversal filter systems, a large number of taps, or the wavelength channels provided by multi-wavelength optical sources, are required to improve their performance. Compared to other multi-wavelength optical sources, optical microcombs can provide significantly increased numbers of wavelength channels by using compact micro-scale resonators, which is critical for improving the processing accuracy of MWP transversal signal processors. Although a range of signal processing functions have been realized, they only used Gaussian input waveforms for demonstrations, while the ability to handle various input signal waveforms is essential for practical applications.

In reference [14], the capability of microcomb-based MWP signal processors for dealing with various input signal waveforms is experimentally demonstrated. The processing accuracy of different input waveforms, including Gaussian, triangle, parabolic, super Gaussian, and nearly square waveforms, is quantified. These results provide guidance for microcomb-based MWP signal processors when processing microwave signals of various waveforms.

3. Conclusions

Although it is far from an exhaustive overview, which would not be possible due to the extensive and broad research activities in nonlinear photonics, this collection of scientific articles presents and discusses several research topics, covering interesting aspects such as new effects, promising materials, and important fields of application. It is my wish that it will serve as a stimulus for students and researchers to further expand the potential of nonlinear photonics devices via fundamental investigations and practical applications.

Acknowledgments: We would like to thank all of the contributing authors of this Special issue for their support. We also thank the reviewers for dedicating their time to ensure the quality of the submitted papers.

Conflicts of Interest: The authors declare no conflicts of interest.

References

1. Krasnodebski, M. Throwing light on photonics: The genealogy of a technological paradigm. *Centaurus* **2018**, *60*, 3–24. [CrossRef]
2. Boyd, R.W. *Nonlinear Optics*; Academic Press: Cambridge, MA, USA, 2003.
3. Righini, G.C.; Sirleto, L. *Advances in Nonlinear Photonics*, 1st ed.; Woodhead Publishing Series in Electronic and Optical Materials; Elsevier: New York, NY, USA, 2023. [CrossRef]
4. Compton, G.D.; Kuzyk, M.G. Deterministic Shaping of Quantum Light Statistics. *Photonics* **2024**, *11*, 287. [CrossRef]
5. Iglesias Martínez, M.E.; Garibo-i-Orts, Ò.; Conejero, J.A. Decoding Anomalous Diffusion Using Higher-Order Spectral Analysis and Multiple Signal Classification. *Photonics* **2025**, *12*, 145. [CrossRef]
6. Tang, D.; Zhan, J.; Wang, D.; Wang, H.; Peng, Y.; Tiu, Z.C.; Zhou, Y. Extra-Cavity Modulating a Soliton Molecule with Chirped Gaussian Pulse Shape. *Photonics* **2024**, *11*, 1098. [CrossRef]

7. Smirnov, Y.; Tikhov, S. The Nonlinear Eigenvalue Problem of Electromagnetic Wave Propagation in a Dielectric Layer Covered with Graphene. *Photonics* **2023**, *10*, 523. [CrossRef]
8. Sidorov, N.; Palatnikov, M.; Teplyakova, N.; Pyatyshev, A.; Skrabatun, A. Some Optical Properties of $LiNbO_3$:Gd^{3+}(0.003):Mg^{2+} (0.65 wt %) Single Crystal: A Promising Material for Laser Radiation Conversion. *Photonics* **2023**, *10*, 921. [CrossRef]
9. Kolmychek, I.A.; Novikov, V.B.; Gusev, N.S.; Pashen'kin, I.Y.; Karashtin, E.A.; Murzina, T.V. Second Harmonic Generation Versus Linear Magneto-Optical Response Studies of Laser-Induced Switching of Pinning Effects in Antiferromagnetic/Ferromagnetic Films. *Photonics* **2023**, *10*, 1303. [CrossRef]
10. El Ghafiani, M.; Noual, A.; Amrani, M.; Moutaouekkil, M.; El Boudouti, E.H. Numerical Investigation of Localized Surface Plasmons in Gold Nano-Ridge Dimer-on-Mirror Structures. *Photonics* **2024**, *11*, 817. [CrossRef]
11. Quandt, A.; Wamwangi, D.; Kumalo, S. The Use of Core-Shell Nanoparticles in Photovoltaics. *Photonics* **2025**, *12*, 555. [CrossRef]
12. Ranjan, R.; Sirleto, L. Stimulated Raman Scattering Microscopy: A Review. *Photonics* **2024**, *11*, 489. [CrossRef]
13. Flannigan, L.G.K.; Atwi, A.; Kashak, T.; Poitras, D.; Xu, C.-Q. Compact Diode-Pumped Solid-State Laser with Intracavity Pump-Enhanced DFG Emitting at ~3.5 Microns. *Photonics* **2023**, *10*, 1164. [CrossRef]
14. Li, Y.; Sun, Y.; Wu, J.; Ren, G.; Corcoran, B.; Xu, X.; Chu, S.T.; Little, B.E.; Morandotti, R.; Mitchell, A.; et al. Processing Accuracy of Microcomb-Based Microwave Photonic Signal Processors for Different Input Signal Waveforms. *Photonics* **2023**, *10*, 1283. [CrossRef]

photonics

Article

Deterministic Shaping of Quantum Light Statistics

Garrett D. Compton * and Mark G. Kuzyk

Department of Physics and Astronomy, Webster Hall, Washington State University,
Pullman, WA 99164-2814, USA; kuz@wsu.edu
* Correspondence: garrett.compton@wsu.edu

Abstract: We propose a theoretical method for the deterministic shaping of quantum light via photon number state selective interactions. Nonclassical states of light are an essential resource for high-precision optical techniques that rely on photon correlations and noise reshaping. Notable techniques include quantum enhanced interferometry, ghost imaging, and generating fault-tolerant codes for continuous variable optical quantum computing. We show that a class of nonlinear-optical resonators can transform many-photon wavefunctions to produce structured states of light with nonclassical noise statistics. The devices, based on parametric down conversion, utilize the Kerr effect to tune photon-number-dependent frequency matching, inducing photon-number-selective interactions. With a high-amplitude coherent pump, the number-selective interaction shapes the noise of a two-mode squeezed cavity state with minimal dephasing, illustrated with simulations. We specify the requisite material properties to build the device and highlight the remaining material degrees of freedom which offer flexible material design.

Keywords: quantum optics; nonlinear optics; quantum state design; quantum control; quantum state engineering; ENZ; EIT; cavity-QED

1. Introduction

Systems that utilize quantum information resources generally take advantage of coherent macroscopic superposition and/or entanglement. The increased information density and sensitivity to measurements of such quantum states, compared to their classical counterparts, is central to their application in modern quantum information technologies like quantum key distribution [1,2], quantum computing [3–5], and metrology at the fundamental noise limit [6–8]. Efforts toward engineering the spatio-temporal profile of quantum states of light have already found great success [9–11]. This article focuses instead on shaping the noise statistics of quantum states of light for such applications. Both macroscopic superposition and entanglement are generically found in states supported by nonlinear systems [12,13]. However, optical nonlinearities are typically weak for closed systems. Much attention has therefore been paid to developing methods for generating quantum states of light which utilize wavefunction collapse as a source of nonlinear evolution, where a particular measurement outcome heralds the desired quantum state [14–18]. Though hugely successful, heralded state preparation is inherently nondeterministic, making sequential preparations low-probability events.

The approach presented here overcomes low-heralding success rates using the paradigm of deterministic generation of quantum states. Examples in the literature can be found where the probabilistic nature of measurement is circumvented [19–31].

One example utilizes the quantum Zeno effect [19–21]. By strongly coupling a specially engineered system to an environment, the state of the system is continuously projected to a subspace of states that are effectively uncoupled from the environment, called a decoherence free subspace (DFS). External driving fields may then evolve states deterministically and coherently within the DFS.

With the advent of epsilon-near-zero materials [22] and polariton-enhanced scattering, commonly facilitated by electronically induced transparency [23–26], large effective nonlinearities become increasingly realistic as well. When realized in a high-Q optical resonator, strong optical nonlinearities with relatively long evolution time and low loss facilitate deterministic quantum optical state engineering. Trapped atom and other quantum emitter systems show great promise for this application [20,21,27–31].

Such systems accomplish universal control of an oscillator state by relying on an array of emitters or by coupling to a single two-level emitter. With an array of quantum emitters, the quantum information content of a state of interest is first written onto the array, typically in a decoherence-free subspace, then mapped to super-radiant light-matter states which couple to an optical waveguide [20,21,27,28]. Alternatively, arbitrary state transformations may be realized in a cavity with a single quantum emitter by repeatedly addressing the emitter with a control beam to introduce a Berry phase to cavity field Fock states and applying coherent displacement operations to the cavity field in sequences prescribed within [29–31]. Notably, Krastanov et al. [30,31] utilize a dispersive coupling between the emitter and a microwave cavity field to realize a selective number-dependent arbitrary phase (SNAP) gate which selectively rotates the phase of each Fock state of the cavity field.

This work introduces an alternative technique for precision shaping of quantum noise statistics without the need for precise emitter engineering and addressing. We consider a material with first-, second-, and third-order electric susceptibilities at optical frequencies (it is straightforward to generalize our approach for nonperturbative constitutive relations) and show that when second- and third-order nonlinearities are comparable in strength, the optical Kerr effect alters the frequency matching condition for optical parametric oscillation such that the interaction between field states is strongly photon-number-dependent. Our analysis separates the phase-rotating interactions from the mode-mixing interactions, providing us a deeper intuition for the dynamics we observe. Specifically, we first solve the combined first- and third-order system exactly and study the second-order mode-mixing in the interaction picture. The general dynamical characteristics are illustrated and we utilize the effect to demonstrate a procedure for the deterministic generation of approximate Fock states, among other interesting states. General material constraints are highlighted and developed into figures of merit to guide the experimental realizations of this method.

2. Background

We begin with a brief discussion of electromagnetic field quantization. Dirac recognized that Hamiltonian mechanics have simple rules for canonical quantization, that Lagrangian mechanics are well suited for relativistic gauge invariant classical field theories, and that they both poorly handle what the other does well. So, to formulate relativistic quantum theory, he generalized Hamiltonian mechanics to handle systems with constraints [32]. His mechanics encode symmetries that are simple to write into a Lagrangian, into constraints on a Hamiltonian, recovering Hamilton's equations of motion with a generalized Poisson bracket called the Dirac bracket. The canonical quantization of Dirac's bracket reveals, in part, Dirac's relativistic theory of quantum mechanics. Born and Infeld adopted Dirac's procedure to show that Maxwell's equations in the Coulomb gauge assume a Hamiltonian formulation [33,34] where the Hamiltonian is the classical field energy

$$H = \int_V dv \int_{-\infty}^{t} dt \, \frac{E \cdot \partial_{t'} D}{2} + \frac{H \cdot \partial_{t'} B}{2} \tag{1}$$

and the fields obey a version of the canonical Dirac bracket relations,

$$\begin{aligned} \{B^j(r), B^k(r')\} &= \{D^j(r), D^k(r')\} = 0 \\ \{D^j(r), B^k(r')\} &= \epsilon^{jlk} \frac{\partial}{\partial x^l} \delta(r - r') \equiv \delta^{tr}(r - r') \, , \end{aligned} \tag{2}$$

with the second equation defining the transverse delta function $\delta^{tr}(r-r')$ (the transverse delta function captures the constraints on the field enforced by the Coulomb gauge. In a medium without free charges or free currents, Maxwell's equations ensure that D and B are always transverse to the direction of radiation propagation, unlike E and H. Furthermore, since both D and B are divergenceless, so must be their commutator. The transverse delta function is consistent with these two properties. It has no divergence and acts as an ordinary delta function on transverse fields). According to Dirac's canonical quantization, the field amplitudes are promoted to operators, and the Dirac brackets become commutators

$$\{D^j(r), B^k(r')\} = \delta^{tr}(r-r') \rightarrow [D^j(r), B^k(r')] = i\hbar\delta^{tr}(r-r') \,. \tag{3}$$

Quesada et al. [35] later showed that, to correctly recover Maxwell's equations in nonlinear dielectric media from the Born–Infeld Dirac brackets, it is imperative to consider the displacement field and magnetic flux density to be the fundamental dynamical variables. Here, 'fundamental' indicates that D and B are Hermitian linear combinations of Bose operators, i.e., they are Boson fields, whereas the electric and magnetic fields E and H are generally higher polynomials of Bose operators. For modern derivations and perspectives of quantum field theory in linear dielectric media, see [36–39].

For a finite linear cavity, the eigenstates of the system are enforced by the dielectric structure and boundary conditions of the cavity. Let us restrict ourselves to one effective spatial dimension, scalar susceptibilities, and a single polarization of the field modes without loss of generality (our analysis depends only on the algebraic structure of the Hilbert space and the dynamics generated by the Hamiltonian, so is easily generalized to richer mode structures). This amounts to a single parameter family of eigenstates. Consider a uniform cavity of length L with periodic boundary conditions, as is the case for a micro-ring resonator [40]. The allowed wavenumbers, k, within the cavity are integer multiples of $\frac{2\pi}{L}$. The fields D and B are physical, therefore real-valued. Each may then be expanded with respect to the basis of eigenmodes in the following form:

$$\begin{aligned} D(z) &= \sum_k D_k(z) = \frac{1}{\sqrt{AL}} \sum_k \alpha_k e^{ikz} + \alpha_k^* e^{-ikz} \\ B(z) &= \sum_k B_k(z) = \frac{1}{\sqrt{AL}} \sum_k \beta_k e^{ikz} + \beta_k^* e^{-ikz} \,, \end{aligned} \tag{4}$$

where A represents the transverse cross-sectional area of the modes in the cavity (if the refractive index is not isotropic, A defines an effective cross-sectional area, discussed in [40]).

The field amplitudes of a single photon, ζ_k—defined later—are determined such that the energy of a photon with well-defined momentum $\hbar k$ agrees with the Einstein energy of a photon, $\hbar\omega(k)$. Suppose a dispersive permitivity such that the Fourier components of the linear electric field and displacement field are related by $\tilde{E}(k) = \frac{1}{\epsilon(k)}\tilde{D}(k)$. If the medium is dispersive, each frequency mode solves a different eigenvalue equation, so normalization requires special care [37]. We rescale the amplitude by the single photon amplitude, $\alpha_k \rightarrow \zeta_k a_k$, and promote the amplitude variable to a Bose operator obeying the canonical commutation relations

$$[a_k, a_{k'}] = 0, \quad [a_k, a_{k'}^\dagger] = \delta_{kk'} \,. \tag{5}$$

The displacement field operator becomes

$$D(z) = \sum_k i\sqrt{\frac{\hbar\omega(k) v_{g,k} n_k^3}{2ALc}} \left(a_k e^{ikz} - a^\dagger e^{-ikr}\right) \equiv \sum_k i\zeta_k \left(a_k e^{ikz} - a^\dagger e^{-ikr}\right) , \tag{6}$$

where we have adopted a common quantum optics phase convention of initial phase $\pi/2$; $v_{g,k}$ is the group velocity, and n_k the refractive index of the mode with wave-vector k. The magnetic flux field operator obtains a similar form [37].

Having properly dealt with linear dispersion, the Hamiltonian for linear quantum optics takes the form of an infinite set of harmonic oscillators

$$H^{(1)} = \sum_k \hbar\omega(k) a_k^\dagger a_k \,. \tag{7}$$

Nonlinear Quantum Optics

We introduce nonlinear optics, assuming that the dispersion of nonlinear susceptibilities is negligible for frequency bands we are interested in. The electric field is related to the displacement field and polarization field, P, via [41]

$$E = D - 4\pi P = \frac{1}{\epsilon} D - 4\pi P^{nl} \,. \tag{8}$$

We can recall from Equation (1) the electric field contribution to the field energy, $\frac{E\cdot D}{2}$. The energy term $\frac{1}{\epsilon} D \cdot D$ is handled in the linear part of the Hamiltonian upon quantization. The remaining interaction $-P^{nl} \cdot D$ relays the effect of the nonlinear polarization P^{nl}.

We assume that the polarization at a point in space depends on the displacement field at that point, i.e., the response function is local. Expanding the nonlinear polarization in a power series of the displacement field amplitudes up to the third order, the part of the Hamiltonian associated with classical nonlinear dynamics is

$$\begin{aligned} H^{(nl)} &= \frac{-P^{nl} \cdot D}{2} = \\ &- 2\pi \sum_{k_0,k_1,k_2} \int dr^3\, \Gamma^{(2)}(\vec{r}; \omega(k_0), \omega(k_1), \omega(k_2))) D_{k_0}(\vec{r},t) D_{k_1}(\vec{r},t) D_{k_2}(\vec{r},t) \\ &- 2\pi \sum_{k_0,k_1,k_2,k_3} \int dr^3\, \Gamma^{(3)}(\vec{r}; \omega(k_0), \omega(k_1), \omega(k_2), \omega(k_3))) \\ &\times D_{k_0}(\vec{r},t) D_{k_1}(\vec{r},t) D_{k_2}(\vec{r},t) D_{k_3}(\vec{r},t) \equiv H^{(2)} + H^{(3)}, \end{aligned} \tag{9}$$

where $\Gamma^{(2)}$ is the second-order displacement field susceptibility that is responsible for optical parametric oscillation (OPO)—often referred to as parametric down conversion. $\Gamma^{(3)}$ is the third-order displacement field susceptibility responsible for four effects: the two optical Kerr effects called self-phase modulation (SPM) and cross-phase modulation (XPM), a mode-mixing process called four-wave mixing (FWM), and a static-field-induced OPO. Both are defined by the power series expansion of the polarization field in terms of displacement field [40].

The ordinary nonlinear susceptibilities are defined analogously for the electric field [42], namely

$$P = \Gamma^{(1)} D + \Gamma^{(2)} D^{\otimes 2} + \Gamma^{(3)} D^{\otimes 3} + \ldots = \chi^{(1)} E + \chi^{(2)} E^{\otimes 2} + \chi^{(3)} E^{\otimes 3} + \ldots \tag{10}$$

Substituting the relation $D = E + 4\pi P(E)$ into Equation (10) and collecting terms with common powers of electric field, ignoring tensor structure and dispersion, yields the relations

$$\Gamma^{(1)} = \frac{\chi^{(1)}}{\epsilon}, \quad \Gamma^{(2)} = \frac{\chi^{(2)}}{\epsilon^3}, \quad \Gamma^{(3)} = \frac{\chi^{(3)}}{\epsilon^4} - 8\pi \frac{(\chi^{(2)})^2}{\epsilon^5} \,. \tag{11}$$

3. Approach

We aim to show how strong Kerr nonlinearities affect OPO. The essential result is that photon-number-dependent frequency shifts from the Kerr effect generate a photon-number-dependent detuning in OPO. The number-dependent detuning is exploited for coherent deterministic number-selective shaping of noise statistics in select modes. To make this effect clear, we constrain our view to the subspace of three optical modes engaging in OPO, called signal, idler, and sum. We first solve the system exactly under the action of the sum of Hamiltonian terms that preserve photon number in each mode. Evolution under this Hamiltonian gives nontrivial structure to the time-dependent displacement field operator; it is scaled by a diagonal unitary operator with a frequency that is a function of the photon number operators. OPO is analyzed in the interaction picture, exhibiting photon-number-selective behavior. The nature of this behavior is discussed, and explored numerically in Section 4.1.

3.1. Eigenstates of Kerr Hamiltonian

The Hamiltonian $H^{(3)}$ from Equation (9) is responsible for two distinct photon number density-preserving processes, SPM and XPM, and a mode-mixing process, FWM. FWM is a rich effect, leading to optical frequency combs and soliton propagation in micro-resonators [43]. However, our aim is to study an entirely different phenomenon in the same environment, so we do not wish to handle these complications presently. We will suppose that the dispersion relation of our system is specified such that no FWM is phase matched for the active modes.

Unlike FWM, SPM and XPM are always phase-matched. These effects, together with the linear Hamiltonian, dominate the phase dynamics of the system. The corresponding Hamiltonian is diagonal in the Fock basis and, equivalently, conserves photon number for each mode. Our present objective is to solve for the time evolution of the field operator under photon number density-preserving Hamiltonian $H_0 \equiv H^{(1)} + H^{(3)}$ such that $H = H_0 + H^{(2)}$, where $H^{(1)}$, $H^{(2)}$, and $H^{(3)}$ are defined in Equations (7) and (9). We later show how this time evolution affects $H^{(2)}$, producing a photon-number-selective interaction under the right conditions. We will consider the dynamics of three modes interacting in the cavity: a sum, signal, and idler mode with frequencies ω_Σ, ω_s, and ω_i, respectively.

We assume permutation invariance of all susceptibilities and abbreviate the frequency dependence of Kerr susceptibilities $\Gamma^{(3)}(-\omega_1, \omega_1, -\omega_2, \omega_2)) = \Gamma^{(3)}_{1,2} = \Gamma^{(3)}_{2,1}$ (permutation invariance typically does not hold near resonance, where we expect our system to reside given the constraints derived later. However, accounting for variability in $\Gamma^{(3)}$ simply amounts to substituting the average $\bar{\Gamma}^{(3)}$ in place of $\Gamma^{(3)}$ and does not qualitatively change the results). Furthermore, we uphold the rotating wave (secular) approximation and assume spatially constant third-order susceptibility. After ordering all permutations of the creation and annihilation operators to be simple powers of the mode number operators $N_j = a_j^\dagger a_j$, the Hamiltonian H_0 pertaining to the sum, signal, and idler modes is

$$
\begin{aligned}
H_0 = &\Big(\hbar\omega_\Sigma - 12\pi L\zeta_\Sigma^2(2\zeta_\Sigma^2\Gamma^{(3)}_{\Sigma,\Sigma} + \zeta_s^2\Gamma^{(3)}_{\Sigma,s} + \zeta_i^2\Gamma^{(3)}_{\Sigma,i})\Big)N_\Sigma \\
&+ \Big(\hbar\omega_s - 12\pi L\zeta_s^2(2\zeta_s^2\Gamma^{(3)}_{s,s} + \zeta_i^2\Gamma^{(3)}_{s,i} + \zeta_\Sigma^2\Gamma^{(3)}_{s,\Sigma})\Big)N_s \\
&+ \Big(\hbar\omega_i - 12\pi L\zeta_i^2(2\zeta_i^2\Gamma^{(3)}_{i,i} + \zeta_\Sigma^2\Gamma^{(3)}_{i,\Sigma} + \zeta_s^2\Gamma^{(3)}_{i,s})\Big)N_i \\
&- 24\pi L\Big(\zeta_\Sigma^4\Gamma^{(3)}_{\Sigma,\Sigma}N_\Sigma^2 + \zeta_s^4\Gamma^{(3)}_{s,s}N_s^2 + \zeta_i^4\Gamma^{(3)}_{i,i}N_i^2 \\
&+ \zeta_\Sigma^2\zeta_s^2\Gamma^{(3)}_{\Sigma,s}N_\Sigma N_s + \zeta_\Sigma^2\zeta_i^2\Gamma^{(3)}_{\Sigma,i}N_\Sigma N_i + \zeta_i^2\zeta_s^2\Gamma^{(3)}_{i,s}N_i N_s\Big)
\end{aligned}
\tag{12}
$$

Heisenberg's equation of motion, $\dot{a} = \frac{1}{i\hbar}[a, H_0]$, tells us the time evolution for the field annihilation operators a_Σ, a_s, and a_i. To condense our notation, let us define the Kerr frequencies

$$
g_{ij} \equiv \frac{24\pi AL}{\hbar}\zeta_i^2\zeta_j^2\Gamma^{(3)}_{ij} \tag{13}
$$

where g_{jj} are the frequencies of self-phase modulation and $g_{i,j\neq i}$ are the frequencies of cross-phase modulation. These frequencies are proportional to the classical Kerr frequency shifts $\delta\omega$ in accordance with the relation $g\langle N\rangle = \delta\omega I$ where I is the classical intensity and

$\langle N \rangle$ is the mean photon number. The time evolution of each field annihilation operator under H_0 is

$$\begin{aligned} a_\Sigma(t) &= \exp\Big\{-i\Big(\omega_\Sigma - (g_{\Sigma\Sigma} + \frac{g_{\Sigma s}}{2} + \frac{g_{\Sigma i}}{2}) - g_{\Sigma\Sigma}(2N_\Sigma + 1) + g_{\Sigma s}N_s + g_{\Sigma i}N_i)\Big)t\Big\}a_\Sigma(0) \\ a_s(t) &= \exp\Big\{-i\Big(\omega_s - (g_{ss} + \frac{g_{s\Sigma}}{2} + \frac{g_{si}}{2}) - g_{ss}(2N_s + 1) + g_{s\Sigma}N_\Sigma + g_{si}N_i)\Big)t\Big\}a_s(0) \\ a_i(t) &= \exp\Big\{-i\Big(\omega_i - (g_{ii} + \frac{g_{i\Sigma}}{2} + \frac{g_{is}}{2}) - g_{ii}(2N_i + 1) + g_{i\Sigma}N_\Sigma + g_{is}N_s)\Big)t\Big\}a_i(0)\,. \end{aligned} \tag{14}$$

It should be noted that the exponentials above do not commute with the $t = 0$ field annihilation operators. To further simplify our notation, we call $a(0) = a$, and use ξ to represent the exponential operators on the right-hand side of Equation (14), such that

$$a_\Sigma(t) = \xi_\Sigma a_\Sigma, \quad a_s(t) = \xi_s a_s, \quad a_i(t) = \xi_i a_i \tag{15}$$

Substituting these operators into Equation (6) yields the time-dependent displacement field operator for the nonlinear field described by H_0,

$$D_k(z,t) = i\zeta_k\Big(\xi_k(t)a_k e^{ikz} - e^{-ikz}a_k^\dagger \xi_k^*(t)\Big)\,. \tag{16}$$

The time-dependent displacement field operator is the essential result of this subsection and the starting point for our analysis in the following Section 3.2. Let us take a brief detour to understand how the ξ operators affect optical pumping and derive our first constraint.

Pump Photon Blockade

The ξ operators neatly quantify the photon blockade effect, reviewed in [26], which affects optical pumping of the resonator. To see this, consider a cavity pumped by a near-monochromatic beam that is near resonance to a cavity mode with wavenumber k_j. As the photon number in the cavity increases from N_j to $N_j + 1$, the frequency shifts by $g_{jj}(2N_j + 1)$. If the frequency of the next excitation shifts out of the pump linewidth, the cavity will no longer be excited, i.e., additional photons are blockaded from entering by the photons already in the cavity. Consequently, obtaining a high cavity intensity requires a pump beam with a wide linewidth in the presence of very strong Kerr nonlinearities. However, a wide pump linewidth yields unwanted excitations, so it is best for the Kerr nonlinearities involving the pump frequency to be small.

If small Kerr frequencies at the pump frequency are challenging to achieve, we consider the following constraint: self-phase modulation by a positive susceptibility redshifts the frequency by $-g_{jj}N_j^2$. If $\Gamma_{jj}^{(3)} > 0$, the largest photon number achievable in the cavity pumped by a beam with a spectral edge at some fraction f of ω_j is

$$N_{MAX} = \sqrt{\frac{(1-f)\omega_j}{g_{jj}}} \tag{17}$$

When considering optical parametric oscillation in the next sections, we drive the sum frequency cavity mode with an external pump. It is most important that the external pump spectrum does not overlap with the signal and idler modes to avoid unwanted excitation. Therefore, it is necessary that $f > 1/2$ and preferable that $f > 3/4$. We discuss later in Section 4.2, and simply note now, that a large coherent state for the sum mode minimizes dephasing and enhances OPO efficiency. Because unwanted excitations and high pump intensity are coincident, it is best for the Kerr nonlinearities involving the pump frequency to be negative, if not small.

3.2. Number-Selective Optical Parametric Oscillation

The previous section solved the Kerr Hamiltonian exactly and introduced the operators ξ to simplify our analysis of optical parametric oscillation. With an exact time evolution for the Kerr photons in hand, we construct the Hamiltonian for number-selective optical parametric oscillation (NSOPO). In the interaction picture, the OPO Hamiltonian evolves via the photon number preserving propagator, $U_0 = \exp\{-iH_0 t/\hbar\}$. From Equation (9), we obtain the time-dependent OPO Hamiltonian in the interaction picture,

$$H_I = U_0^\dagger H^{(2)} U_0 = -2\pi \sum_{k_0,k_1,k_2} \int dr^3\, \Gamma^{(2)}(\vec{r};\omega(k_0),\omega(k_1),\omega(k_2)))D_{k_0}(\vec{r},t)D_{k_1}(\vec{r},t)D_{k_2}(\vec{r},t)\,. \tag{18}$$

We assume the susceptibility $\Gamma^{(2)}$ is spatially local, spatially constant, and frequency permutation invariant (as is the case for $\Gamma^{(3)}$, permutation invariance is not valid for the strong nonlinearities required of NSOPO. However, accounting for variability simply amounts to substituting the average $\bar{\Gamma}^{(2)}$ in place of $\Gamma^{(2)}$ and does not qualitatively change the result). The time-dependent displacement field operators, Equation (16), are substituted into Equation (18). Terms of the form $a_\Sigma^\dagger a_i^\dagger a_s^\dagger$ and $a_\Sigma a_s a_i$ are neglected by assuming the rotating wave approximation for the three modes, but the time dependence from the Kerr-induced frequency mismatch is preserved. All possible permutations of the three relevant mode contributions to the field operators are reordered to a common tractable form, accounting for the noncommutativity of a and ξ. The simplified interaction Hamiltonian is thus

$$H_I = -i12\pi AL\zeta_\Sigma\zeta_s\zeta_i\left(\Gamma^{(2)}_{\Sigma,s,i}a_\Sigma^\dagger a_s a_i \xi_\Sigma^* \xi_s \xi_i e^{-i(2g_{ss}+2g_{ii}+g_{si}-g_{\Sigma s}-g_{\Sigma i})t} - h.c.\right) \tag{19}$$

where *h.c.* denotes a Hermitian conjugate.

We suppose that the sum frequency mode is a high-amplitude coherent state $|\alpha_\Sigma\rangle$ such that $\bar{N}_\Sigma = |\alpha_\Sigma|^2 \gg \bar{N}_s, \bar{N}_i$. Then, we can reasonably replace the operator a_Σ with the c-number α_Σ without worrying much about dephasing effects. For example, if the shaping of signal and idler modes is done in a range of m photons, the purity of the manipulated quantum state is of the order of $e^{-|\frac{m}{2\alpha_\Sigma}|^2}$ (See Section 4.2).

To simplify notation, we define the NSOPO complex squeezing parameter Ξ and OPO resonance detuning frequency $\Delta(N_\Sigma, N_s, N_i)$ such that

$$\Xi e^{i\Delta t} \equiv \left(\frac{12\pi AL}{\hbar}\zeta_\Sigma\zeta_s\zeta_i\Gamma^{(2)}_{\Sigma,s,i}\alpha_\Sigma^*\right)\left(\xi_\Sigma^*\xi_s\xi_i e^{-i(2g_{ss}+2g_{ii}+g_{si}-g_{\Sigma s}-g_{\Sigma i})t}\right) \tag{20}$$

where

$$\Xi \equiv \frac{12\pi AL}{\hbar}\zeta_\Sigma\zeta_s\zeta_i\Gamma^{(2)}_{\Sigma,s,i}\alpha_\Sigma^*\,, \tag{21}$$

$$\Delta \equiv \delta + G_\Sigma N_\Sigma + G_S(N_s+1) + G_i(N_i+1)\,, \tag{22}$$

and

$$\begin{aligned}
\delta &\equiv \omega_\Sigma - \omega_s - \omega_i + 2(g_{ss}+g_{ii}-g_{\Sigma\Sigma}) \\
G_\Sigma &\equiv g_{\Sigma s} + g_{\Sigma i} - 2g_{\Sigma\Sigma} \\
G_s &\equiv 2g_{ss} + g_{is} - g_{\Sigma s} \\
G_i &\equiv 2g_{ii} + g_{si} - g_{\Sigma i}
\end{aligned} \tag{23}$$

are the set $\{G\}$ of detuning weights for each photon number, as well as a static detuning δ.

The interaction Hamiltonian is now expressed compactly as

$$H_I = -i\hbar\left(a_s a_i \Xi e^{i\Delta t} - e^{-i\Delta t}\Xi^* a_i^\dagger a_s^\dagger\right) \tag{24}$$

Without Kerr nonlinearities, non-degenerate parametric down-conversion generates two-mode squeezing between the signal and idler modes [44]. With Kerr nonlinearities, the squeezing parameter hosts a detuning oscillation with a photon-number-dependent frequency, making the two-mode squeezing amplitude dependent. Let us explore precisely how this feature affects the interaction.

The OPO detuning $\Delta(N_\Sigma, N_s, N_i)$ governs how number states experience two-mode squeezing. A simple, and very useful, case to study is when a large-α_Σ coherent state $|\alpha_\Sigma\rangle$ is injected into the nonlinear cavity in the presence of a two-mode squeezed state of signal and idler frequencies. We choose to study NSOPO on a two-mode squeezed state rather than a single-mode state to highlight how noise shaping is identically present in the entangled signal and idler modes, though NSOPO may be realized with a single mode as well. Experimentally, NSOPO of a two-mode squeezed state would require a method for coherently injecting the state into the nonlinear cavity initially in the vacuum state. Acknowledging the technological challenge (this may be realized, for example, with an acousto-optic modulator), we will assume it is possible and investigate its consequences.

We assume an initial two-mode squeezed state—a state which is a superposition of all equal populations of signal and idler photons. Evident from the interaction operator $a_\Sigma^\dagger a_s a_i$, the difference in signal and idler photon numbers is a conserved quantity, so the state remains a superposition of equal photon numbers for all times, much less than the cavity decay time. When $G_\Sigma = 0$, the OPO detuning Δ is independent of the sum frequency photon number and NSOPO is coherently driven by the sum frequency beam. Each OPO interaction event will decrement or increment the signal and idler photon numbers by one, transporting probability amplitude along the space of states. Adjacent states see a difference in detuning of

$$\gamma \equiv G_s + G_i = 2(g_{ss} + g_{ii} + g_{is} - g_{\Sigma\Sigma}) \tag{25}$$

Let us suppose there is a photon number, $N_s = N_i = n'$, such that the detuning is minimal. We call n' the shape center. Then, the OPO Hamiltonian in Dirac notation takes the form

$$H_I = -i\hbar \sum_{m=0}^{\infty} (m+1)\left(\Xi e^{i(\gamma(n'-m)+\Delta(n'))t}|m\rangle\langle m+1| - \Xi^* e^{-i(\gamma(n'-m)+\Delta(n'))t}|m+1\rangle\langle m|\right). \tag{26}$$

This Hamiltonian generates a particularly interesting evolution when Ξ and γ are of similar magnitude; however, it is challenging to describe analytically. We explore this behavior in the following section.

4. Results and Discussions

In the previous section, we developed a theoretical method for directly targeting photon number states to coherently shape the quantum statistics of an optical field in a cavity. In the following section, we numerically simulate the model and illustrate how the parameters can be controlled to generate quantum states with useful structure. The constraints established here and in Section 3 are elaborated on and brought together to determine what regions of parameter space are viable for experimental realization.

4.1. Simulations

The time-dependent Schrödinger equation corresponding to the interaction Hamiltonian, Equation (26), is numerically evolved with a Crank–Nicolson method, which preserves unitarity. Figure 1 reveals the characteristic behavior of NSOPO. It shows that probability amplitude tends to coalesce around number states separated by a spacing

$$s = \frac{2\pi}{\gamma t} \tag{27}$$

where states see constructive interference from two-mode squeezing. These peaks rapidly fall off in amplitude when near the shape center, but fall off more slowly when away from the shape center maintaining a wide tail in their height distribution. The peak heights increase monotonically with $|\Xi|$ until reaching a maximum, after which the peak begins to wind itself and develops a complicated structure.

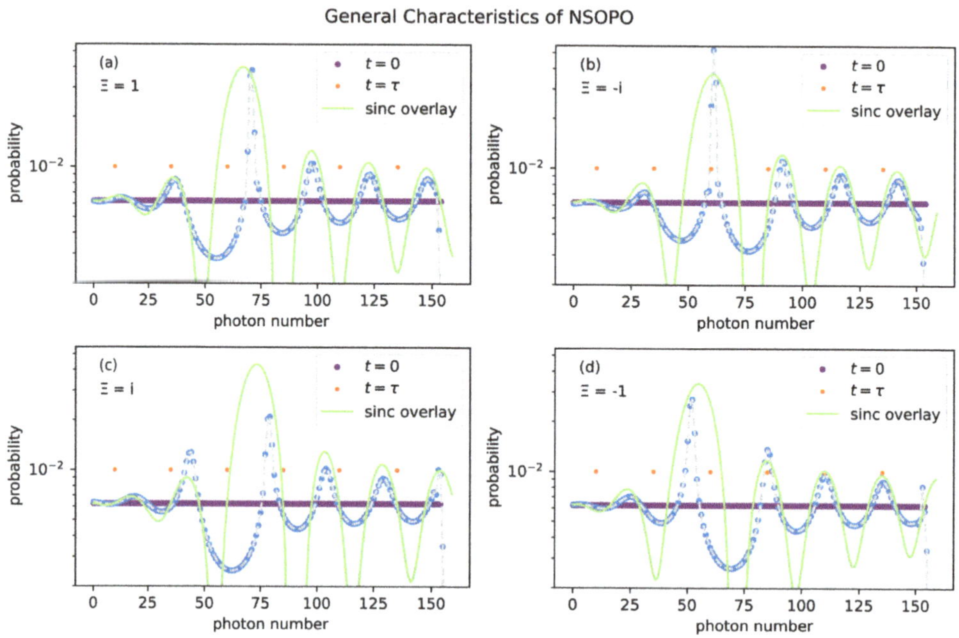

Figure 1. General characteristics of NSOPO. In each plot, the state is initially a uniform real-valued distribution. The plots differ only in the phase of the NSOPO complex squeezing parameters Ξ. For all plots, the gouge center is $n' = 60$, the Kerr detuning $\gamma = \pi$, and the evolution time $\tau = 0.08$. Orange dots are spaced by the predicted peak spacing, $s = \frac{2\pi}{\gamma\tau}$, and aligned with the shaping center. The green curve shows a first-order approximation of shaping that follows the peak positions and magnitude closely. The blue curve is the shaped photon probability distribution resulting from an initial uniform distribution.

The peak nearest the shape center deviates from periodic spacing, consequently shifting the pattern for states with a higher photon number. The location of the central peak depends on the phase of the NSOPO squeezing parameter Ξ. For $\Xi = -i$, the peak is centered on the shape center. For $\Xi = i$, there is a deep trough at the shape center with the nearest peaks equidistant to the left and right. For $\Xi = 1(-1)$, the nearest peak and trough are to the right and left (left and right) of the shape center. The plots in Figure 1 underlay the function

$$\frac{1}{N}\left[\mathrm{sinc}\Big(\gamma(n' - m)\tau + arg(\Xi) + \pi/2\Big)(m+1)\tau + 1\right] \tag{28}$$

which approximately replicates the height and location of each peak, breaking accuracy at the central peak. This function is obtained by integrating the Hamiltonian over time τ and acts as a more tractable guideline for engineering states with shaping.

These shaping principles can be applied in several ways. Setting $arg(\Xi) = -\pi/2$ such that the shape center is aligned with the peak of a state and the peak spacing s to be greater than the width of the state, the photon number probability will coalesce into an approximate Fock state. Figure 2a illustrates this process, generating a state with signal photon number probabilities $P(n = 52) \approx 45\%$, $P(n = 54, 53, 51, 50) \approx 11\%$. Alternatively, if the shape center is set far from the concentration of initial state probabilities, clean periodic oscillations in photon number probabilities, with a period s, are introduced into the state. Figure 2b–d illustrate this process: (b) shows precision shaping of a coherent state with $s = 2$; (c) is the same as (b) but acting on a two-mode squeezed vacuum; (d) shows shaping with $s = 6$.

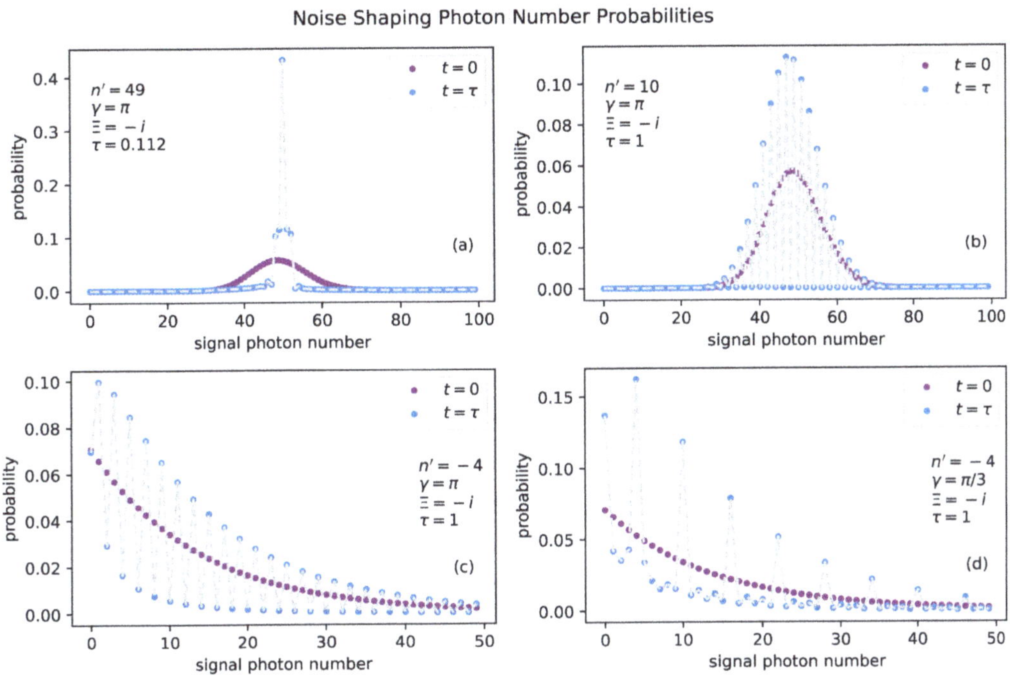

Figure 2. Examples of photon probability shaping offered by NSOPO. (**a**) An approximate Fock state of $n = 52$ accomplished by placing the shaping center on the peak of an initial coherent state with an amplitude of $\alpha_{s,i} = 7$. (**b**) A precision shaping of the same coherent state. Parameter γt is chosen such that peak spacing s is two photons. (**c**) A two-mode squeezed vacuum state with $s = 2$. (**d**) A two-mode squeezed vacuum with $s = 6$.

In tandem with photon number probabilities, NSOPO offers control over states in optical phase space. The following demonstrates an interesting case of this control: generating a family of displaced Schrödinger's cats with a single operation. It is important to remember that, in the Schrödinger picture, the modes experience self-phase modulation and cross-phase modulation via the Kerr effect. Introducing the condition $G_\Sigma = 0$ into Equation (12), we see that the term of the Hamiltonian

$$H^{Kerr} = -\hbar\left(g_{ss}N_s^2 + g_{ii}N_i^2 + g_{is}N_iN_s\right), \tag{29}$$

generates nonlinear dynamics for the signal and idler modes. Noting that N is an integer, we see that states of a single mode are periodic in time t when each Kerr frequency satisfies $gt = 2\pi m$, with m an integer. For simple rational values of m, the evolution maps coherent states to generalized coherent states [12].

We extend this argument to the two-mode squeezed state, recalling $N_i = N_s$. We choose a preferred final state by selecting m to satisfy

$$(g_{ss} + g_{ii} + g_{si})t = 2\pi m = (\gamma/2 + g_{\Sigma\Sigma})t\,. \tag{30}$$

Equation (27) provides a similar constraint regarding NSOPO peak spacing. If we suppose $g_{\Sigma\Sigma} = 0$, then $(g_{ss} + g_{ii} + g_{si}) = \gamma/2$, making Equations (30) and (27) the same constraint with $2m = 1/s$. Setting $s = 4$, we obtain a four-phase Schrödinger cat state. Figure 3 shows the action of NSOPO on a coherent state, (a), that is split into a generalized four-phase Schrödinger cat state, (b). Depending on the phase $arg(\Xi)$, NSOPO restricts the amplitude of the cat to two of four phases, shown in (c–f). The axes in each plot are independent phase space quadratures of the displacement field. Not shown in Figure 3, shifting the shape center by one rotates the action of NSOPO among the operations shown in (c–f). Namely, under $n' \to n' + 1$, the plots map to each other according to $c, d, e, f \to f, c, d, e$. This behavior alludes to efficient quantum logic gates for macroscopic quantum states, though further investigation is needed to understand the range of this capability.

Figure 4 shows the effect of incoherent loss on the generation of an approximate Fock state, similar to that shown in Figure 2a. In this instance, an initial product state $|\psi\rangle$ of a signal/idler two-mode coherent state with amplitude $\alpha_{s,i} = 2$ and a large coherent pump state is evolved according to the local Lindblad master equation

$$\begin{aligned}\frac{d\rho(t)}{dt} &= \frac{1}{i\hbar}[H^{(1)} + H^{(2)} + H^{(3)}, \rho(t)] \\ &+ \frac{\Gamma_{cav}}{2}\Big(2a_s\rho(t)a_s^\dagger - \rho(t)a_s^\dagger a - a_s^\dagger a_s\rho(t) + 2a_i\rho(t)a_i^\dagger - \rho(t)a_i^\dagger a - a_i^\dagger a_i\rho(t)\Big)\end{aligned} \tag{31}$$

where $\rho(0) = |\psi\rangle\langle\psi|$ and Γ_{cav} represents the incoherent decay rate of the cavity at the signal and idler frequencies.

To characterize the viability of the system for generating useful and pure quantum states, we introduce a parameter that is similar in spirit to cooperativity [45]. Cooperativity is defined for light-matter systems with specific quantum models of the material and quantifies the ratio of the light-matter interaction rate to the loss rate of the system. Our theory is derived from a constitutive relation, Equation (10), so it does not offer the detail required to provide the exact cooperativity. However the strength-to-loss ratio γ/Γ_{cav} of the adjacent state detuning γ, i.e., the Kerr frequency of the signal-idler state, and the incoherent decay rate of the cavity at the signal and idler frequencies Γ_{cav}, serves a similar role.

Plot (a) shows the growth of quantum infidelity $1 - F$ with respect to reciprocal NSOPO peak spacing $\gamma t = \frac{2\pi}{s}$. Each curve represents a different value of the strength-to-loss ratio γ/Γ_{cav}. For $\gamma t < 1$, infidelity scales linearly with γt. For $|\Xi|t = 0.3$, $n' = 4$, and $\gamma t = 0.9$, we find the numerical result

$$\Gamma_{cav}/\gamma \approx \frac{1-F}{3} \,. \tag{32}$$

This inverse proportionality between the strength-to-loss ratio and infidelity is substantial. Interestingly, generating approximate Fock states of higher photon numbers require a smaller γt, such that the spacing is larger, thus high Fock state generation demands less of the strength-to-loss ratio for the same quantum fidelity.

4.2. Pump Dephasing

A key challenge to overcome for preserving the quantum coherence of noise-shaped states of light is to mitigate the dephasing effects of the pump beam. The two criteria to satisfy are a strong pump beam, quantified below, and a vanishing detuning weight for the pump frequency, $G_\Sigma = 0$. Specifically, we demand that the detuning weights satisfy

$$G_\Sigma N_\Sigma \approx |2g_{\Sigma\Sigma} - g_{\Sigma s} - g_{\Sigma i}||\alpha_\Sigma|^2 \ll G_s N_s + G_i N_i = \gamma N_s \,. \tag{33}$$

A vanishing pump detuning weight ensures that the desired OPO detuning Δ is independent of pump photon number, rendering NSOPO not number-selective for the pump field. This condition is important because a pump photon-number-selective process implies that far fewer states in the entire Hilbert space contribute to OPO, dramatically decreasing the efficiency of the process. Furthermore, if the interactions were number-selective for the pump, the state of the pump field would vary dramatically for different possible interaction histories, resulting in strong dephasing.

When NSOPO is not number-selective for the pump, the OPO interaction is simply a displacement operation, which minimally disturbs the structure of the pump field. To illustrate, let us consider the state $|\alpha_\Sigma, n_s, n_i\rangle$ time-evolved under the OPO interaction H_I with $G_\Sigma = 0$. The result is a superposition of products of shifted coherent states in the sum frequency and incremented number states in the signal and idler. The coherent states are, at most, shifted by a displacement corresponding to the maximum photon number change of the signal and idler. For large amplitude coherent states, shifts in the coherent states on the scale of shifts in the signal and idler yield minimal deviations in the overlap between the coherent pump states.

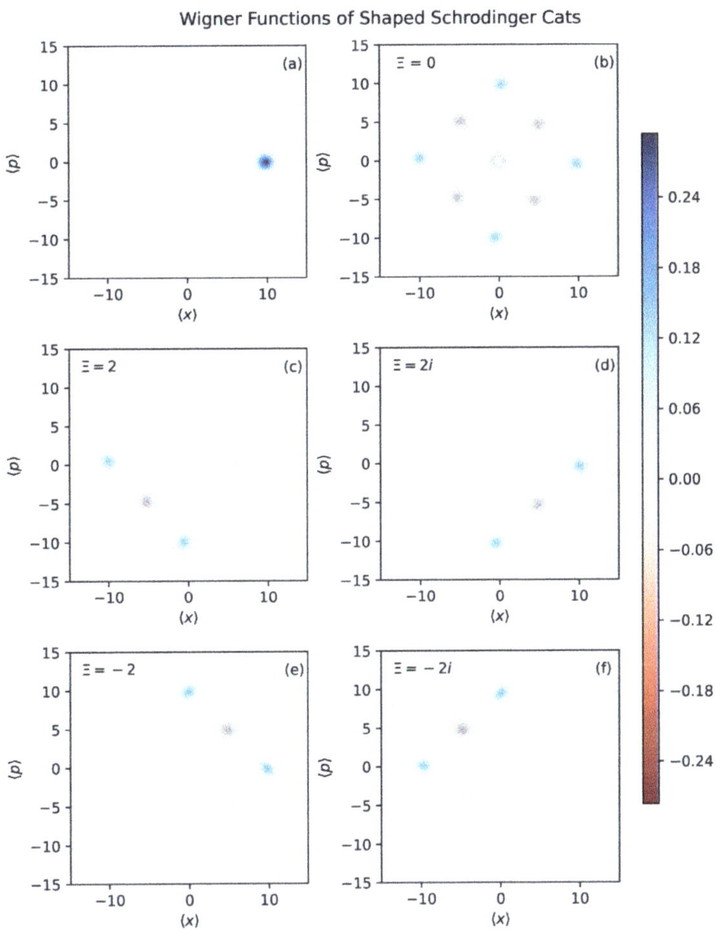

Figure 3. (**a**) Wigner function of an initial coherent state of amplitude $\alpha = 7$. (**b**) The coherent state subject to Kerr rotation without any NSOPO. (**c–f**) The state shaped with NSOPO complex squeezing parameter $\pm 2, \pm 2i$. In all plots, blue regions are positively valued and red regions are negative. Every state is shaped with parameters $\gamma = \pi/2, \tau = 1$, and $n' = 8$. As the phase of Ξ is varied by increments of $\pi/2$, number-selective shaping changes predictably.

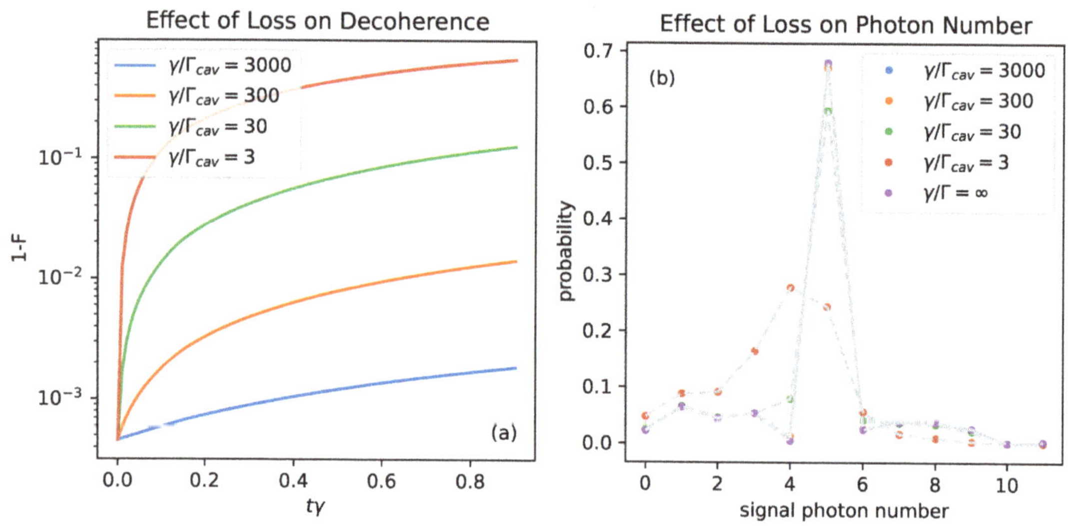

Figure 4. Both plots present data from a simulation generating an approximate N=5 Fock state in the presence of incoherent loss. The state is evolved from an initial two-mode squeezed coherent state of amplitude $\alpha = 2$. The Hamiltonian parameters are $|\Xi|\tau = 0.3$, $\gamma\tau = 0.9$, and $n' = 4$. Plot (**a**) shows the growth of quantum infidelity $1 - F$ with respect to reciprocal NSOPO peak spacing $\gamma t = \frac{2\pi}{s}$. Each curve represents a different value of the strength-to-loss ratio γ/Γ_{cav} where γ is the the adjacent state detuning and Γ_{cav} is the decay rate of the cavity at the signal and idler frequencies. Plot (**b**) shows the effect of loss on corresponding photon number probabilities.

To obtain a generic measure of state purity, we note that the histories of Fock state amplitudes are carried along with NSOPO peaks. The pump beam coherent states that are entangled with either peak acquire dephasing contributions from pump beam coherent states with a mean shifted by the interval of photon numbers the NSOPO peak has traveled over its evolution. This entanglement between the pump, signal, and idler contributes a dephasing in the signal and idler density matrix. The purity of the signal/idler density matrix, $Tr(\rho^2_{s/i})$, is well approximated by the squared overlap of the shifted and unshifted pump beam states. Thus, for a pump beam in a coherent state $|\alpha_\sigma\rangle$, signal/idler in initial two-mode squeezed coherent state $|\alpha_{s,i}\rangle$, and spacing s, the purity measure is

$$\gamma_{purity} = \exp\left\{ - \left|\frac{\alpha_{s,i}}{4s\alpha_\Sigma}\right|^2 \right\}. \tag{34}$$

4.3. Summary of Constraints and Figures of Merit

Five conditions must be met in order to realize coherent NSOPO. The following subsections summarize each condition, roughly ordered from simplest to most difficult to satisfy:

1. In the presence of strong Kerr nonlinearities, exciting a cavity pump field to high intensity requires a broad external pump linewidth. However, a broad external pump will excite superfluous modes in the system. Let us suppose the pump beam is a broad spectrum coherent state $|\alpha_\Sigma\rangle$. To mitigate unintended interactions from modes above the signal and idler frequency, we restrict the magnitude of the sum mode self-phase modulation frequency by Equation (17) to obtain

$$g_{\Sigma\Sigma} < \frac{\omega_\Sigma}{4|\alpha_\Sigma|^4}. \tag{35}$$

Thus, it is essential for $g_{\Sigma\Sigma}$ to be small or negative.

2. The purity measure Equation (34) indicates the constraint

$$\frac{\alpha_{s,i}}{4\alpha_\Sigma} \ll s\,. \tag{36}$$

The precision of noise-shaping is limited by the relative uncertainty of the signal/idler photon number noise with respect to the pump photon number noise and is generally easy to control.

3. Numerical simulations place useful values of $\frac{\Xi}{\gamma}$ in the range from 0.1 to 10. For an order of magnitude estimate of this constraint, let us consider a case where $g_{ss} = g_{ii} = g_{is}$ and $g_{\Sigma\Sigma} = 0$. Using the definitions of γ, Ξ, and g from Equations (6), (13), (21), and (25),

$$\frac{\Xi}{\gamma} = \frac{\Xi}{6g} \approx \frac{\zeta_\Sigma \Gamma^{(2)} \alpha_\Sigma^*}{12\zeta_s\zeta_i\Gamma^{(3)}}\,. \tag{37}$$

The quantity ζ is the single photon displacement field amplitude in Equation (6). Careful use of Equation (11) relates the displacement field susceptibilities to the more commonly measured electric field susceptibilities. Expanding the ratio with the definitions of each frequency yields

$$\begin{aligned}\frac{\Xi}{\gamma} &\approx \sqrt{\frac{LA}{18\hbar\omega}}\sqrt{\frac{cv_{g,\Sigma}n_\Sigma^3}{v_{g,s}v_{g,i}n_s^3n_i^3}}\frac{\Gamma^{(2)}\alpha_\Sigma^*}{\Gamma^{(3)}} \\ &\approx \sqrt{\frac{LA}{18\hbar\omega}}\sqrt{\frac{cv_{g,\Sigma}}{v_{g,s}v_{g,i}}}\sqrt{\frac{1}{n_\Sigma n_s n_i}}\frac{\chi^{(2)}\alpha_\Sigma^*}{\chi^{(3)}} \in [0.1, 10]\end{aligned} \tag{38}$$

This criterion is straightforward to satisfy in real materials. For however large the magnitude of $\chi^{(3)}$ is, the strength of OPO can be increased with the intensity of the pump beam and the quantization volume.

4. As discussed in Subsection 4.2 above, shaping maintains quantum coherence if the pump field detuning weight vanishes, $G_\Sigma = 0$. Equation (33) provides a qualitative expression of the requisite smallness of G_Σ. The upshot is that greater pump power requires higher precision in the proximity of G_Σ to zero. This condition is easiest to satisfy if Kerr frequencies related to the pump are independently small, which is challenging if the other Kerr frequencies are to be incredibly large. Appendix A discusses a class of systems that can satisfy exactly this scenario by generating nonlinearities on resonance with the signal and idler, and off-resonance with the pump.

5. To apply the theory developed in this paper, all shaping should occur on a time scale τ shorter than that defined by the acceptable infidelity $1 - F_{acc}$ and desired spacing according to

$$\frac{\Gamma}{\gamma}\frac{2\pi}{s} = \frac{\tau}{\tau_{cav}} < \frac{1 - F_{acc}}{s}\,. \tag{39}$$

where $\tau_{cav} = 1/\Gamma_{cav}$ is the cavity decay time. This constrains the lower bound on the Kerr frequencies g_{ss}, g_{ii}, and g_{si}. Let us assume $g_{ss} = g_{ii} = g_{si} = g$ and $g_{\Sigma,\Sigma} = 0$ for simplicity. Restating the numerical result for the high-fidelity regime Equation (32), but substituting in the cavity lifetime $\tau_{cav} = \frac{2Q}{\omega}$, we obtain

$$\frac{3}{1-F} \approx \frac{\gamma}{\Gamma_{cav}} = \frac{12gQ}{\omega}\,.$$

Substituting ζ from Equation (6) into the definition of the Kerr frequencies g from Equation (13), and relating $\Gamma^{(3)}$ to $\chi^{(3)}$ with Equation (11), we find that the strength-to-loss ratio is

$$\frac{\gamma}{\Gamma_{cav}} = 72\pi Q\frac{\hbar\omega}{V}\frac{v_g^2}{c^2}\frac{\chi^{(3)}}{n^2} \tag{40}$$

Finally, substituting in the classical form of the nonlinear refractive index $n_2 = \frac{24\pi^2\chi^{(3)}}{cn^2}$, we obtain a figure of merit $\mathcal{F}$ for NSOPO that is proportional to the strength-to-loss ratio,

$$\frac{1}{1-F} \approx \frac{Q\hbar\omega v_g^2 n_2}{\pi V c} \equiv \mathcal{F} \tag{41}$$

This figure of merit commands our attention; it is the most challenging constraint to practically satisfy among those in this list. Candidate systems will have large $\mathcal{F}$ at the signal and idler frequencies and small $\mathcal{F}$ at the sum frequency. In Appendix A, we show that compliance is likely attainable by addressing the state of the art for achieving strong nonlinearities with minimal loss and controllable group velocity and showing that these systems perform well within the boundary of the constraints.

4.4. Control Parameters

The shape center n' is controlled by modulating the linear refractive index of the material, and therefore the static detuning parameter δ, in time. To understand the extent of modulation we may ask, what are the possible rates of change for the shape center with respect to refractive index shifts? Let us recall the expression for δ, (23), which contains the frequency difference $\omega_\Sigma - \omega s - \omega_i$. In the simplest case, the frequencies are related to the fixed wavenumbers of each mode by

$$\omega = \frac{kc}{n}. \tag{42}$$

where $\frac{d\omega}{dn}$ is maximal, so is $\frac{d\delta}{dn}$. Therefore, it is desirable for the refractive index of each mode to be near zero. This conclusion is agreeable with the figures of merit derived above. One can change the refractive index of the material by, for example, introducing a static field, or by utilizing acoustic oscillations in the cavity material.

The NSOPO squeezing parameter, (21), is primarily controlled by modulating the amplitude and phase of the pump beam. In practice, one should be aware that changing the shape center affects the strength of NSOPO in three ways:

1. Equation (26) shows that OPO strength is scaled by $(m+1)$, where m is the signal photon number of the state being acted upon. Thus, the OPO strength is scaled by the shape center.
2. In some materials, e.g., organic materials, changing a static field to modulate the refractive index will reorient molecules affecting $\Gamma^{(2)}$.
3. Static fields will also introduce a $\Gamma^{(3)}$ contribution to OPO strength Ξ.

Ultimately, determining the parameters for preparing a particular state will be a matter of empirical tuning guided by rough numerical estimates.

A final note on state control: if the static field can be modulated on a time scale faster than the state preparation time τ, more complicated shaping of the field noise statistics can be accomplished, beyond fixed shaping parameters.

5. Conclusions

We have proposed a novel method for the deterministic manipulation of quantum states of light. The underlying theoretical tools required for its implementation are developed and the physical constraints for its practical implementation described. The method utilizes strong Kerr nonlinearities to introduce photon number dependence into the frequency matching condition for optical parametric oscillation, rendering the OPO interaction photon-number-selective. Number-selective interactions expand the capabilities of current CV quantum optical state control and preparation techniques by introducing targeted deterministic manipulations to subspaces of a light field's Fock space.

Simulations indicate the generation of highly nonclassical quantum states of light. We establish five necessary conditions to realize NSOPO in practice, which will aid the search for candidate materials and optical configurations. The constraints are consistent with a common figure of merit, $\mathcal{F}$, expressed in Equation (41). The best materials for NSOPO will have large $\mathcal{F}$ at the signal and idler frequencies, and small $\mathcal{F}$ at the sum frequency. The constraints can likely be realized in specialized systems that are driven into electronically induced transparency and spontaneously generated coherence.

Future research will aim to realize the requisite constraints in detailed models of highly resonant cavity light coupled with matter systems, with special attention paid to dark resonances. From specific models, the approach derived in this paper will be adapted to accommodate Hamiltonians beyond the simple cubic and quartic interactions treated above. These new models will be used to guide experimental design and to search for novel sculpted states. Future research will also investigate generation of entanglement between two uncorrelated signal and idler states, and study how noise shaping can be generalized to pulses of light, in particular to solitons in optical resonators with attention paid to mitigating the dephasing effects of quantum soliton propagation.

Author Contributions: Conceptualization, G.D.C. and M.G.K.; Data curation, G.D.C.; Formal analysis, G.D.C. and M.G.K.; Funding acquisition, M.G.K.; Investigation, G.D.C. and M.G.K.; Methodology, G.D.C. and M.G.K.; Project administration, M.G.K.; Resources, M.G.K.; Software, G.D.C.; Supervision, M.G.K.; Validation, G.D.C.; Visualization, G.D.C.; Writing—original draft, G.D.C.; Writing—review and editing, G.D.C. and M.G.K. All authors have read and agreed to the published version of the manuscript.

Funding: This research received no external funding.

Institutional Review Board Statement: Not applicable.

Informed Consent Statement: Not applicable.

Data Availability Statement: The code for simulations and original data presented in the study are openly available in the GitHub repository https://github.com/GarrettCompton/NSOPO.git.

Acknowledgments: We thank Ned and Nancy Wogman for the Wogman Physics Fellowship, the William J. Shelton Scholarship, and the donors to the Innovative Physics Fund and the Department of Physics for support.

Conflicts of Interest: The authors declare no conflicts of interest. The funders had no role in the design of the study; in the collection, analyses, or interpretation of data; in the writing of the manuscript; or in the decision to publish the results.

Abbreviations

The following abbreviations are used in this manuscript:

OPO	Optical parametric oscillation
SPM	Self-phase modulation
XPM	Cross-phase modulation
FWM	Four-wave mixing
NSOPO	Number-selective optical parametric oscillation
TMSV	Two-mode squeezed vacuum
ENZ	Electronically induced transparency
CV	Continuous variable
DFS	Decoherence-free subspace
SGC	Spontaneously generated coherence

Appendix A

The present theoretical paper describes how one might achieve the deterministic shaping of a Fock state. This appendix gives a plausibility argument that the required figure of merit is achievable. However, meeting the constraints discussed in Section 4.3 may require new materials and device architectures. Here, we individually consider the measured properties of known material systems to estimate a reasonable range of the figure of merit.

EIT is an example of a process that exhibits the requisite properties for NSOPO. Schmidt and Imamoglu [26] showed that a four-level system can be configured to obtain a giant cross-phase modulation strength. With the additional presence of spontaneously generated coherence (SGC), which can be accomplished in a five-level system [46], the EIT window can be shifted such that the Kerr nonlinearity is maximal when absorption is near zero. Furthermore, the five-level system proposed offers direct control over the group velocity, which ranges from highly sub- to super-luminal and with photon-number-dependent behavior [47].

Obtaining giant nonlinearities in near resonant systems with EIT gives us two other desirable features for NSOPO. The peaks of the linear and nonlinear susceptibilities are sufficiently sharp near resonance to make FWM minimal due to a rapidly increasing phase mismatch and rapidly decaying interaction strength. In addition, off-resonant nonlinearities will be comparatively weak. Such a system can be tuned so that the sum frequency is off resonance with the material while the signal and idler are on resonance. Then, Constraint (33) is satisfied by virtue of small Kerr nonlinearities at the pump frequency and FWM can be reliably neglected for all modes.

Bang and Doai, [46,47], report values of self and/or cross-phase modulating nonlinear index in a gas of ^{85}Rb with density 10^8 atoms/cm^3 on the order of

$$n_2 = \frac{24\pi^2\chi^{(3)}}{cn^2} \approx 5 \times 10^{-13}\frac{s \cdot cm^2}{erg} \tag{A1}$$

In principle, the atom density can be made orders of magnitude greater in a solid-state device. With modest concentrations of the active molecule on the order of 10^{18} molecules/cm^3 [48], the third-order susceptibility for non-interacting molecules is enhanced in proportion to the molecular density, increasing the nonlinear index to

$$n_2 \approx 5 \times 10^{-3}\frac{s \cdot cm^2}{erg}\,. \tag{A2}$$

Supposing that the signal and idler wavelengths are on the order of 1 μm, the resonator cross-sectional area is 1 μm × 1 μm, and the group velocity of the signal and idler are both c. Then, the figure of merit $\mathcal{F}$ is to within an order of magnitude given by

$$\mathcal{F} = \frac{Q}{L}10^3\,. \tag{A3}$$

For a length on the order of 10λ and cavity quality factor on the order 10^4 [49,50], the infidelity $\mathcal{F}^{-1}$ is 10^{-10}. This scenario is the upper boundary of feasible materials of this type, so we do not expect to reach it in practice, but it offers several orders of magnitude of leeway for trade-offs in designing materials and devices that shape photon statistics.

While this rough calculation shows the plausibility of deterministic shaping given known materials' properties, an important question remains as to the plausibility of making a solid-state waveguide device. Realizing the benefits of EIT and SGC in a solid-state waveguide device would require a material that, in analogy, is made with atoms that do not strongly interact with each other. A possible material system is a dye-doped polymer, which can be fabricated into waveguides, and whose linear and nonlinear optical properties can be controlled with atom/molecule doping.

One way to incorporate the active atom/molecule into a solid is to use a polymer, sol-gel host, or other host—which would lead to higher dopant concentration, thus a larger nonlinearity [51]. Such materials can be made into thin films and fibers [52] and photolithographic techniques can be used to make waveguide resonators [53]. Interactions between the host material and the atom/molecule can be minimized by using molecular pincers—called chelates—to isolate them. Chelated rare-earth atoms are commonly used in amplifiers [54]. The ability to get higher densities of active molecules to increase nonlinearity, the flexibility of choosing molecules for their spectral properties, the shielding from the environment to mitigate dephasing through chelation, and the reduction of inhomogeneous broadening by cooling the material all provide the means for optimization.

Each of these approaches presents challenges, but designing a device is far beyond the scope of this paper. However, the numbers suggest that making a device for the deterministic shaping of quantum light statistics is possible and our work defines the material requirements that will guide future experimental efforts.

References

1. Sangouard, N.; Simon, C.; Gisin, N.; Laurat, J.; Tualle-Brouri, R.; Grangier, P. Quantum repeaters with entangled coherent states. *J. Opt. Soc. Am. B* **2010**, *27*, A137. [CrossRef]
2. Brask, J.B.; Rigas, I.; Polzik, E.S.; Andersen, U.L.; Sørensen, A.S. Hybrid Long-Distance Entanglement Distribution Protocol. *Phys. Rev. Lett.* **2010**, *105*, 160501. [CrossRef]
3. Sakaguchi, A.; Konno, S.; Hanamura, F.; Asavanant, W.; Takase, K.; Ogawa, H.; Marek, P.; Filip, R.; Yoshikawa, J.I.; Huntington, E.; et al. Nonlinear feedforward enabling quantum computation. *Nat. Commun.* **2023**, *14*, 3817. [CrossRef]
4. Gottesman, D.; Kitaev, A.; Preskill, J. Encoding a qubit in an oscillator. *Phys. Rev. A* **2001**, *64*, 012310. [CrossRef]
5. Lidar, D.A.; Chuang, I.L.; Whaley, K.B. Decoherence-Free Subspaces for Quantum Computation. *Phys. Rev. Lett.* **1998**, *81*, 2594–2597. [CrossRef]
6. Polino, E.; Valeri, M.; Spagnolo, N.; Sciarrino, F. Photonic quantum metrology. *AVS Quantum Sci.* **2020**, *2*, 024703. [CrossRef]
7. Oszmaniec, M.; Augusiak, R.; Gogolin, C.; Kołodyński, J.; Acín, A.; Lewenstein, M. Random Bosonic States for Robust Quantum Metrology. *Phys. Rev. X* **2016**, *6*, 041044. [CrossRef]
8. Unternährer, M.; Bessire, B.; Gasparini, L.; Perenzoni, M.; Stefanov, A. Super-resolution quantum imaging at the Heisenberg limit. *Optica* **2018**, *5*, 1150–1154. [CrossRef]
9. Vasilev, G.S.; Ljunggren, D.; Kuhn, A. Single photons made-to-measure. *New J. Phys.* **2010**, *12*, 063024. [CrossRef]
10. Morin, O.; Körber, M.; Langenfeld, S.; Rempe, G. Deterministic Shaping and Reshaping of Single-Photon Temporal Wave Functions. *Phys. Rev. Lett.* **2019**, *123*, 133602. [CrossRef]
11. Huang, Y.; Qi, Z.; Yang, Y.; Li, Y.; Sun, Y.; Tang, Y.; Ni, F.; Li, L.; Zheng, Y.; Chen, X. Frequency-insensitive spatiotemporal shaping of single photon in multiuser quantum network. *npj Quantum Inf.* **2023**, *9*, 83. [CrossRef]
12. Yurke, B.; Stoler, D. Generating quantum mechanical superpositions of macroscopically distinguishable states via amplitude dispersion. *Phys. Rev. Lett.* **1986**, *57*, 13–16. [CrossRef]
13. Kolesnikow, X.C.; Bomantara, R.W.; Doherty, A.C.; Grimsmo, A.L. Gottesman-Kitaev-Preskill state preparation using periodic driving. *arXiv* **2023**, arXiv:2303.03541.
14. Asavanant, W.; Nakashima, K.; Shiozawa, Y.; Yoshikawa, J.I.; Furusawa, A. Generation of highly pure Schrödinger's cat states and real-time quadrature measurements via optical filtering. *Opt. Express* **2017**, *25*, 32227. [CrossRef]
15. Takase, K.; Yoshikawa, J.I.; Asavanant, W.; Endo, M.; Furusawa, A. Generation of optical Schrödinger cat states by generalized photon subtraction. *Phys. Rev. A* **2021**, *103*, 013710. [CrossRef]
16. Chen, Y.R.; Hsieh, H.Y.; Ning, J.; Wu, H.C.; Chen, H.L.; Shi, Z.H.; Yang, P.; Steuernagel, O.; Wu, C.M.; Lee, R.K. Generation of heralded optical 'Schroedinger cat' states by photon-addition. *arXiv* **2023**, arXiv:2306.13011.
17. Fukui, K.; Endo, M.; Asavanant, W.; Sakaguchi, A.; Yoshikawa, J.I.; Furusawa, A. Generating the Gottesman-Kitaev-Preskill qubit using a cross-Kerr interaction between squeezed light and Fock states in optics. *Phys. Rev. A* **2022**, *105*, 022436. [CrossRef]
18. Ben Hayun, A.; Reinhardt, O.; Nemirovsky, J.; Karnieli, A.; Rivera, N.; Kaminer, I. Shaping quantum photonic states using free electrons. *Sci. Adv.* **2021**, *7*, eabe4270. [CrossRef]
19. Beige, A.; Braun, D.; Tregenna, B.; Knight, P.L. Quantum computing using dissipation to remain in a decoherence-free subspace. *Phys. Rev. Lett.* **2000**, *85*, 1762–1765. [CrossRef]
20. González-Tudela, A.; Paulisch, V.; Chang, D.E.; Kimble, H.J.; Cirac, J.I. Deterministic Generation of Arbitrary Photonic States Assisted by Dissipation. *Phys. Rev. Lett.* **2015**, *115*, 163603. [CrossRef]
21. Verstraete, F.; Wolf, M.M.; Ignacio Cirac, J. Quantum computation and quantum-state engineering driven by dissipation. *Nat. Phys.* **2009**, *5*, 633–636. [CrossRef]
22. Vertchenko, L.; Nikitin, M.; Lavrinenko, A. Near-zero-index platform in photonics: Tutorial. *J. Opt. Soc. Am. B* **2023**, *40*, 1467. [CrossRef]
23. Camacho-Guardian, A.; Bastarrachea-Magnani, M.; Pohl, T.; Bruun, G.M. Strong photon interactions from weakly interacting particles. *Phys. Rev. B.* **2022**, *106*, L081302. [CrossRef]
24. Ribeiro, R.F.; Campos-Gonzalez-Angulo, J.A.; Giebink, N.C.; Xiong, W.; Yuen-Zhou, J. Enhanced optical nonlinearities under collective strong light-matter coupling. *Phys. Rev. A* **2021**, *103*, 063111. [CrossRef]
25. Hu, X.X.; Zhao, C.L.; Wang, Z.B.; Zhang, Y.L.; Zou, X.B.; Dong, C.H.; Tang, H.X.; Guo, G.C.; Zou, C.L. Cavity-enhanced optical controlling based on three-wave mixing in cavity-atom ensemble system. *Opt. Express* **2019**, *27*, 6660–6671. [CrossRef]
26. Imamoḡlu, A.; Schmidt, H.; Woods, G.; Deutsch, M. Strongly interacting photons in a nonlinear cavity. *Phys. Rev. Lett.* **1997**, *79*, 1467–1470. [CrossRef]
27. Porras, D.; Cirac, J.I. Collective generation of quantum states of light by entangled atoms. *Phys. Rev. A* **2008**, *78*, 053816. [CrossRef]

28. Perarnau-Llobet, M.; González-Tudela, A.; Cirac, J.I. Multimode Fock states with large photon number: Effective descriptions and applications in quantum metrology. *Quantum Sci. Technol.* **2020**, *5*, 025003. [CrossRef]
29. Law, C.K.; Eberly, J.H. Arbitrary Control of a Quantum Electromagnetic Field. *Phys. Rev. Lett.* **1996**, *76*, 1055–1058. [CrossRef]
30. Krastanov, S.; Albert, V.V.; Shen, C.; Zou, C.L.; Heeres, R.W.; Vlastakis, B.; Schoelkopf, R.J.; Jiang, L. Universal control of an oscillator with dispersive coupling to a qubit. *Phys. Rev. A* **2015**, *92*, 040303. [CrossRef]
31. Heeres, R.W.; Vlastakis, B.; Holland, E.; Krastanov, S.; Albert, V.V.; Frunzio, L.; Jiang, L.; Schoelkopf, R.J. Cavity State Manipulation Using Photon-Number Selective Phase Gates. *Phys. Rev. Lett.* **2015**, *115*, 137002. [CrossRef]
32. Dirac, P.A.M. Generalized Hamiltonian Dynamics. *Can. J. Math.* **1950**, *2*, 129–148. [CrossRef]
33. Born, M.; Infeld, L. On the quantization of the new field equations I. *Proc. R. Soc. Lond.* **1934**, *147*, 522–546.
34. Born, M.; Infeld, L. Foundations of the new field theory. *Proc. R. Soc. Lond. A Math. Phys. Sci.* **1934**, *144*, 425–451. [CrossRef]
35. Quesada, N.; Sipe, J.E. Why you should not use the electric field to quantize in nonlinear optics. *Opt. Lett.* **2017**, *42*, 3443. [CrossRef]
36. Bhat, N.A.R.; Sipe, J.E. Hamiltonian treatment of the electromagnetic field in dispersive and absorptive structured media. *Phys. Rev. A* **2006**, *73*, 063808. [CrossRef]
37. Raymer, M.G. Quantum theory of light in a dispersive structured linear dielectric: a macroscopic Hamiltonian tutorial treatment. *J. Mod. Opt.* **2020**, *67*, 196–212. [CrossRef]
38. Hillery, M.; Mlodinow, L.D. Quantization of electrodynamics in nonlinear dielectric media. *Phys. Rev. A Gen. Phys.* **1984**, *30*, 1860–1865. [CrossRef]
39. Drummond, P.D.; Hillery, M. *The Quantum Theory of Nonlinear Optics*; Cambridge University Press: Cambridge, UK, 2014.
40. Quesada, N.; Helt, L.; Menotti, M.; Liscidini, M.; Sipe, J. Beyond photon pairs: Nonlinear quantum photonics in the high-gain regime. *Adv. Opt. Photonics* **2022**, *14*, 291–403. [CrossRef]
41. Jackson, J.D. Electrodynamics, Classical. In *Digital Encyclopedia of Applied Physics*; Wiley-VCH Verlag GmbH & Co. KGaA: Weinheim, Germany, 2003.
42. Boyd, R.W. *Nonlinear Optics*, 4th ed.; Academic Press: San Diego, CA, USA, 2020.
43. Pasquazi, A.; Peccianti, M.; Razzari, L.; Moss, D.J.; Coen, S.; Erkintalo, M.; Chembo, Y.K.; Hansson, T.; Wabnitz, S.; Del'Haye, P.; et al. Micro-combs: A novel generation of optical sources. *Phys. Rep.* **2018**, *729*, 1–81. [CrossRef]
44. Alam, M.; Mandal, S.; Wahiddin, M.R. Squeezing, mixed mode squeezing, amplitude squared squeezing and principal squeezing in a non-degenerate parametric oscillator. *Optik* **2018**, *157*, 1035–1052. [CrossRef]
45. Kroeze, R.M.; Marsh, B.P.; Lin, K.Y.; Keeling, J.; Lev, B.L. High Cooperativity Using a Confocal-Cavity–QED Microscope. *PRX Quantum* **2023**, *4*, 020326. [CrossRef]
46. Bang, N.H.; Van Doai, L. Colossal Kerr nonlinearity without absorption in a five-level atomic medium. *Sci. Rep.* **2024**, *14*, 1554. [CrossRef]
47. Van Doai, L. The effect of giant Kerr nonlinearity on group velocity in a six-level inverted-Y atomic system. *Phys. Scr.* **2020**, *95*, 035104. [CrossRef]
48. Singer, K.D.; Kuzyk, M.G.; Holland, W.R.; Sohn, J.E.; Lalama, S.J.; Comizzoli, R.B.; Katz, H.E.; Schilling, M.L. Electro-optic phase modulation and optical second-harmonic generation in corona-poled polymer films. *Appl. Phys. Lett.* **1988**, *53*, 1800–1802. [CrossRef]
49. Cheng, X.; Hong, J.; Spring, A.M.; Yokoyama, S. Fabrication of a high-Q factor ring resonator using LSCVD deposited Si3N4 film. *Opt. Mater. Express* **2017**, *7*, 2182–2187. [CrossRef]
50. Ling, T.; Chen, S.L.; Guo, L.J. Fabrication and characterization of High Q polymer micro-ring resonator and its application as a sensitive ultrasonic detector. *Opt. Express* **2011**, *19*, 861–869. [CrossRef]
51. Dirk, C.W.; Kuzyk, M.G. Squarylium dye-doped polymer systems as quadratic electrooptic materials. *Chem. Mater.* **1990**, *2*, 4–6. [CrossRef]
52. Welker, D.J.; Tostenrude, J.; Garvey, D.W.; Canfield, B.K.; Kuzyk, M.G. Fabrication and characterization of single-mode electro-optic polymer optical fiber. *Opt. Lett.* **1998**, *23*, 1826–1828. [CrossRef]
53. Zheng, J.; Lu, M.; Wagner, C.J.; Cunningham, B.T.; Eden, J.G. Spectral characteristics of single and coupled microresonator lasers comprising a replica-molded Bragg grating and dye-doped polymer. *J. Opt. Soc. Am. B* **2012**, *29*, 209. [CrossRef]
54. Koeppen, C.; Yamada, S.; Jiang, G.; Garito, A.F.; Dalton, L.R. Rare-earth organic complexes for amplification in polymer optical fibers and waveguides. *J. Opt. Soc. Am. B* **1997**, *14*, 155. [CrossRef]

photonics

MDPI

Article

Decoding Anomalous Diffusion Using Higher-Order Spectral Analysis and Multiple Signal Classification

Miguel E. Iglesias Martínez [1], Òscar Garibo-i-Orts [1,2] and J. Alberto Conejero [1,*]

1 Instituto Universitario de Matemática Pura y Aplicada, Universitat Politècnica de València, 46022 València, Spain; miigmar@upv.es (M.E.I.M.); oscar.garibo@professor.universidadviu.com (Ò.G.-i.-O.)
2 GRID—Grupo de Investigación en Ciencia de Datos, Valencian International University—VIU, 46002 València, Spain
* Correspondence: aconejero@upv.es

Abstract: Anomalous diffusion is characterized by nonlinear growth in the mean square displacement of a trajectory. Recent advances using statistical methods and recurrent neural networks have made it possible to detect such phenomena, even in noisy conditions. In this work, we explore feature extraction through parametric and non-parametric spectral analysis methods to decode anomalously diffusing trajectories, achieving reduced computational costs compared with other approaches that require additional data or prior training. Specifically, we propose the use of higher-order statistics, such as the bispectrum, and a hybrid algorithm that combines kurtosis with a multiple-signal classification technique. Our results demonstrate that the type of trajectory can be identified based on amplitude and kurtosis values. The proposed methods deliver accurate results, even with short trajectories and in the presence of noise.

Keywords: anomalous diffusion; subdiffusion; superdiffusion; higher-order spectral analysis; multiple signal classification

1. Introduction

Anomalous diffusion appears in many physical, chemical, biological, and human phenomena. Examples include molecular encounters in reactions [1–3], cellular signaling [4,5], the foraging of animals [6–8] the spread of diseases [9], and trends in financial markets and climate records. In disordered photonic systems, we can find examples in which light propagation exhibits a subdiffusive behavior [10]. This is the case when photons remain trapped for longer periods due to multiple scattering events, but we can also find examples of superdiffusive behavior, where photons propagate more rapidly due to collective phenomena or localized excitations [11].

Deviations from Brownian diffusion, known as anomalous diffusion (AnDi), occur when the mean square displacement (MSD) growth with time has an exponent other than one. The classic example of AnDi is Brownian motion, which describes the motion of a microscopic particle in a fluid as a consequence of thermal forces [12,13].

Several methods have been used in the classification of anomalous diffusion trajectories. Lately, neural networks have been the most widely used method as they can efficiently characterize the anomalous diffusion by determining the exponent of a single short path exceeding the standard estimate based on MSD when the available data are limited, as is often the case in experiments [14–20].

Recent advancements in machine learning, specifically in deep learning and neural networks, have demonstrated considerable success in analyzing complex patterns and predicting events in noisy data environments [21,22]. Recurrent neural networks (RNNs), in particular, have shown promise for identifying temporal dependencies and sequential patterns, which are essential for characterizing anomalous diffusion processes [23,24]. The integration of statistical methods with RNNs has enabled the development of robust frameworks that can detect and predict the occurrence of anomalous diffusion, even in the presence of significant noise [25,26]. This approach aligns with recent works in the field of deep learning, where neural networks are increasingly applied to complex, non-linear systems to uncover underlying behaviors and trends [21,22]. By leveraging these techniques, we aim to improve accuracy for identifying diffusion types without the need for extensive preprocessing or noise filtering. This paper, therefore, contributes to the broader application of a combination of higher-order statistics and artificial intelligence in analyzing and predicting phenomena within stochastic and noisy environments, positioning our work within the current trend of using deep learning to solve complex analytical challenges.

Regarding other classification methods, several works use spectral analysis [27–29], where criteria based on the power spectrum are used for Fractional Brownian Motion (FBM) trajectories and other related work using Bayesian inference [30–32].

Some works have been conducted using moving displacement arithmetic based on first and second-order statistics, such as the mean value and the variance to analyze anomalous diffusion [33], as well as in specific applications, such as to characterize the random movement patterns of animals [34], and in the classification of particle trajectories in living cells [35].

Although a punctual classification of the studied sequence is obtained with the above methods, they are not immune to interference. In processes where noise is also present, good results are not obtained. In general, using estimators [36] that do not involve prior information about the data, such as prior training, can speed up the computational process as these methods only depend on a successful non-linear combination that identifies the analyzed processes.

Parametric and non-parametric high-resolution spectral analysis can effectively analyze anomalous diffusion as an estimator of the characteristics and descriptive patterns that random trajectories may present, differentiating the helpful information from noise within the anomalous diffusion series. The use of high-order spectral analysis provides certain advantages over classical processing methods like power spectrum because the statistics of random processes of orders greater than two are theoretically zero [37,38], as well as over high-resolution spectral analysis [39,40], which is based on the calculation of the spectrum using the vector subspaces of the eigenvalue matrix of the information to be analyzed [41,42].

In this work, we propose a non-parametric estimation approach using the signal bispectrum in combination with first-order statistics, such as the mean value, to characterize anomalous diffusion sequences. We also propose combining high-resolution spectral analysis with fourth-order statistics such as kurtosis to decode anomalous diffusion trajectories for a reduced data set. By now, our methods cannot yet be compared with the huge data sets used for training the most successful methods in the Andi Challenge. In this regard, we have focused on presenting an alternative perspective for studying anomalous diffusion using signal analysis tools and a more understandable interpretation of these models' behavior.

The trajectories used in the experiments presented in this work follow one of these five types of anomalous diffusion [43,44]: Annealed Transit Time Model (ATTM), Continuous

Time Random Walk (CTRW), Fractional Brownian Motion (FBM), Lévy Walk (LW), and Scaled Brownian Motion (SBM).

The paper is organized as follows. Section 2 provides a brief theoretical description of the fundamentals of anomalous diffusion involving each trajectory. Section 3 provides the theoretical foundations that support the analysis based on higher-order spectra, and Section 4 shows the experimental description of each trajectory. Section 5 describes the obtained results using the first proposed method based on the mean value of the bispectrum. Section 6 describes the results using the analysis of the proposed hybrid algorithm based on multiple signal classification and kurtosis. Finally, we discuss the results in Section 7 and show the conclusions in Section 8.

2. Fundamentals of Anomalous Diffusion

The process that gives rise to diffusion is stochastic or random, clearly illustrated by the paradigmatic example of Brownian motion. This motion consists of the irregular and unpredictable displacement of small particles suspended on the surface of a fluid and was explained in [45,46]. Unlike a normal diffusion process, in which the squared displacement is a linear function of time, anomalous diffusion is a diffusion process that involves a non-linear relationship between the mean square displacement (MSD) σ_r^2 and time t [47]. That is, the probability $P(x,t)$ of encountering a particle at time t and position x is described using a power law $\sigma_r^2 \sim Dt^{\alpha}$ where D is the diffusion coefficient. There are several models and derivations of anomalous diffusion trajectories. The following is a theoretical description of the models discussed in this work.

2.1. Annealed Transit Time Model (ATTM)

In this model, the initial point of the trajectory starts at $x = 0$ for $t = 0$, and we follow Langevin's equation [48] at several instances. The random motion of the particle diffuses with a coefficient D_1 for a time τ_1. Then, successive pairs are generated in the form $[D_n, \tau_n]$. The length of the interval τ_n varies with D_n, such that the Probabilistic Density Function (PDF) has a mean of $E[\tau_n, D_n] = D^{-\gamma}$. For $\gamma > 0$, we expect to find larger periods of low dispersal Brownian motion combined with shorter periods of high dispersal [49,50].

2.2. Continuous Time Random Walk (CTRW)

The CTRW model is a generalization of the classical random walk, where a particle can be described as a sequence of movements sampled from a Gaussian distribution with zero mean. Besides, there are also waiting times between displacements that are sampled from a power-law distribution $\Psi(t) = t^{-\sigma}$. This long waiting time can lead to anomalous diffusion [51].

2.3. Fractional Brownian Motion (FBM)

FBM is a generalization of classical Brownian motion in which the jumps are not independent. FBM is a continuous-time Gaussian process $B_H(t)$ defined on $[0, T]$, with $E[B_H(t)] = 0$ for all $t \geq 0$, and a covariance function given by [52]:

$$E[B_H(t)B_H(s)] = \frac{1}{2}\left[|t|^{2H} + |s|^{2H} - |t-s|^{2H}\right] \quad (1)$$

where H is the Hurst exponent, $0 \leq H \leq 1$. If $H = \frac{1}{2}$, the process reduces to standard Brownian motion. For $H > \frac{1}{2}$, increments are positively correlated (superdiffusion), while for $H < \frac{1}{2}$, increments are negatively correlated (subdifussion), see [53,54].

2.4. Lévy Walk (LW)

Lévy walks (LWs) are a particular case of CTRW where dispersal lengths correlate with waiting times. The PDF describing random time intervals between consecutive jumps is a power-law function $\psi(t) = t^{-\sigma-1}$ (as in the CTRW). Besides, the probability of a Δx dispersal length is given by $\Psi(\Delta x, t) = \frac{1}{2}\delta(|\Delta x) - vt)\psi(t)$ [44,55,56].

2.5. Scaled Brownian Motion: SBM

This model is derived from the Langevin equation. However, here, the diffusivity changes with time, even in the Gaussian noise case [57–59].

3. Higher-Order Spectral Analysis: Bispectrum

After briefly showing the theoretical background of the anomalous diffusion trajectories, let us briefly introduce the higher-order spectral analysis. The *bispectrum* is defined as the Fourier Transform of the third-order cumulant, which is given by (2):

$$B(f_1, f_2) = \sum_{\tau_1=-N+1}^{N-1} \sum_{\tau_2=-N+1}^{N-1} C_{3x}(\tau_1, \tau_2) \cdot e^{-2\pi f_1 \tau_1} \cdot e^{-2\pi f_2 \tau_2} = \frac{1}{N^2} X(f_1, f_2) \cdot X(f_1) \cdot X(f_2) \tag{2}$$

where $X(f)$ is the Fourier Transform of the sequence $\{x(n)\}_{n=0}^{N-1}$, C_{3x} is the third-order cumulant of the input sequence, and sets f_1 and f_2 are integers from the matrix of bispectral frequencies obtained from the bispectrum calculation.

In the spectral domain, the bispectrum can also be computed through convolution in the spectral domains, as shown in (3). The convolution theorem for bispectrum analysis extends the principles of the Fourier transform to higher-order spectral analysis. Specifically, in the context of the bispectrum, which is a second-order frequency domain representation involving two frequency components, the theorem states that the bispectrum of the convolution of two signals in the time domain is equal to the product of their bispectra in the frequency domain.

$$B(f_1, f_2) = X(f_1 + f_2) \cdot X(f_1) \cdot X(f_2) \tag{3}$$

We used the absolute value of the bispectrum signal as described in (4), which will be computed from (2), where N is the number of rows of the square matrix $N \times N$ obtained from the bispectrum. The obtained result is a complex $N \times N$ matrix that contains the frequency values of the amplitude bispectrum matrix of the data set.

$$B_a(f_1, f_2) = |B(f_1, f_2)|_i \forall i = 1, \ldots, N \tag{4}$$

Once Equation (4) has been obtained and in order to characterize each trajectory, in this work, we use an indicator based on the mean value of the bispectrum (Equation (5)), which is defined as follows:

$$M(B_a(f_1, f_2)) = \frac{1}{N} \sum_{i=1}^{N} B_a(f_1, f_2). \tag{5}$$

3.1. Hybrid Algorithm: Multiple Signal Classification and Kurtosis

The proposed method is divided into two phases to ensure accurate frequency estimation and analysis. First, we obtain the pseudospectrum of the signal through the MUSIC algorithm, and then we calculate the bispectrum to capture non-linear interactions and phase coupling between different frequency components. This two-phase approach ensures that the method combines high-resolution frequency estimation (from the MUSIC algorithm) with a deeper analysis of the frequency interactions (from bispectral or higher-order techniques), leading to more robust and accurate signal analysis.

3.1.1. First Phase: Pseudospectrum Estimation Using the MUSIC Algorithm

In this phase, we apply the multiple signal classification (MUSIC) algorithm to estimate the pseudospectrum of a given trajectory using Schmidt's eigenspace analysis method [60]. The MUSIC algorithm performs an eigenspace analysis of the data's correlation matrix, separating the signal subspace from the noise subspace. The pseudospectrum $Pm(f)$, computed from the covariance of the eigenvalues of the data for each trajectory, is expressed as follows:

$$Pm(f) = \frac{1}{\sum_{k=1}^{p}\left(1 - \left|v_k^H e(f)\right|^2\right)} \tag{6}$$

where v_k^H is the Hermitian transpose (or conjugate transpose) of the eigenvector v_k, and $e(f)$ is the steering vector corresponding to frequency f. Specifically:

- v_k^H denotes the Hermitian transpose of the eigenvector v_k. This operation involves taking the transpose of the matrix and applying the complex conjugate to each element. It is used in signal processing to perform orthogonal analysis of the signal components.
- $e(f)$ represents the steering vector at frequency f, which describes the response of the system to a signal at that particular frequency. In the MUSIC algorithm, the steering vector helps evaluate how well the signal aligns with the subspace spanned by the eigenvectors.

The term $\left|v_k^H e(f)\right|^2$ represents the squared magnitude of the projection of the steering vector $e(f)$ onto the eigenvector v_k and is used to determine the contribution of the signal in the corresponding eigenspace.

3.1.2. Second Phase: Bispectral Analysis

This phase enhances the frequency characterization by detecting phase relationships not captured by second-order methods like the MUSIC algorithm alone. We compute the bispectrum of the signal and then we evaluate the kurtosis to identify the trajectory patterns according to the following relation:

$$Pk = \frac{E\left(Pm(f) - v^4\right)}{\sigma^4} \tag{7}$$

where v and σ are the mean and the standard deviation of $Pm(f)$, and E represents the expected value. This provides the characterization of each trajectory.

4. Experimental Description

We study five anomalous diffusion trajectory models: ATTM, CTRW, FBM, LW, and SBM. We have generated several trajectories using the same procedure as in [43,44]. Each trajectory has a standard deviation σ, with $0.5 \leq \sigma \leq 1.0$; a variable exponent α, with $0.1 \leq \alpha \leq 0.9$; and a length of 2^N, where $5 \leq N \leq 10$.

The sequence length is chosen to be a power of two because it is advantageous for the computation of the bispectrum. This choice optimizes the computational efficiency, particularly when applying Fast Fourier Transforms (FFTs), which operate most efficiently when the length of the sequence is a power of two. As the bispectrum involves evaluating phase relationships between multiple frequencies, a sequence length that is a power of two ensures compatibility with FFT algorithms and enhances the precision of the frequency domain analysis. Additionally, using window lengths that are multiples of two for the bispectrum computation ensures uniform segmentation and avoids data loss at the edges of the windows. Figure 1 is an example of the bispectrum contour of an anomalous diffusion trajectory, in this case, for an ATTM sequence. As can be seen, a bispectral pattern can be unique for each sequence and can also be observed when identifying the characteristics in the trajectory analyzed.

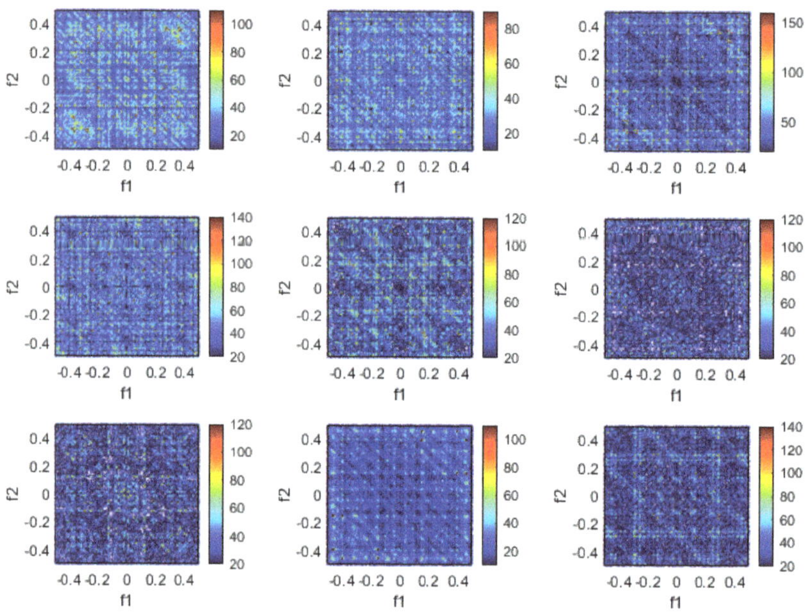

Figure 1. An example of the bispectrum of an ATTM trajectory. f_1 and f_2 contain the frequencies of the Fourier transform of the third-order cumulant of the data. The axes represent the bispectral content of each frequency that generates the anomalous diffusion sequence.

The bispectrum results in a matrix containing the frequencies of the Fourier transform of the third-order cumulant of the data. The axes represent the bispectral content of each frequency in the harmonics that generate the anomalous diffusion sequence.

We briefly describe the data set of trajectories:

1. For the ATTM and CTRW models, we considered 108 trajectories that represent the set of combinations of length, 2^N with $5 \leq N \leq 10$, $\sigma = 0.5, 1$, and $\alpha = 0.1, 0.2, \ldots, 0.9$.
2. For the LW model, we considered 108 trajectories that represent the set of combinations of length, 2^N with $5 \leq N \leq 10$, $\sigma = 0.5, 1$, and $\alpha = 1.1, 1.2, \ldots, 1.9$.
3. For the FBM and SBM models, we considered 210 trajectories that represent the set of combinations of length, 2^N with $5 \leq N \leq 10$, $\sigma = 0.5, 1$, and $\alpha = 0.05, \ldots, 1.95$.

5. Results Based on the Mean Value of Bispectrum

To obtain a classification method based on the previous results, we propose using methods based on first-order statistics combined with bispectrum analysis, see Equation (5). This classification type aims to characterize each trajectory so that each of the sequences studied can be quantitatively differentiated. Initially, the bispectrum of each sequence is taken as a basis for a specific trajectory, as aforementioned in Section 4.

Figure 2 shows the results obtained using the proposed indicator based on the bispectrum mean. For ATTM and CTRW trajectories, it can be seen that there are significant differences between the sequences for $\alpha \geq 0.2$, with the mean value increasing as long as the trajectory length does. There are no marked differences between the sequences of dimensions 32 and 64, respectively, whose mean value does not oscillate and remains almost unchanged over time.

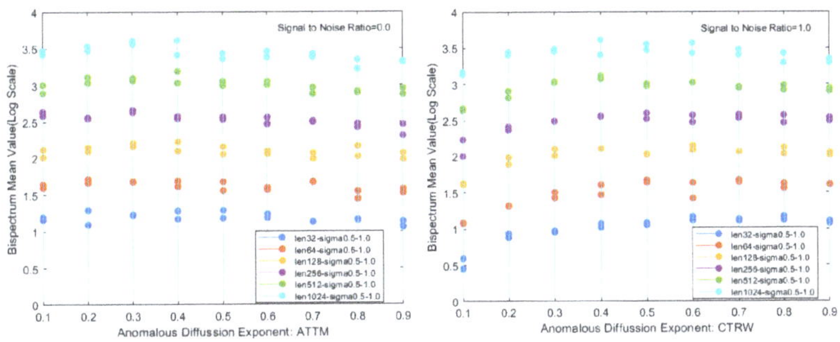

(**a**) (**b**)

Figure 2. Mean value of the bispectrum for the (**a**) ATTM and (**b**) CTRW trajectories.

It can also be seen in Figure 3 that the proposed algorithm can characterize the FBM trajectories. In this case, the result of the mean values of the bispectrum provides a representative curve as a function of the length of each sequence. In Figure 3, it can be noticed, analogous to Figure 2, that the average bispectrum value curve increases in amplitude as the sequence becomes longer, but with a more clear separation according to the sequence length. In the case of FBM, we see clear increasing trends depending on the length and the anomalous diffusion exponent. In contrast, LW shows a more homogeneous behavior for each sequence length, but with a wider separation between different lengths. Figure 4 shows the results obtained for SBM, which are closer to the ATTM model in the subdiffusion regime. For SBM, the bispectrum's mean values show a decreasing trend as the sequence length shortens, demonstrating a level of regularity not seen in LW trajectories.

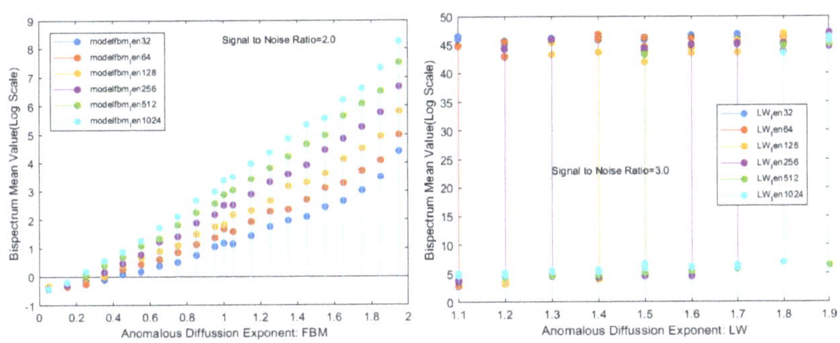

(**a**) (**b**)

Figure 3. Mean value of the bispectrum for: (**a**) FBM and (**b**) LW trajectories.

Finally, in Figure 5, we show the overall behavior of the bispectral amplitude of the five analyzed models. It can be seen that the behavior of ATTM, CTRW, and SBM follows a similar trend, especially for the lowest lengths of 32, 64, and 12, increasing the amplitude as long as the trajectory length increases. This also happens for the FBM trajectory; the trend with the amplitude–length variation is also increasing, as we have already indicated, but mainly for long trajectories. Finally, the behavior of the LW trajectory is quite erratic and non-stationary, where no specific pattern can be distinguished, in contrast with the other cases.

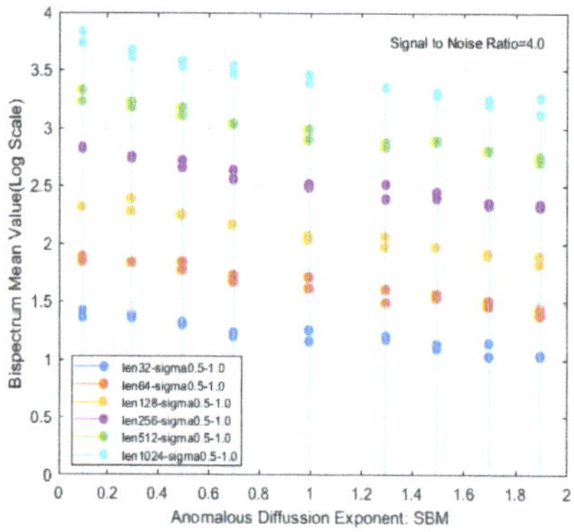

Figure 4. Mean value of the bispectrum for SBM trajectories.

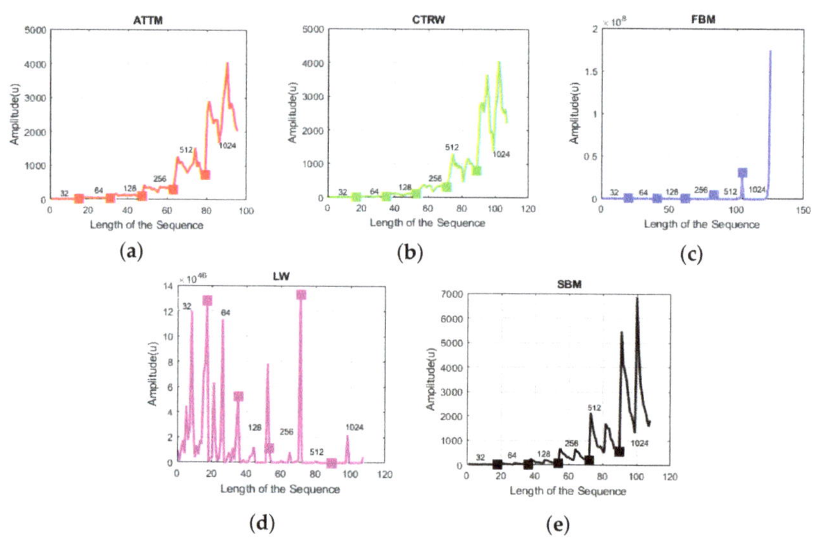

Figure 5. Overall behavior of the bispectral amplitude according to sequence length: (**a**) ATTM, (**b**) CTRW, (**c**) FBM, (**d**) LW, and (**e**) SBM.

Histogram Analysis

To decode each of the previously analyzed models in terms of the bispectrum and to evaluate how the amplitude is probabilistically distributed, we study the bispectrum amplitude probability histogram of each model to characterize each model. The obtained results are shown in Figure 6. We can see that all models have an asymmetric bispectrum probability distribution, and the highest amplitude probability is found at the lowest amplitudes. We also noticed that all models have different amplitude scale values as well as probability frequencies, except for the ATTM and CTRW trajectories, which have the same amplitude scale but slight differences.

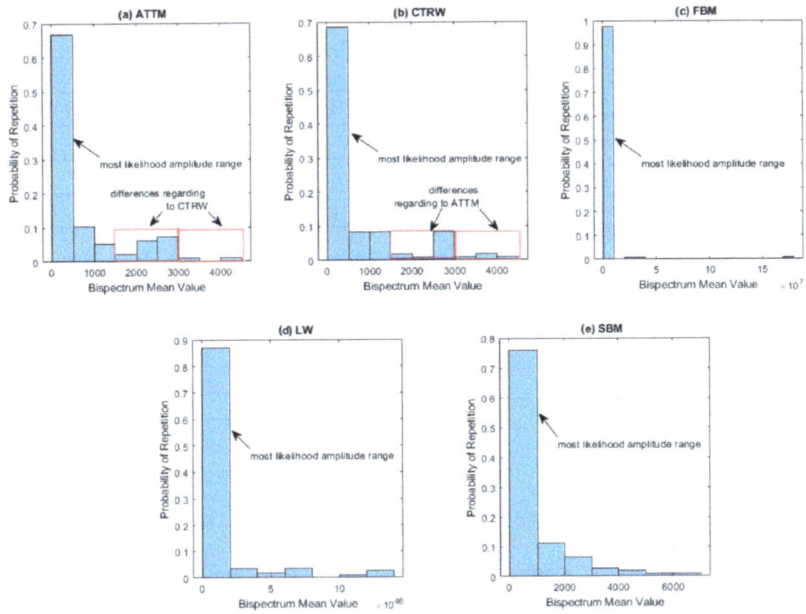

Figure 6. Probability histogram distribution for each trajectory: (**a**) ATTM trajectory, (**b**) CTRW trajectory, (**c**) FBM, (**d**) LW, and (**e**) SBM.

To show the quantitative differences in the probability histograms of each type of trajectory, a correlation analysis is performed using Pearson's coefficient. The obtained results are shown in Table 1, where it is possible to appreciate the coherence of the correlation values with the similarities described above for the ATTM and CTRW sequences, respectively. Likewise, significant correlation values are also observed in the FBM–SBM and LW–SBM pairs, respectively, although the differences in the amplitude scale between the FBM, LW, and SBM trajectories are remarkable. This makes it easy to differentiate between the type of trajectory.

Table 1. Correlation matrix of the five trajectories used in the experiments.

	ATTM	CTRW	FBM	LW	SBM
ATTM	1	0	0	0	0
CTRW	0.7499	1	0	0	0
FBM	0.1016	0.1958	1	0	0
LW	−0.00616	−0.2267	−0.2390	1	0
SBM	−0.2261	0.1144	0.8255	0.7841	1

6. Results Using the Hybrid Algorithm: Multiple Signal Classification and Kurtosis

This section shows the results using a high-resolution spectral analysis method based on the covariance data matrix eigenvalues decomposition, in combination with a fourth-order statistic such as kurtosis, to achieve the identification of each model. The method reduces the computational cost and enhances noise immunity by combining two non-parametric estimators where the noise is zero. The proposed method is evaluated for each of the analyzed sequences. The results are shown in Figure 7.

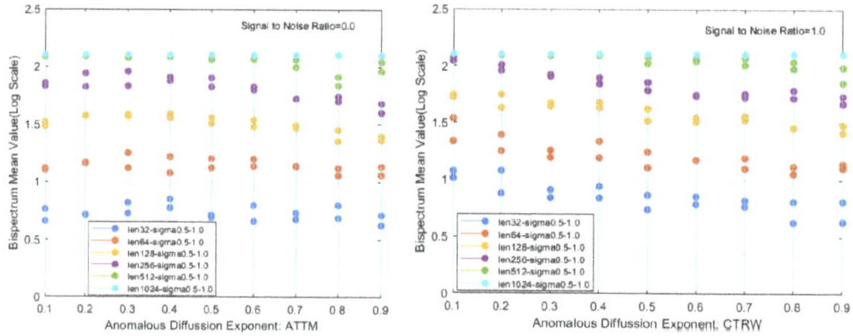

(**a**) (**b**)

Figure 7. Results using the MUSIC–kurtosis hybrid method: (**a**) ATTM trajectory (**b**) CTRW trajectory.

In Figures 7–9, it can be seen that the same pattern is obtained using the mean value of the bispectrum of each sequence. In this case, the kurtosis value of each trajectory decreases as the length of the sequence decreases. On the one hand, in all cases, it is possible to visualize with greater clarity the pattern of the short-length sequences 32–64 and 128, respectively. Likewise, the pattern of the LW trajectory is better defined. On the other hand, Figure 10 shows the results of the behavior of the kurtosis obtained from Equation (7) for each of the analyzed trajectories.

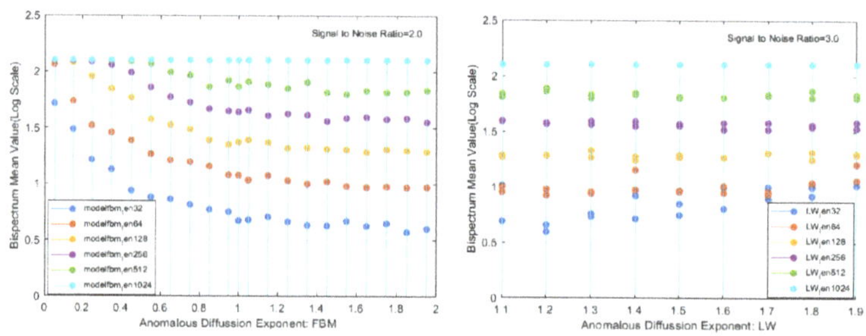

(**a**) (**b**)

Figure 8. Results using the MUSIC–kurtosis hybrid method: (**a**) FBM trajectory (**b**) LW trajectory.

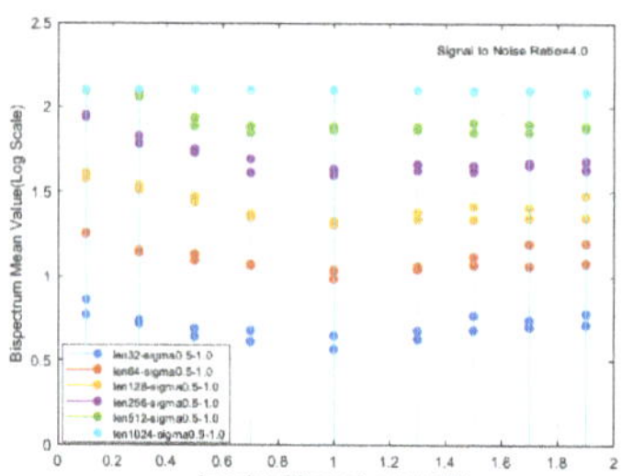

Figure 9. Results using the MUSIC–kurtosis hybrid method for: SBM trajectory.

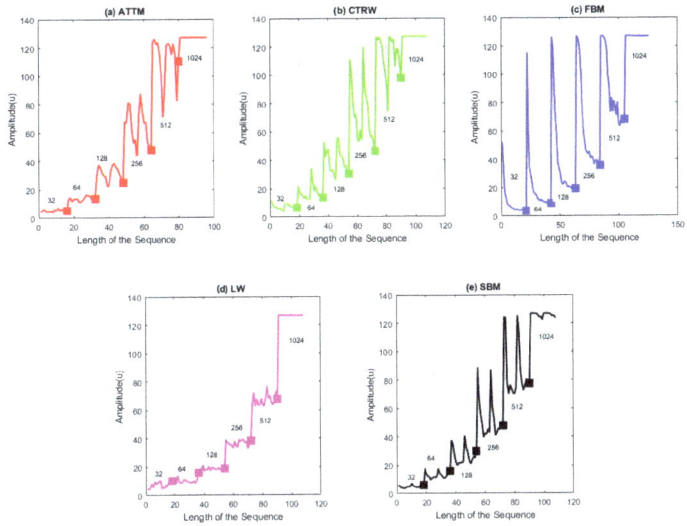

Figure 10. Overall behavior of the MUSIC–kurtosis according to sequence length: (**a**) ATTM, (**b**) CTRW, (**c**) FBM, (**d**) LW, and (**e**) SBM.

An upward-trending behavior in amplitude can be noticed as the sequence length increases. It can also be observed in all cases that the kurtosis of the sequence of length 1024 remains constant along the trajectory, with zero slope, except in the case of the SBM model. This feature identifies this model's respect for others.

The histogram for each of the trajectories was also obtained. Figure 11 shows the obtained results, and in Table 2 we show the correlation matrix for each model, where it can be seen that there are no notable differences in each of the histograms, given that all the correlation values are greater than 0.5.

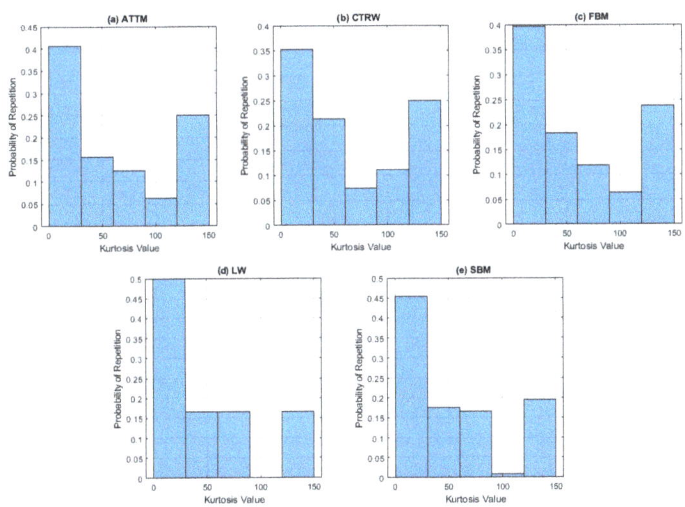

Figure 11. Probability histogram distribution for each trajectory using the MUSIC–kurtosis algorithm: (**a**) ATTM trajectory, (**b**) CTRW trajectory, (**c**) FBM (**d**) LW, and (**e**) SBM.

Table 2. Correlation matrix of the five trajectories used in the experiments: MUSIC–kurtosis algorithm.

	ATTM	CTRW	FBM	LW	SBM
ATTM	1	0	0	0	0
CTRW	0.8860	1	0	0	0
FBM	0.5676	0.6199	1	0	0
LW	0.5327	0.8880	0.8432	1	0
SBM	0.9426	0.5144	0.9553	0.8566	1

Likewise, there is no discernible pattern of asymmetry in each distribution that could differentiate each trajectory.

7. Discussion

Having obtained the experimental results of the two analysis proposals to identify the anomalous diffusion trajectories, it can be said that:

- With the use of both the bispectrum-based method and the method using multiple signal classification and kurtosis, it is possible to identify each of the anomalous diffusion trajectories analyzed.
- The method based on multiple signal classification and kurtosis gives better results than the method based on the mean value of the bispectrum in the task of differentiating the type of trajectory according to its sequence length.
- With the bispectrum-based method, better results are obtained to identify the type of trajectory based on the probability histogram or the distribution of the amplitude values. As a quantitative indicator of this, the obtained results from the correlation matrix in both cases are shown.
- The average amplitude values of the bispectrum follow an asymmetric probability distribution, which is common for all the trajectories analyzed in the experiments.

8. Conclusions

This work evaluates the bispectrum as a tool for characterizing anomalous diffusion trajectories. Five types of anomalous diffusion trajectories were analyzed, with varying numbers of sequences and exponents. The results demonstrate that, from a graphical standpoint, distinct features can be extracted, which may serve as classification patterns to identify each of the studied trajectories.

In all cases, except for the LW trajectory, the proposed algorithm—based on the mean values of the bispectrum—shows a decrease in performance as the length of the trajectory decreases. For sequences of lengths 256, 512, and 1024, the classification indicators are quantitatively more prominent compared with those of shorter sequences, such as lengths 128, 64, and 32.

Additionally, an analysis based on the histogram of the mean bispectrum value for each sequence was performed to identify each trajectory in terms of the probability distribution of bispectrum amplitudes. The results reveal that the histogram for each trajectory is distinguishable from the others, except the ATTM and CTRW trajectories, which exhibit similarities in amplitude scale, although with appreciable differences.

In the second part of the study, a high-resolution spectral analysis method, MUSIC (Multiple Signal Classification), combined with the kurtosis of the resulting spectrum, was applied. This method improves the identification of shorter sequences (lengths 32, 64, and 128) compared to the bispectrum-based method, but does not improve the analysis of probability distributions using histograms.

This work represents a first step towards characterizing anomalous diffusion through parametric and non-parametric spectral feature estimation based on vector subspaces. We describe the mean value and histogram description of the signals depending on the diffusion type and classification method. It is not our goal, which facilitates the understanding of the different movements from the perspective of signal theory.

Author Contributions: Conceptualization, M.E.I.M. and J.A.C.; Software, M.E.I.M. and Ò.G.-i.-O.; Investigation, M.E.I.M., Ò.G.-i.-O. and J.A.C.; Writing—original draft, M.E.I.M. and J.A.C.; Writing—review & editing, M.E.I.M., Ò.G.-i.-O. and J.A.C.; Visualization, M.E.I.M. All authors have read and agreed to the published version of the manuscript.

Funding: M.E.I.M. was funded by the postdoctoral research scholarship "Ayudas para la recualificación del sistema universitario español 2021–2023. Modalidad: Margarita Salas", UPV, Ministerio de Universidades, Plan de Recuperación, Transformación y Resiliencia, Spain, funded by the European Union-Next Generation EU. Ò.G.-i.-O. and J.A.C. are supported by by European Union—NextGenerationEU, ANDHI project CPP2021-008994.

Institutional Review Board Statement: Not applicable.

Informed Consent Statement: Not applicable.

Data Availability Statement: Dataset available on request from the authors.

Acknowledgments: The authors appreciate the facilities provided by the Galician Supercomputing Center (CESGA: https://www.cesga.es/ (accessed from May to June 2024) for the development of modeling calculations.

Conflicts of Interest: The authors declare no conflict of interest.

References

1. Guigas, G.; Weiss, M. Sampling the cell with anomalous diffusion—The discovery of slowness. *Biophys. J.* **2008**, *94*, 90–94. [CrossRef]
2. Klafter, J.; Sokolov, I.M. Anomalous diffusion spreads its wings. *Phys. World* **2005**, *18*, 29. [CrossRef]
3. Shen, Z.; Plouraboué, F.; Lintuvuori, J.S.; Zhang, H.; Abbasi, M.; Misbah, C. Anomalous diffusion of deformable particles in a honeycomb network. *Phys. Rev. Lett.* **2023**, *130*, 014001. [CrossRef] [PubMed]
4. Qiang, Y. Numerical Simulation of Anomalous Diffusion with Application to Medical Imaging. Ph.D. Thesis, Queensland University of Technology, Brisbane, Australia, 2013.
5. Manzo, C.; Garcia-Parajo, M.F. A review of progress in single particle tracking: From methods to biophysical insights. *Rep. Prog. Phys.* **2015**, *78*, 124601.
[CrossRef]
6. Viswanathan, G.M.; Afanasyev, V.; Buldyrev, S.V.; Murphy, E.J.; Prince, P.A.; Stanley, H.E. Lévy flight search patterns of wandering albatrosses. *Nature* **1996**, *381*, 413–415. [CrossRef]
7. Viswanathan, G.M.; Buldyrev, S.V.; Havlin, S.; da Luz, M.G.E.; Raposo, E.P.; Stanley, H.E. Optimizing the success of random searches. *Nature* **1999**, *401*, 911–914. [CrossRef]
8. Sims, D.W.; Southall, E.J.; Humphries, N.E.; Hays, G.C.; Bradshaw, C.J.; Pitchford, J.W.; James, A.; Ahmed, M.Z.; Brierley, A.S.; Hindell, M.A.; et al. Scaling laws of marine predator search behaviour. *Nature* **2008**, *451*, 1098–1102. [CrossRef]
9. Goiriz, L.; Ruiz, R.; Garibo-i Orts, Ò.; Conejero, J.A.; Rodrigo, G. A variant-dependent molecular clock with anomalous diffusion models SARS-CoV-2 evolution in humans. *Proc. Natl. Acad. Sci. USA* **2023**, *120*, e2303578120. [CrossRef]
10. Longhi, S. Subdiffusive dynamics in photonic random walks probed with classical light. *arXiv* **2024**, arXiv:2410.02287. [CrossRef]
11. Alves-Pereira, A.R.; Nunes-Pereira, E.J.; Martinho, J.M.G.; Berberan-Santos, M.N. Photonic superdiffusive motion in resonance line radiation trapping partial frequency redistribution effects. *J. Chem. Phys.* **2007**, *126*, 154505. [CrossRef]
12. Bouchaud, J.-P.; Georges, A. Anomalous diffusion in disordered media: Statistical mechanisms, models and physical applications. *Phys. Rep.-Rev. Sec. Phys. Lett.* **1990**, *195*, 127–293. [CrossRef]
13. Sokolov, I.M.; Klafter, J. From diffusion to anomalous diffusion: A century after Einstein's Brownian motion. *Chaos* **2005**, *15*, 26103. [CrossRef]
14. dos Santos, M.A.F. Analytic approaches of the anomalous diffusion: A review. *Chaos Solitons Fractals* **2019**, *124*, 86–96. [CrossRef]

15. Bo, S.; Schmidt, F.; Eichhorn, R.; Volpe, G. Measurement of anomalous diffusion using recurrent neural networks. *Phys. Rev. E* **2019**, *100*, 010102. [CrossRef]
16. Firbas, N.; Garibo-i Orts, Ò.; Garcia-March, M.A.; Conejero, J.A. Characterization of anomalous diffusion through convolutional transformers. *J. Phys. A-Math. Theor.* **2023**, *56*, 014001. [CrossRef]
17. Garibo-i Orts, Ò.; Baeza-Bosca, A.; Garcia-March, M.A.; Conejero, J.A. Efficient recurrent neural network methods for anomalously diffusing single particle short and noisy trajectories. *J. Phys. A-Math. Theor.* **2021**, *54*, 504002. [CrossRef]
18. Garibo-i Orts, Ò.; Firbas, N.; Sebastiá, L.; Conejero, J.A. Gramian angular fields for leveraging pretrained computer vision models with anomalous diffusion trajectories. *Phy. Rev. E* **2023**, *107*, 034138. [CrossRef]
19. Gentili, A.; Volpe, G. Characterization of anomalous diffusion classical statistics powered by deep learning (CONDOR). *J. Phys. A-Math. Theor.* **2021**, *54*, 314003. [CrossRef]
20. Muñoz-Gil, G.; Garcia-March, M.A.; Manzo, C.; Martín-Guerrero, J.D.; Lewenstein, M. Single trajectory characterization via machine learning. *New J. Phys.* **2020**, *22*, 013010. [CrossRef]
21. LeCun, Y.; Bengio, Y.; Hinton, G. Deep learning. *Nature* **2015**, *521*, 436–444. [CrossRef]
22. Goodfellow, I.; Bengio, Y.; Courville, A. *Deep Learning*; MIT Press: Cambridge, MA, USA, 2016.
23. Hochreiter, S.; Schmidhuber, J. Long short-term memory. *Neural Comput.* **1997**, *9*, 1735–1780. [CrossRef] [PubMed]
24. Cho, K.; van Merrienboer, B.; Gulcehre, C.; Bahdanau, D.; Bougares, F.; Schwenk, H.; Bengio, Y. Learning phrase representations using RNN encoder-decoder for statistical machine translation. In Proceedings of the 2014 Conference on Empirical Methods in Natural Language Processing (EMNLP), Doha, Qatar, 25–29 October 2014; pp. 1724–1734.
25. Rumelhart, D.E.; Hinton, G.E.; Williams, R.J. Learning representations by back-propagating errors. *Nature* **1986**, *323*, 533–536. [CrossRef]
26. Graves, A.; Mohamed, A.; Hinton, G. Speech recognition with deep recurrent neural networks. In Proceedings of the 2013 IEEE International Conference on Acoustics, Speech and Signal Processing, Vancouver, BC, Canada, 26–31 May 2013; pp. 6645–6649.
27. Krapf, D.; Marinari, E.; Metzler, R.; Oshanin, G.; Xu, X.; Squarcini, A. Power spectral density of a single Brownian trajectory: What one can and cannot learn from it. *New J. Phys.* **2018**, *20*, 023029. [CrossRef]
28. Metzler, R. Brownian motion and beyond: First-passage, power spectrum, non-Gaussianity, and anomalous diffusion. *J. Stat. Mech.-Theory Exp.* **2019**, *2019*, 114003. [CrossRef]
29. Sposini, V.; Metzler, R.; Oshanin, G. Single-trajectory spectral analysis of scaled Brownian motion. *New J. Phys.* **2019**, *21*, 073043. [CrossRef]
30. Hinsen, K.; Kneller, G.R. Communication: A multiscale Bayesian inference approach to analyzing subdiffusion in particle trajectories. *J. Chem. Phys.* **2016**, *145*, 151101. [CrossRef]
31. Thapa, S.; Lomholt, M.A.; Krog, J.; Cherstvy, A.G.; Metzler, R. Bayesian analysis of single-particle tracking data using the nested-sampling algorithm: Maximum-likelihood model selection applied to stochastic-diffusivity data. *Phys. Chem. Chem. Phys.* **2018**, *20*, 29018–29037. [CrossRef]
32. Salzenstein, F.; Boudraa, A. Multidimensional directional derivatives and AM-FM dual-band image demodulation by higher order Teager-Kaiser operators. *Digit. Signal Process* **2022**, *129*, 103641. [CrossRef]
33. Maraj, K.; Szarek, D.; Sikora, G.; Balcerek, M.; Wyłomańska, A.; Jabłoński, I. Measurement instrumentation and selected signal processing techniques for anomalous diffusion analysis. *Meas. Sens.* **2020**, *7*, 100017. [CrossRef]
34. Vilk, O.; Aghion, E.; Nathan, R.; Toledo, S.; Metzler, R.; Assaf, M. Classification of anomalous diffusion in animal movement data using power spectral analysis. *J. Phys. A-Math. Theor.* **2022**, *55*, 334004. [CrossRef]
35. Janczura, J.; Kowalek, P.; Loch-Olszewska, H.; Szwabiński, J.; Weron, A. Classification of particle trajectories in living cells: Machine learning versus statistical testing hypothesis for fractional anomalous diffusion. *Phys. Rev. E* **2020**, *102*, 032402. [CrossRef]
36. Bellman, R. A markovian decision process. *J. Math. Mech.* **1957**, *6*, 679–684. [CrossRef]
37. Iglesias-Martínez, M.E.; Antonino-Daviu, J.A.; Fernández de Córdoba, P.; Conejero, J.A. Higher-order spectral analysis of stray flux signals for faults detection in induction motors. *Appl. Math. Nonlinear Sci.* **2020**, *5*, 1–14. [CrossRef]
38. Iglesias-Martínez, M.E.; Garcia-March, M.A.; Milián, C.; Fernández de Córdoba, P. *Algorithms for Noise Reduction in Signals*; IOP Publishing: Bristol, UK, 2022; pp. 2053–2563.
39. Yang, X.; Wang, K.; Zhou, P.; Xu, L.; Liu, J.; Sun, P.; Su, Z. Ameliorated-multiple signal classification (Am-MUSIC) for damage imaging using a sparse sensor network. *Mech. Syst. Signal Proc.* **2022**, *163*, 108154. [CrossRef]
40. Yang, X.; Wang, K.; Zhou, P.; Xu, L.; Su, Z. Imaging damage in plate waveguides using frequency-domain multiple signal classification (F-MUSIC). *Ultrasonics* **2022**, *119*, 106607. [CrossRef]
41. Dokhanchi, S.H.; Mysore, B.S.; Mishra, K.V.; Ottersten, B. A mmWave automotive joint radar-communications system. *IEEE Trans. Aerosp. Electron. Syst.* **2019**, *55*, 1241–1260. [CrossRef]

42. Joanes, D.N.; Gill, C.A. Comparing measures of sample skewness and kurtosis. *J. R. Stat. Soc. Ser. D Stat.* **1998**, *47*, 183–189. [CrossRef]
43. Muñoz-Gil, G.; Volpe, M.A.; García-March, G.; Metzler, R.; Lewenstein, M.; Manzo, C. The anomalous diffusion challenge: Single trajectory characterisation as a competition. In *Emerging Topics in Artificial Intelligence 2020*; SPIE: Bellingham, WA, USA, 2020; Volume 11469, pp. 42–51.
44. Muñoz-Gil, G.; Volpe, G.; Garcia-March, M.A.; Aghion, E.; Argun, A.; Hong, C.B.; Bland, T.; Bo, S.; Conejero, J.A.; Firbas, N.; et al. Objective comparison of methods to decode anomalous diffusion. *Nat. Commun.* **2021**, *12*, 6253. [CrossRef] [PubMed]
45. Islam, M.A. Einstein–Smoluchowski diffusion equation: A discussion. *Phys. Scr.* **2004**, *70*, 120. [CrossRef]
46. Islam, M.A. Fickian diffusion equation—An unsolved problem. *Phys. Scr.* **2004**, *70*, 114. [CrossRef]
47. Sanvee, B.A.; Lohmann, R.; Horbach, J. Normal and anomalous diffusion in the disordered wind-tree model. *Phys. Rev. E* **2022**, *106*, 024104. [CrossRef]
48. Glatt-Holtz, N.E.; Herzog, D.P.; McKinley, S.A.; Nguyen, H.D. The generalized Langevin equation with power-law memory in a nonlinear potential well. *Nonlinearity* **2020**, *33*, 2820. [CrossRef]
49. Akimoto, T.; Yamamoto, E. Distributional behavior of diffusion coefficients obtained by single trajectories in annealed transit time model. *J. Stat. Mech.-Theory Exp.* **2016**, *2016*, 123201. [CrossRef]
50. Massignan, P.; Manzo, C.; Torreno-Pina, J.A.; García-Parajo, M.F.; Lewenstein, M.; Lapeyre, G.J., Jr. Nonergodic subdiffusion from Brownian motion in an inhomogeneous medium. *Phys. Rev. Lett.* **2014**, *112*, 150603. [CrossRef]
51. Scher, H.; Montroll, E.W. Anomalous transit-time dispersion in amorphous solids. *Phys. Rev. B* **1975**, *12*, 2455. [CrossRef]
52. Napolitano, A. Cyclostationarity: Limits and generalizations. *Signal Process* **2016**, *120*, 323–347. [CrossRef]
53. Mandelbrot, B.; Ness, J.W.V. Fractional Brownian motions, fractional noises, and applications. *SIAM Rev.* **1968**, *10*, 422. [CrossRef]
54. Jeon, J.H.; Metzler, R. Fractional Brownian motion and motion governed by Langevin equation in confined geometries. *Phys. Rev. E* **2010**, *81*, 021103. [CrossRef] [PubMed]
55. Klafter, J.; Zumofen, G. Lévy statistics in a Hamiltonian system. *Phys. Rev. E* **1975**, *49*, 4873. [CrossRef] [PubMed]
56. Zaburdaev, V.; Denisov, S.; Klafter, J. Lévy walks. *Rev. Mod. Phys.* **2015**, *87*, 483. [CrossRef]
57. S.C. Lim; Muniandy, S.V. Self-similar Gaussian processes for modeling anomalous diffusion. *Phys. Rev. E* **2002**, *66.2*, 021114. [CrossRef]
58. Fuliński, A. Anomalous diffusion and weak nonergodicity. *Phys. Rev. E* **2011**, *83*, 061140. [CrossRef]
59. Jeon, J.-H.; Chechkin, A.V.; Metzler, R. Scaled Brownian motion: A paradoxical process with a time dependent diffusivity for the description of anomalous diffusion. *Phys. Chem. Chem. Phys.* **2014**, *16*, 15811–15817. [CrossRef]
60. Marple, S.L., Jr. *Digital Spectral Analysis with Applications*; Prentice-Hall, Englewood Cliffs, NJ, USA, 1987.

hv **photonics**

MDPI

Communication

Extra-Cavity Modulating a Soliton Molecule with Chirped Gaussian Pulse Shape

Daqian Tang [1], Junxiao Zhan [1], Dayu Wang [1], Haoming Wang [1], Yangyang Peng [1], Zian Cheak Tiu [2] and Yan Zhou [1,*]

[1] School of Science, Shanghai Institute of Technology, Shanghai 201418, China; 246182108@mail.sit.edu.cn (D.T.); 246182102@mail.sit.edu.cn (J.Z.); 236182112@mail.sit.edu.cn (D.W.); 236182113@mail.sit.edu.cn (H.W.); 2012231208@mail.sit.edu.cn (Y.P.)

[2] Faculty of Engineering & Quantity Surveying, INTI International University, Nilai 71800, Negeri Sembilan, Malaysia; ziancheak.tiu@newinti.edu.my

* Correspondence: yzhou@sit.edu.cn

Abstract: In this work, we theoretically simulate the modulation of a soliton molecule that has an initial chirped Gaussian pulse shape in a 1 μm extra-cavity optical fiber modulation system. Different soliton parameters in orthogonal polarizations are applied to achieve controllable optical solitons' output with specific properties in the time/frequency domain. For instance, when the phase difference is changed, both pulse shapes' and corresponding optical spectra's peak intensities will have a sudden change when the orthogonal phase difference is $\pi/2$. These simulation results provide a beneficial reference value for extra-cavity shaping of different solitons that come from nonlinear optical systems. Optimally, the reported results could pave the groundwork for industrial growth in ultrafast laser design.

Keywords: modulation; soliton molecule; chirped Gaussian; group velocity dispersion; optical fiber

1. Introduction

As a kind of localized nonlinear wave, optical solitons have been widely studied in various kinds of nonlinear optical systems [1–5]. Among these systems, ultrafast fiber laser systems (especially mode-locked fiber laser oscillators) are often employed to explore soliton dynamic characteristics [6,7], such as soliton generation, soliton collision, and the soliton's real-time evolution. Ultrafast fiber lasers have already been extensively applied in optical fiber communication, medical treatment, and aerospace. The theoretical research about optical solitons that exist in fiber laser oscillators can be developed through solving Ginzburg–Landau equations (scalar solitons) and coupled Ginzburg–Landau equations (vector solitons). The difference between a scalar soliton and a vector soliton is that a scalar soliton has degenerate modes in orthogonal polarizations, whereas a vector soliton has non-degenerate modes in orthogonal polarizations. While in a mode-locked fiber laser experiment, an optical soliton can be generated through managing different intra-cavity parameters to achieve the mode-locking operation. These parameters include gain (gain fiber length and ion doping concentration), dispersion (such as second-order dispersion, third-order dispersion, and fourth-order dispersion), nonlinear absorption (saturable power/energy, modulation depth, nonlinear response bandwidth, and so on), nonlinearity (pump power, fiber length, and fiber nonlinearity), birefringence (linear and nonlinear), relaxation time of saturable absorber, band-pass filtering effect, and some other factors. In some previous works about intra-cavity generating optical solitons, Zhang et al. experimentally generated noise-like pulses, in a bidirectional mode-locked fiber laser, with the use of step-index multimode fiber [8]. Li et al. employed a small-period long-period fiber grating in a normal dispersion mode-locked fiber laser and finally achieved soliton pulse output at L-band [9]. Chen et al. reported the observation of multi-state solitons in an Er-doped fiber laser, using γ-MnO_2-deposited photonic crystal fiber [10]. Zhao et al.

theoretically investigated the properties of vector quartic solitons in a quartic-dispersion-dominant mode-locked fiber laser [11]. The peak intensities/single pulse energies of soliton pulses emitted from ultrafast fiber laser oscillators can be further amplified with the use of optical fiber laser amplifiers. It is worth mentioning that the integration of multimode fibers in fiber laser cavities can effectively improve the soliton pulse energies/peak powers. In recent years, temporal-spatial mode-locking techniques in ultrafast fiber lasers have been widely investigated by photonics communities. For instance, Lou et al. reported an all-fiber, temporal-spatial, mode-locked fiber laser based on an InP quantum dot saturable absorber [12]. In a fiber laser system, linear optical fiber birefringence influences the polarization state of the generated optical soliton. And scalar soliton will transform to vector soliton [13–19] when scalar approximation is not negligible. These vector solitons include group-velocity-locked, polarization-rotated, and polarization-locked. Different vector solitons have their own features, such as polarization states, orthogonal components' central wavelengths, and optical spectra. The generation and dynamic evolution properties of vector solitons have been widely explored by some research groups. Until now, the optical soliton's working wavelength has covered the visible wavelength band to the mid-infrared wavelength band. It should be noted that spatial solitons have also been explored in some previous works, such as four-wave mixing dipole solitons [20] and photonic graphene edge solitons in atomic or atomic-like ensembles [21].

Except for the intra-cavity generation and power/energy scaling of optical soliton pulses, the extra-cavity modulating optical solitons have also been conducted theoretically and experimentally in some previous papers [22,23]. For instance, Wang et al. theoretically investigated extra-cavity modulation of group-velocity locked vector dissipative solitons [24]. Jin et al. reported the theoretical and experimental results of extra-cavity modulating group-velocity locked vector solitons at 1.55 μm [25]. As compared with intra-cavity modulation, extra-cavity modulation is more flexible, and the adjustment range is larger. So, the extra-cavity modulation system plays a significant role in the field of optical soliton modulation.

However, there are limited works reporting the modulation of chirped Gaussian soliton molecules in extra-cavity optical fiber modulation systems. A soliton molecule is a kind of soliton that can be formed when pulse splitting (restricted by the soliton area theorem) occurs under the balance between gain, loss, nonlinearity, dispersion, and so on. A soliton molecule consists of two or more solitons, and these solitons have fixed phase differences and temporal pulse intervals between adjacent solitons. Soliton molecules can be efficiently generated in fiber laser oscillators through increasing pump power or decreasing saturable power/energy of intra-cavity saturable absorbers. The possible experimental realization techniques of soliton molecules with chirped Gaussian pulses are the constructions of passively mode-locked fiber lasers, accompanied by intra-cavity management of pump power, group-velocity dispersion, saturable absorbers, and so on. In this work, extra-cavity modulating a soliton molecule with a chirped Gaussian pulse shape is developed in a 1 μm optical fiber system. The optical fiber modulation system can efficiently modulate temporal and frequency properties of optical soliton molecules and their orthogonal components. These results further explore modulating optical solitons that are unique characters in an extra-cavity optical fiber modulation system. And the modulation system can be further expanded to modulate optical solitons at other wavelengths by replacing the optical elements (such as wave plates and single-mode fiber) that are working at corresponding wavelengths.

2. Theoretical Model

Figure 1 shows the proposed extra-cavity chirped Gaussian soliton molecule modulation system in our simulation. The chirped Gaussian soliton molecule comes from a fiber laser source. ISO is an optical fiber isolator, which is used for isolating reflected optical signals. The initial chirped Gaussian soliton molecule is collimated by the first collimator (COL1) and collected by the second collimator (COL2). There is one λ/2 wave plate and two λ/4 wave plates between COL1 and COL2, which are used to flexibly alter the soliton pulse's polarization state through rotating along the propagation direction axis. Then, the elliptically polarized pulse is split and combined by two optical fiber polarization beam splitters (please see PBS1 and PBS2 in Figure 1). In the two branches between PBS1 and PBS2, amplitude modulators (AM1 and AM2) can independently change electric amplitudes in orthogonal directions. Chirped fiber Bragg gratings (CFBG1 and CFBG2) can independently alter the pulse chirps in orthogonal directions. Cirl and Cir2 are two optical fiber circulators, ensuring unidirectional propagation of optical pulses. Time delay lines (DL1 and DL2) are used to alter time delay between orthogonal directions. After combination with PBS2, the polarization state can be further varied by a polarization controller (PC). After that, the modulated chirped Gaussian soliton molecule is collimated by the third collimator (COL3). Finally, the parallel laser beam is secondarily polarization-split by PBS3, and orthogonal electric field components will transmit or reflect by PBS3 and then be detected by the two detectors (D1 and D2).

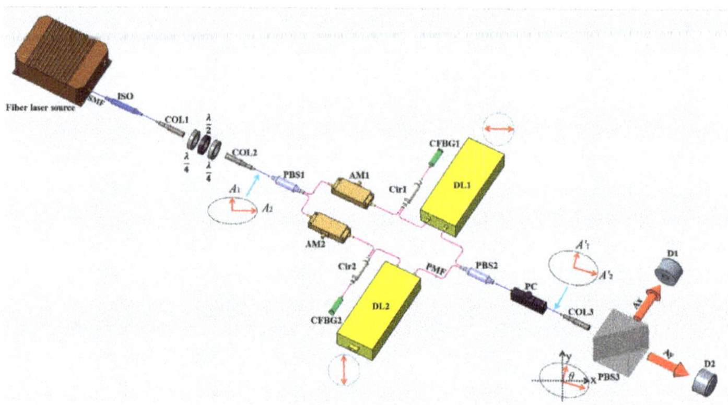

Figure 1. Chirped Gaussian soliton molecule modulation system.

The fundamental of this simulation is initiated as follows:

$$A_1'(z,t) = A_1 e^{i\frac{2\pi ct}{\lambda}} \frac{\exp\left\{-\frac{(1+iC_1)[(t-\Delta T/2)/T_0]^2}{2[1-iz_1(1+iC_1)/L_D]}\right\}}{\sqrt{1-iz_1(1+iC_1)/L_D}} + \frac{A_1}{\sqrt{2}} e^{i\frac{2\pi ct}{\lambda}} \sum_{k=-1,+1} \frac{\exp\left\{-\frac{(1+iC_1)\left[(t-\Delta T/2+kT_{delay})/T_0\right]^2}{2[1-iz_1(1+iC_1)/L_D]}\right\}}{\sqrt{1-iz_1(1+iC_1)/L_D}}, \quad (1)$$

$$A_2'(z,t) = A_2 e^{i(\frac{2\pi ct}{\lambda}+\Delta\varphi)} \frac{\exp\left\{-\frac{(1+iC_2)[(t+\Delta T/2)/T_0]^2}{2[1-iz_2(1+iC_2)/L_D]}\right\}}{\sqrt{1-iz_2(1+iC_2)/L_D}} + \frac{A_2}{\sqrt{2}} e^{i(\frac{2\pi ct}{\lambda}+\Delta\varphi)} \sum_{k=-1,+1} \frac{\exp\left\{-\frac{(1+iC_2)\left[(t+\Delta T/2+kT_{delay})/T_0\right]^2}{2[1-iz_2(1+iC_2)/L_D]}\right\}}{\sqrt{1-iz_2(1+iC_2)/L_D}}. \quad (2)$$

In Equations (1) and (2), $A_1'(z, t)$ and $A_2'(z, t)$ are orthogonal electric amplitudes of chirped Gaussian soliton molecules in the region between PC and COL3. A_1 and A_2 are constant amplitudes. λ is the central wavelength. c is a constant, meaning light transmission speed in a vacuum. t is time. $\Delta\varphi$ is the phase difference in orthogonal electric amplitudes. C_1 and C_2 are pulse chirps, whereas z_1 and z_2 are propagation distances. L_D is dispersion length. T_0 is initial pulse width. T_{delay} means time delay between three solitons in horizontal/vertical directions. While ΔT is the time delay of orthogonal components. k = −1, +1 means there are two solitons with different time delays in one polarization direction. Here, it is assumed that the three solitons in the same polarization have the same phase. And of course, their phases can also be different (such as π phase difference) in some situations.

The orthogonal electric fields' expressions after splitting by PBS3 are written as follows:

$$\begin{bmatrix} A_x(z,t) \\ A_y(z,t) \end{bmatrix} = \begin{bmatrix} \cos\theta & \sin\theta \\ \sin\theta & -\cos\theta \end{bmatrix} \begin{bmatrix} A'_1(z,t) \\ A'_2(z,t) \end{bmatrix}. \tag{3}$$

In Equation (3), $A_x(z, t)$ and $A_y(z, t)$ stand for time- and position-varying electric fields in the x and y directions. x means horizontal and y means vertical. θ is the projection angle when the modulated chirped Gaussian soliton molecule is incident on PBS3.

Then, the time and frequency intensities are written as follows:

$$I_j(z,t) = |A_j(z,t)|^2 \ (j = x \, or \, y), \tag{4}$$

$$I_j(z,\omega) = |A_j(z,\omega)|^2 = \left| \int_{-\infty}^{\infty} A_j(z,t) e^{-i\omega t} dt \right|^2 \ (j = x \, or \, y). \tag{5}$$

In Equations (4) and (5), $I_j(z, t)$ and $I_j(z, \omega)$ are intensities of modulated chirped Gaussian soliton molecules, the time domain, and frequency domain, respectively.

3. Simulation Results

In the simulation, some initial variables are set as follows: λ = 1064 nm, T_0 = 5 ps, and T_{delay} = 25 ps. And then, other soliton parameters will be altered for simulation successively; these variables include A_2/A_1, θ, ΔT, $\Delta\varphi$, C_1, C_2, z_1, and z_2.

Figure 2 gives the results of the modulated chirped Gaussian soliton molecule after PBS3 by changing A_2/A_1. Here, A_2/A_1 = 1/1.3, 1/1.15, 1/1, 1.15/1, and 1.3/1. Others are θ = 5°, ΔT = 0 ps, $\Delta\varphi$ = 0, $C_1 = C_2$ = 0.5, and $z_1/L_D = z_2/L_D$ = 0.05. In Figure 2a–e, the amplitudes in orthogonal directions change alternatively with the increase of A_2/A_1. Orthogonal pulse shapes have different peak intensities even when A_2/A_1 = 1/1, because θ is not equal to zero. In Figure 2f–j, the optical spectra's peak intensities also vary with A_2/A_1, and they have a similar peak intensity variation trend compared with the corresponding pulse shapes.

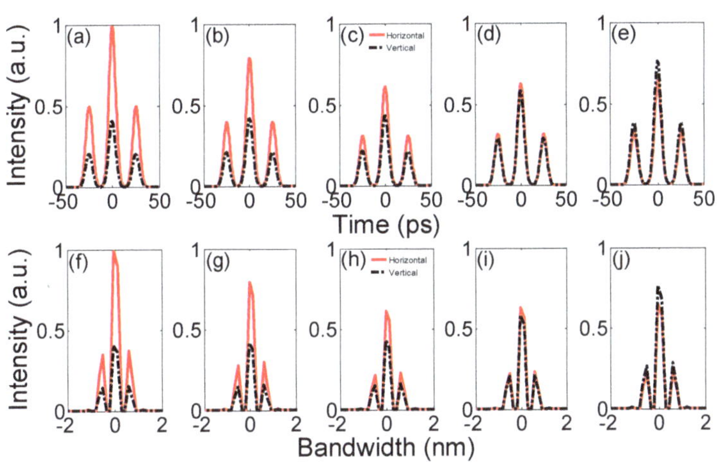

Figure 2. Simulation results in the time domain (**a–e**) and frequency domain (**f–j**) after PBS3, by changing A_2/A_1.

Figure 3 gives the results of the modulated chirped Gaussian soliton molecule after PBS3 by changing θ. Here, θ = 3°, 6°, 9°, 12°, and 15°. Others are A_2/A_1 = 1/1, ΔT = 0 ps, $\Delta\varphi$ = 0, $C_1 = C_2$ = 0.5, and $z_1/L_D = z_2/L_D$ = 0.05. In Figure 3a–e, the amplitude of the horizontal/vertical component of the modulated chirped Gaussian soliton molecule will increase/decrease as θ increases. Because the change in θ through rotating PBS3 will induce soliton pulse energy exchange between orthogonal components of the chirped Gaussian soliton molecule. The corresponding optical spectra in Figure 3f–j have similar peak intensity variation trends.

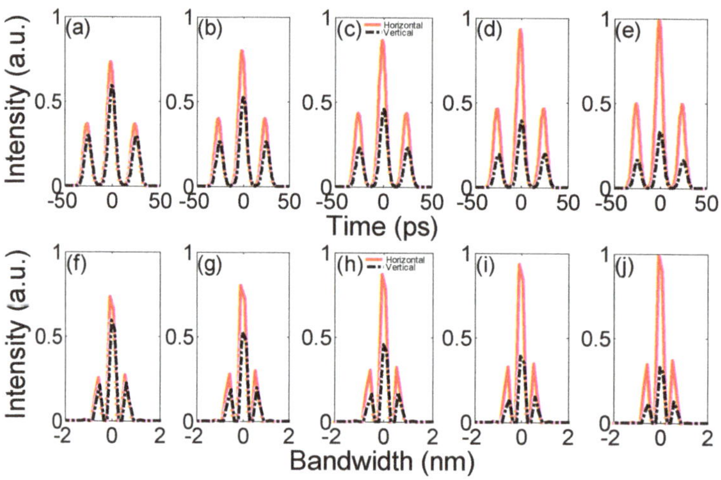

Figure 3. Simulation results in the time domain (**a–e**) and frequency domain (**f–j**) after PBS3, by changing θ.

Figure 4 gives the results of the modulated chirped Gaussian soliton molecule after PBS3 by changing ΔT. Here, ΔT = 1 ps, 2 ps, 3 ps, 4 ps, and 5 ps. Others are $A_2/A_1 = 1/1$, $\theta = 0°$, $\Delta\varphi = 0$, $C_1 = C_2 = 0.5$, and $z_1/L_D = z_2/L_D = 0.05$. In Figure 4, as ΔT increases, the time delay of modulated orthogonal components will also increase (see Figure 4a–e). While the optical spectra demonstrated in Figure 4f–j show that orthogonal optical spectra are always overlapped, meaning group velocity mismatch is always zero for orthogonal components, although the orthogonal phase velocity difference is changed.

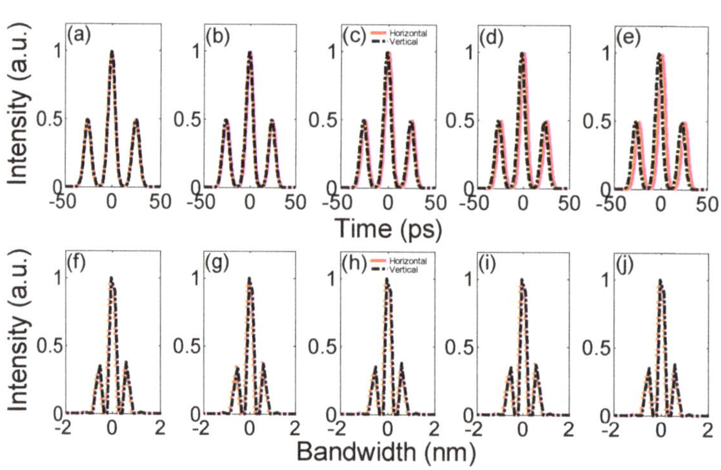

Figure 4. Simulation results in the time domain (**a–e**) and frequency domain (**f–j**) after PBS3, by changing ΔT.

Figure 5 gives the results of the modulated chirped Gaussian soliton molecule after PBS3, by changing $\Delta\varphi$. Here, $\Delta\varphi = \pi/10, \pi/8, \pi/6, \pi/4$, and $\pi/2$. Others are $A_2/A_1 = 1/1$, $\theta = 5°$, $\Delta T = 0$ ps, $C_1 = C_2 = 0.5$, and $z_1/L_D = z_2/L_D = 0.05$. In Figure 5a–e, the modulated pulse shapes in the output port have no obvious changes when $\Delta\varphi = \pi/10, \pi/8, \pi/6$, and $\pi/4$ in both directions. However, pulse peak intensity in horizontal/vertical polarization direction suddenly decreases/increases when $\Delta\varphi = \pi/2$ (in the state of polarization locked), which induces the temporal overlap of waveforms in orthogonal directions. For the corresponding optical spectra, their peak intensities in orthogonal polarizations have the same variation trend. And the orthogonal optical spectra will also be overlapped when $\Delta\varphi = \pi/2$.

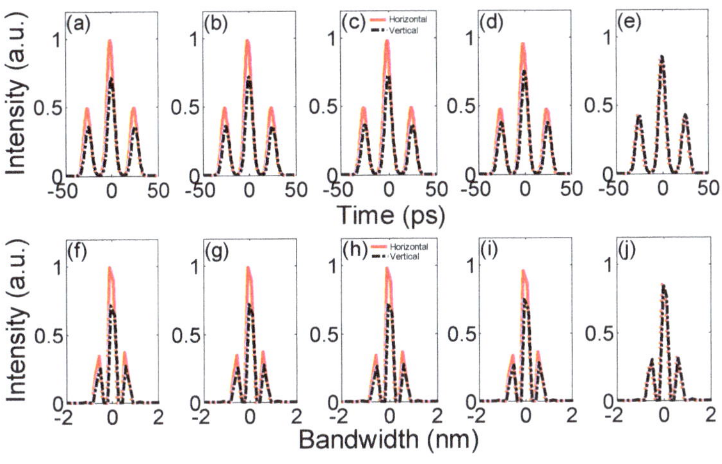

Figure 5. Simulation results in the time domain (**a–e**) and frequency domain (**f–j**) after PBS3, by changing $\Delta\varphi$.

Figure 6 gives the results of the modulated chirped Gaussian soliton molecule after PBS3, by changing C_1 or C_2 (C_1 is equal to C_2). Here, $C_1 = C_2 = 1, 2, 3, 4$, and 5. Others are $A_2/A_1 = 1/1$, $\theta = 5°$, $\Delta T = 0$ ps, $\Delta\varphi = 0$, and $z_1/L_D = z_2/L_D = 0.05$. In Figure 6a–e, peak intensities of both orthogonal soliton molecules will decrease when C_1 or C_2 increases. Pulse shapes keep their smooth property, and pulse widths have no obvious changes, as pulse chirps vary. In Figure 6f–j that shows frequency characters, optical spectra widths will increase and peak intensities will decrease in orthogonal directions as C_1 or C_2 increase. In addition, frequency oscillation occurs for both orthogonal components due to relatively large pulse chirps.

Figure 7 gives the results of the modulated chirped Gaussian soliton molecule after PBS3, by changing C_2 (C_1 is different from C_2). Here, $C_1 = 0.5$, $C_2 = 1, 2, 3, 4$, and 5. Others are $A_2/A_1 = 1/1$, $\theta = 5°$, $\Delta T = 0$ ps, $\Delta\varphi = 0$, and $z_1/L_D = z_2/L_D = 0.05$. In Figure 7a–e, the modulated orthogonal waveforms have no obvious changes when C_2 varies, which is different from the results that are being shown in Figure 6a–e. For the corresponding frequency domain results that are demonstrated in Figure 7f–j, the horizontal component is almost unchanged with the change of C_2. While for the vertical component, its peak intensity decreases and is accompanied by an increased pulse width when C_2 increases.

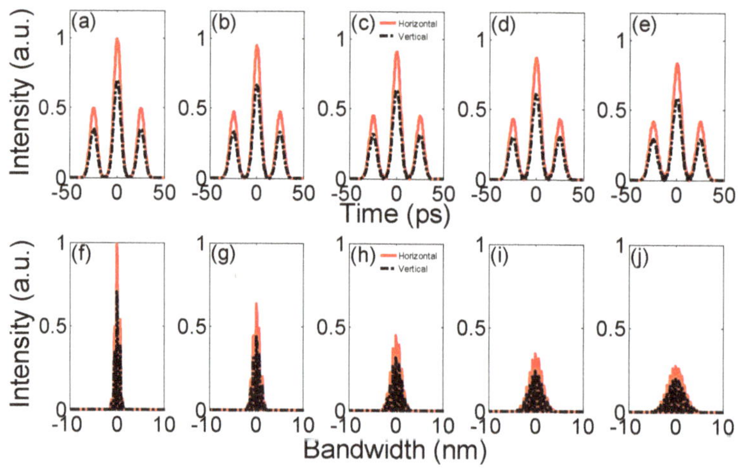

Figure 6. Simulation results in the time domain (**a–e**) and frequency domain (**f–j**) after PBS3, by changing C_1 or C_2 ($C_1 = C_2$).

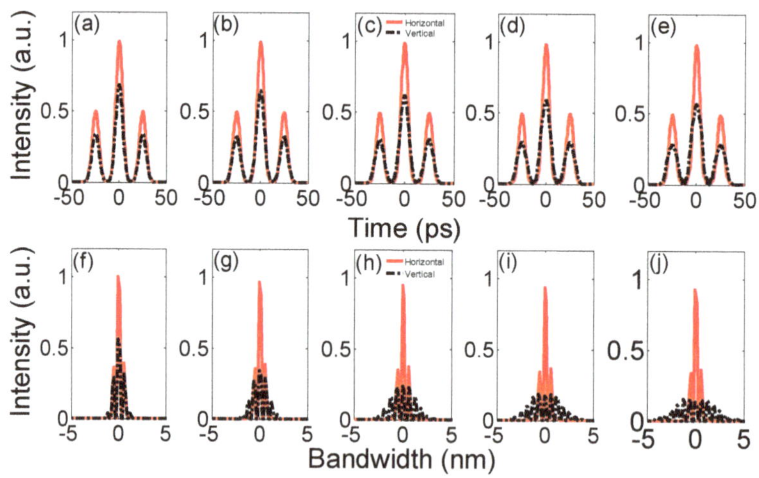

Figure 7. Simulation results in the time domain (**a–e**) and frequency domain (**f–j**) after PBS3, by changing C_2 ($C_1 \neq C_2$).

Figure 8 gives the results of the modulated chirped Gaussian soliton molecule after PBS3, by changing z_1/L_D or z_2/L_D ($z_1/L_D = z_2/L_D$). Here, $z_1/L_D = z_2/L_D = 0.1$, 0.2, 0.3, 0.4, and 0.5. Others are $A_2/A_1 = 1/1$, $\theta = 5°$, $\Delta T = 0$ ps, $\Delta\varphi = 0$, and $C_1 = C_2 = 0.5$. In Figure 8a–e, peak intensities will decrease when propagation distances increase for both orthogonal modes. The values of central solitons' side valleys start not equal to zero when z_1/L_D or z_2/L_D reaches 0.2. In Figure 8f–j, the optical spectra have no obvious changes in orthogonal directions. And horizontal mode always has a higher peak intensity than the vertical mode because of the non-zero θ value.

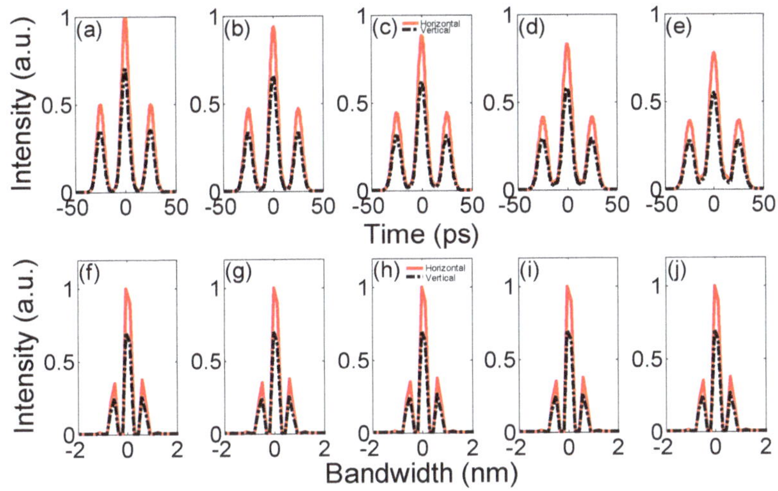

Figure 8. Simulation results in the time domain (**a–e**) and frequency domain (**f–j**) after PBS3, by changing z_1/L_D or z_2/L_D ($z_1/L_D = z_2/L_D$).

Figure 9 gives the results of the modulated chirped Gaussian soliton molecule after PBS3, by changing z_2/L_D ($z_1/L_D \neq z_2/L_D$). Here, z_1/L_D = 0.05, z_2/L_D = 0.1, 0.2, 0.3, 0.4, and 0.5. Others are $A_2/A_1 = 1/1$, $\theta = 5°$, ΔT = 0 ps, $\Delta\varphi = 0$, and $C_1 = C_2 = 0.5$. In Figure 9a–e, the horizontal component is almost unchanged under the variation of z_2/L_D. For the vertical component, its peak intensity slightly decreases as z_2/L_D increases. And temporal mismatch becomes obvious when z_2/L_D = 0.5. In Figure 9f–j, the orthogonal optical spectra have no observable changes under the variation of z_2/L_D.

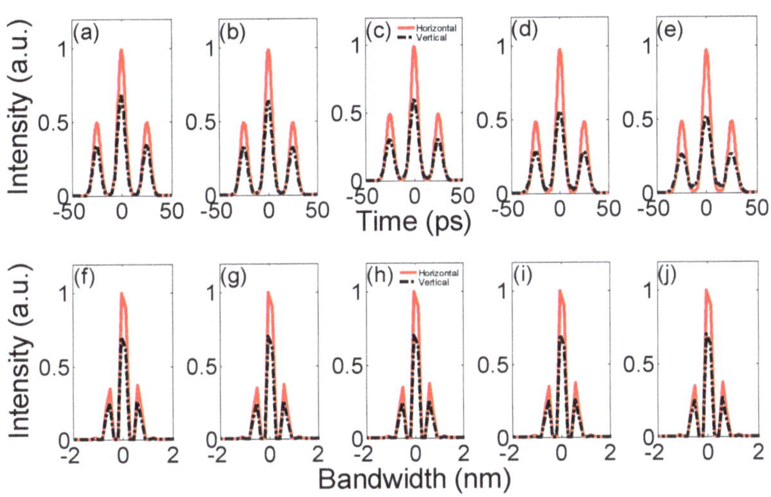

Figure 9. Simulation results in the time domain (**a–e**) and frequency domain (**f–j**) after PBS3, by changing z_2/L_D ($z_1/L_D \neq z_2/L_D$).

4. Conclusions

In this work, we have theoretically simulated the extra-cavity modulation of a soliton molecule with a chirped Gaussian pulse shape through changing different soliton variables. The modulated 1-μm chirped Gaussian soliton molecule can be designed with specific time/frequency properties in orthogonal polarizations by alteration and optimization of nonlinear propagation properties. For the future work, the single-mode fiber modulation system can be further expanded to a multimode fiber setup and to explore the temporal-spatial modulation of different optical solitons.

Author Contributions: Conceptualization, Y.Z.; methodology, D.T.; software, D.W., H.W. and Y.P.; validation, D.T.; formal analysis, J.Z.; investigation, D.T.; data curation, Y.Z.; writing—original draft preparation, D.T.; writing—review and editing, Y.Z. and Z.C.T.; supervision, Y.Z. and Z.C.T.; project administration, Y.Z.; funding acquisition, Y.Z. All authors have read and agreed to the published version of the manuscript.

Funding: This research was funded by the National Natural Science Foundation of China (NSFC), grant number 62105208; the Collaborative Innovation Fund of Shanghai Institute of Technology, grant number XTCX2022-32; and College students' innovation and entrepreneurship project of Shanghai Institute of Technology, grant numbers S202410259563, S202410259567, X2024102591490, X2024102591501, and X2024102591502.

Institutional Review Board Statement: Not applicable.

Informed Consent Statement: Not applicable.

Data Availability Statement: Data are available on request.

Conflicts of Interest: The authors declare no conflicts of interest.

References

1. Kbashi, H.J.; Sergyev, S.V. Optical rogue waves in fiber lasers. *Photonics* **2024**, *11*, 657. [CrossRef]
2. Yang, Y.T.; Zeng, Q.; Yang, Y.Z.; Du, G.G.; Ji, J.H.; Song, Y.F.; Wang, Z.H.; Wang, K. Generation and dynamics of multiple pulses in an ultrafast fiber laser with a single-mode fiber-graded-index multimode fiber-single-mode fiber-based saturable absorber. *Photonics* **2024**, *11*, 52. [CrossRef]
3. Triki, H.; Kruglov, V.I. Generation of solitons and periodic wave trains in birefringent optical fibers. *Chaos Soliton. Fract.* **2024**, *186*, 115300. [CrossRef]
4. Sukiasyan, M.; Avetisyan, V.; Kutuzyan, A. Spectral self-compression of chirp-free pulses in anomalously dispersive optical fibers. *Photonics* **2023**, *10*, 1207. [CrossRef]
5. González-Gaxiola, O.; Biswas, A.; Moraru, L.; Moldovanu, S. Highly dispersive optical solitons in absence of self-phase modulation by Laplace-Adomian decomposition. *Photonics* **2023**, *10*, 114. [CrossRef]
6. Guo, X.X.; Huang, X.W.; Li, X.H.; Luo, P.T.; Gao, C.X.; Wang, R.H.; Wang, Y.S.; Xi, F.; Yin, X.Q.; Zhang, K. 1.9 μm ultra-narrow spectral width mode-locked pulsed laser based on femtosecond laser inscribed FBG. *Opt. Laser Eng.* **2024**, *181*, 108441. [CrossRef]
7. Wang, S.; Li, X.; Li, X.H.; Li, Y.; Qiu, J.T.; Song, B.L.; Dou, X.H.; Qiao, F.H.; Lv, Z.G. Route to L-band GHz nonlinear multimodal interference mode-locked fiber laser. *J. Light. Technol.* **2024**, *42*, 5330–5336. [CrossRef]
8. Zheng, Y.G.; Wang, A.S.; Cai, Y.; Zhang, F.; Zhang, X.H. A bidirectional mode-locked fiber laser for asynchronous noise-like pulse generation. *Photonics* **2024**, *11*, 886. [CrossRef]
9. Li, Q.Y.; Cheng, P.Y.; Zhao, R.; Shu, X.W. Femtosecond laser-written small-period long-period fiber grating for an L-band normal dispersion mode-locked fiber laser. *Photonics* **2024**, *11*, 693. [CrossRef]
10. Chen, E.C.; Lu, P.; Li, X.H.; Huang, X.Z.; Han, Y.H.; Liu, D.M.; Zhang, J.S. γ-MnO_2-deposited photonic crystal fiber: Novel photonic devices for multi-state solitons generation. *Opt. Laser Technol.* **2023**, *159*, 108940. [CrossRef]
11. Zhao, K.J.; Gao, C.X.; Xiao, X.S.; Yang, C.X. Vector quartic solitons in birefringent fibers. *Opt. Lett.* **2021**, *46*, 761–764. [CrossRef] [PubMed]
12. Lou, Y.J.; Ge, B.Y.; Cai, Y.; Zhang, X.H. Spatiotemporal mode-locking of an all-fiber laser based on InP quantum dot saturable absorber. *Opt. Express* **2023**, *31*, 42400–42412. [CrossRef]
13. Cui, Y.D.; Zhang, Y.S.; Huang, L.; Liu, X.M. Dual-state vector soliton in mode-locked fiber laser. *Opt. Laser Technol.* **2024**, *157*, 108674. [CrossRef]
14. He, C.J.; Zhu, Z.W.; Yang, S.; Wang, N.; Yang, Y.Y.; Lin, X.C. Buildup and synchronization regimes of a vector pure-quartic soliton molecule in a fiber laser cavity. *Opt. Express* **2024**, *32*, 11895–11906. [CrossRef]
15. Yang, H.L.; He, S.L. Widely tunable 1.7 μm vector dissipative soliton all-fiber thulium laser. *J. Light. Technol.* **2024**, *42*, 347–353. [CrossRef]
16. Huang, L.; Li, B.X.; Jia, D.F.; Ge, C.F. Generation and real-time dynamics of vector square waves. *Opt. Commun.* **2024**, *569*, 130823. [CrossRef]
17. He, Z.W.; Du, Y.Q.; Zeng, C.; Tian, W.; Zhao, J.L.; Mao, D. Linear coupling-related pulse splitting in fiber lasers. *Opt. Lett.* **2023**, *48*, 5535–5538. [CrossRef] [PubMed]

18. Stalin, S.; Ramakrishnan, R.; Lakshmanan, M. Nondegenerate bright solitons in coupled nonlinear Schrödinger systems: Recent developments on optical vector solitons. *Photonics* **2021**, *8*, 258. [CrossRef]
19. Hu, X.; Guo, J.; Li, L.; Yoo, S.; Tang, D.Y. Evolution from periodic intensity modulations to dissipative vector solitons in a single-mode fiber laser. *Photonics* **2020**, *7*, 103. [CrossRef]
20. Zhang, Y.P.; Wang, Z.G.; Nie, Z.Q.; Li, C.B.; Chen, H.X.; Lu, K.Q.; Xiao, M. Four-wave mixing dipole soliton in laser-induced atomic gratings. *Phys. Rev. Lett.* **2011**, *106*, 093904. [CrossRef]
21. Zhang, Z.Y.; Wang, R.; Zhang, Y.Q.; Kartashov, Y.V.; Li, F.; Zhong, H.; Guan, H.; Gao, K.L.; Li, F.L.; Zhang, Y.P.; et al. Observation of edge solitons in photonic grapheme. *Nat. Commun.* **2020**, *11*, 1902. [CrossRef] [PubMed]
22. Zhou, Y.; Zhang, K.Y.; Liao, M.S.; Zhao, G.Y.; Fang, G.Y. Operating vector solitons with chirped sech pulse shapes. *Photonics* **2022**, *9*, 143. [CrossRef]
23. Zhou, Y.; Li, Y.F.; Fang, Y.Z.; Liao, M.S. Generation of pseudo-high-order polarization-locked vector solitons. *J. Technol.* **2020**, *20*, 196–201.
24. Wang, X.; Li, L.; Geng, Y.; Wang, H.X.; Su, L.; Zhao, L.M. Decomposition of group-velocity-locked-vector-dissipative solitons and formation of the high-order soliton structure by the product of their recombination. *Appl. Opt.* **2018**, *57*, 746–751. [CrossRef]
25. Jin, X.X.; Wu, Z.C.; Li, L.; Zhang, Q.; Tang, D.Y.; Shen, D.Y.; Fu, S.N.; Liu, D.M.; Zhao, L.M. Manipulation of group-velocity-locked vector solitons from fiber lasers. *IEEE Photonics J.* **2016**, *8*, 1501206. [CrossRef]

photonics

MDPI

Article

The Nonlinear Eigenvalue Problem of Electromagnetic Wave Propagation in a Dielectric Layer Covered with Graphene

Yury Smirnov * and Stanislav Tikhov

Mathematics and Supercomputing, Penza State University, 440026 Penza, Russia; tik.stanislav2015@yandex.ru
* Correspondence: smirnovyug@mail.ru

Abstract: The paper focuses on the problem of a monochromatic terahertz TE-polarized wave propagation in a plane dielectric layer filled with a homogeneous isotropic medium; one of the boundaries of the waveguide is covered with a layer of graphene. In fact, the paper aims to find the eigenwaves of the described waveguiding structure. On the one hand, in the study, energy losses both in the dielectric layer and in the graphene layer are neglected; the latter assumption is reasonable in the terahertz range of electromagnetic radiation (on which the paper focuses), where graphene has a strong plasmonic response and much less loss. On the other hand, this study takes into account the significant third-order nonlinearity resulting from the interaction of the electromagnetic wave with the charge carriers in the graphene layer. The paper aims to study the guiding properties of the above structure using primarily an analytical approach. The wave propagation problem is reduced to an eigenvalue problem, where one of the boundary conditions is nonlinear with respect to the sought-for function. The main result of the paper is a dispersion equation allowing for a waveguide of a given thickness to determine a set of its propagation constants and, consequently, a set of its eigenwaves. It is worth noting that the dispersion equation being written in an explicit form can be used to obtain deep qualitative results related to the solvability of the problem and the properties of its solutions. For example, in the paper, the existence of several propagation constants (and, consequently, the eigenwaves) of the studied waveguiding structure is proved under some conditions. Besides studying the problem analytically, the paper presents some numerical results as well. In particular, the presented figures demonstrate how the nonlinearity in graphene affects the propagation constants and eigenwaves, providing the dispersion curves and eigenwaves for nonlinear graphene as well as for the linear one.

Keywords: Maxwell's equations; electromagnetic waves; plane dielectric layer; graphene; nonlinear surface electric conductivity; eigenvalue problem; boundary value problem

1. Introduction

Today, much attention is being paid to 2D materials and 2D electronic components; among the two-dimensional materials, graphene, discovered in 2004 by Geim and Novoselov [1], occupies a special place. Graphene is a hexagonal crystal lattice of carbon atoms; due to its special structure, graphene has a number of unique physical properties. In particular, graphene has properties that make it very promising for various applications in photonics and optoelectronics: combining graphene with other materials, one can create a variety of waveguiding devices with better characteristics compared to waveguides made of «traditional» materials. In [2], there is an overview of waveguides of different configurations based on the use of graphene; such waveguides can serve as components for various optical devices such as photodetectors, modulators, polarizers, etc., significantly increasing their efficiency. In [3–5], waveguiding structures with graphene are shown to be efficient in many fields, such as biosensors, photoelectric detection, and so on; in [6], the authors considered graphene oxide (GO) as a potential absorbent for environmental cleanup and suggested an approach, which enables making full use of GO for this purpose.

The above applications use only the «linear» optical and electrical properties of graphene. In papers [7,8], it was theoretically predicted, and in [9], it was experimentally confirmed that graphene has an extremely strong third-order nonlinearity compared to the commonly used dielectrics and metals; this nonlinearity results from the interaction of the charge carriers in graphene with strong electromagnetic radiation. The discovery of the strong nonlinear response of graphene has led to significant efforts to realize a new generation of practical nonlinear photonic guided wave devices for all optical applications [10].

This paper focuses on the propagation of a monochromatic terahertz TE-polarized electromagnetic wave in a plane dielectric waveguide, one of the surfaces of which is covered with a layer of graphene. In fact, in this paper, we consider the problem of finding the eigenmodes of the waveguide, i.e., such electromagnetic waves that can propagate in the waveguide without obtaining energy from any external source. In eigenwave problems, energy losses in the waveguiding structure are often neglected, assuming that the energy losses are small, and the propagation length of the found eigenwaves is sufficient in practical applications. Such an approach is also applied in this paper: we neglect the energy losses in the dielectric as well as in the graphene layer. We stress that although in the general case graphene absorbs some of the incident electromagnetic wave, in the terahertz range, the absorption is small compared to the strong plasmonic response of graphene. We understand that neglecting the absorption properties of graphene might seem to be a serious drawback of this study. However, we stress that this requirement is unavoidable in our statement of the problem. Section 4 deals with this issue in more detail.

For applications in photonics and optoelectronics, a convenient parameter characterizing graphene is its surface electrical conductivity σ_g; it is very important that the conductivity of graphene can be controlled by an electric field or chemical doping without changing the size and structure of the sample. In accordance with [7–9,11], the conductivity σ_g of graphene depends on the strength of the electric field interacting with graphene. Due to the centrally symmetric structure of graphene, this dependence has the form $\sigma_g = \sigma^{(1)} + \sigma^{(3)}|\mathbf{E}|^2$, where $\sigma^{(1)}$ and $\sigma^{(3)}$ are quantities depending on the wave frequency. As the electric conductivity differs from zero, there arises a surface current. Since these phenomena are localized on the surface, it is essential to take them into account in the boundary conditions of the problem. Taking into account the nonlinear effects arising in the graphene ($\sigma_3 \neq 0$), one obtains the boundary conditions, which are nonlinear with respect to the sought-for function, which, in turn, brings additional complexity to the problem.

The guiding properties of graphene structures play an important role for practical applications. The ability to propagate TM- and TE-polarized waves localized on a graphene monolayer with the dispersion in the terahertz range of electromagnetic radiation is shown in [12–14]. The ability of the structure formed by two graphene layers and a separating thin dielectric layer to hold the localized plasmon modes has been studied in many papers [15–17]. In [18], the authors investigated the features of guided TE wave modes in a plane structure consisting of a set of alternating layers of dielectric and graphene. We stress that the novelty of this paper is that the guiding properties of the considered waveguiding structure (plane dielectric waveguide covered with a graphene layer) are studied taking into account the abovementioned third-order nonlinearity of graphene. The obtained results show that this nonlinearity affects the wave propagation process significantly (for strong electromagnetic radiation) and for this reason cannot be neglected.

2. Materials and Methods

The main physical model we deal with in this study is the system of Maxwell's equations describing the dynamics of an electromagnetic field. Since we assume that everywhere there is no bulk current (the medium in the waveguide is not conducting) and an uncompensated electric charge, and the electromagnetic field harmonically depends on time, then the system of Maxwell's equations can be written in the harmonic mode, see Formula (3). One of the boundaries of the considered waveguiding structure is covered with graphene, which implies a surface current when an electromagnetic wave propagates.

Since this phenomena is localized on the surface, it is essential to take it into account in the boundary conditions of the problem. Graphene is characterized by its surface electric conductivity σ_g. In this study, we take into account the nonlinear optical properties of graphene supposing that σ_g depends on the electric field strength, see Formula (5).

The main goal of the paper is to study the guiding properties of the considered waveguiding structure analytically. To do this, the problem on electromagnetic wave propagation is reduced to an eigenvalue problem, where the eigenvalues correspond to the propagation constants. Solving the main differential equation and taking into account the boundary conditions, we obtain the dispersion equation of the problem written in an explicit form. Studying this equation analytically, we obtain results related to the solvability of the problem and the properties of its solutions.

For solving the eigenvalue problem $\mathcal{P}$, see Section 3.2, numerically, we use mainly the so-called «shooting method». The main schema of the method is following. We fix some segment on γ, say $\gamma \in [\gamma', \gamma'']$, and generate a grid with nodes $\gamma' \leqslant \gamma_1 < \gamma_2 < \ldots < \gamma_n \leqslant \gamma''$. For each $\gamma = \gamma_i$, we solve the Cauchy problem for equation $u''(x) = -(\varepsilon_2 - \gamma^2)u(x)$ with the initial conditions $u(0) = A$, $u'(0) - \sqrt{\gamma^2 - \varepsilon_1} \cdot u(0) = 0$ and evaluate its solution $u \equiv u(x; \gamma_i)$ and the first derivative $u'(x; \gamma_i)$ of this solution at point $x = h$. Then, going through all γ_i, we look for the condition $F(\gamma_i) \cdot F(\gamma_{i+1}) < 0$, where

$$F(\gamma) = u'(h;\gamma) + \sqrt{\gamma^2 - \varepsilon_3} \cdot u(h;\gamma) + 120\pi i\left(\sigma^{(1)} + \sigma^{(3)} u^2(h;\gamma)\right) u(h;\gamma),$$

is true; if it is true, then segment $\gamma \in [\gamma_i, \gamma_{i+1}]$ definitely contains a solution to problem $\mathcal{P}$. The numerical results are presented in Section 3.5. All the numerical methods are implemented with the package «Maple».

3. Results

3.1. Electrodynamic Statement of the Problem

A monochromatic TE-polarized electromagnetic wave

$$(\mathbf{E}, \mathbf{H})e^{-i\omega t}, \tag{1}$$

where ω is a circular frequency,

$$\mathbf{E} = \left(0, \mathrm{E}_y(x), 0\right)e^{i\gamma z}, \quad \mathbf{H} = \left(\mathrm{H}_x(x), 0, \mathrm{H}_z(x)\right)e^{i\gamma z} \tag{2}$$

are complex amplitudes, and γ is (unknown) real spectral parameter, propagates in a plane dielectric layer

$$\Sigma = \{(x,z) \in \mathbb{R}^2 : 0 \leqslant x \leqslant h\}$$

sandwiched between two half-spaces $x < 0$ and $x > h$. The boundary $x = 0$ is open, and at the boundary $x = h$, there is a graphene layer, which causes a surface current at this boundary.

Waveguide Σ is filled with a homogeneous isotropic medium characterized by a constant permittivity ε_2; half-spaces $x < 0$ and $x > h$ are filled with homogeneous isotropic mediums characterized by constant permittivities ε_1 and ε_3, respectively, such that $1 \leqslant \varepsilon_1 \leqslant \varepsilon_3 \leqslant \varepsilon_2$. Everywhere, permeability $\mu = \mu_0$, where μ_0 is the magnetic constant. The geometry of the problem is presented in Figure 1.

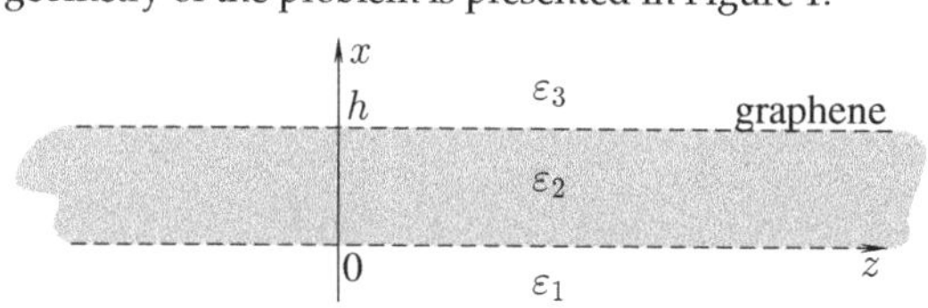

Figure 1. Geometry of the problem.

Field (1) satisfies Maxwell's equations

$$\operatorname{rot}\mathbf{H} = -i\omega\varepsilon_0\varepsilon\mathbf{E}, \quad \operatorname{rot}\mathbf{E} = i\omega\mu\mathbf{H}, \tag{3}$$

where ε_0 is the dielectric constant, and

$$\varepsilon = \begin{cases} \varepsilon_1, & x < 0, \\ \varepsilon_2, & 0 < x < h, \\ \varepsilon_3, & x > h. \end{cases}$$

Since the boundaries of the waveguide are open, and the electromagnetic field penetrates into half-spaces $x < 0$ and $x > h$, it is essential to impose on (1) the condition of decaying at $x \to \pm\infty$. The tangential component E_y of the electric field is continuous at both boundaries, due to the absence of a surface charge there. The tangential component of the magnetic field is also continuous at the boundary $x = 0$, due to the absence of a surface current there; however, it undergoes a jump at the boundary $x = h$, due to the surface current of charge carried in graphene (the current is induced by the electromagnetic wave), and the jump is equal to surface current density $\mathbf{j}$. Thus, at $x = h$, the discussed component satisfies the following condition

$$[\mathrm{n}, \mathbf{H}^+ - \mathbf{H}^-] = \mathbf{j} = \sigma_g\mathbf{E}, \tag{4}$$

where n is a unit vector of the normal directed along the x axis, $\mathbf{H}^+$ and $\mathbf{H}^-$ are the values of magnetic field above and below the surface $x = h$, respectively, $[*,*]$ is the vector product, and σ_g is the surface conductivity of graphene.

As was previously noted, the electric conductivity σ_g of graphene depends on the electric field coupling to the charge carriers in graphene; due to the central symmetric structure of graphene, this dependence has the form

$$\sigma_g = \sigma^{(1)} + \sigma^{(3)}|\mathbf{E}|^2, \tag{5}$$

where $\sigma^{(1)}$ and $\sigma^{(3)}$ are generally some complex numbers depending on frequency ω [7–9,19,20].

The linear part $\sigma^{(1)}$ of graphene's electric conductivity is determined by formulas [21,22]. We assume that $\Re\sigma^{(1)} = 0$ and $\Im\sigma^{(1)} > 0$, i.e.,

$$\sigma^{(1)} = i\sigma', \quad \sigma' > 0. \tag{6}$$

Such a restriction on σ_1 is fair in THz range, where graphene has a strong plasmonic response and much less loss [23,24]. For determining $\sigma^{(3)}$ in (5), there exist several formulas [7,19,20]; for example, one can apply the formula in [7]; in accordance with this formula, quantity $\sigma^{(3)}$ is purely imaginary and $\Im\sigma^{(3)} < 0$, i.e.,

$$\sigma^{(3)} = -i\sigma'', \quad \sigma'' > 0. \tag{7}$$

The main problem — we call it problem $\mathcal{P}'$ — is to find such values of parameter $\gamma = \gamma'$, where there exists field (1) satisfying Maxwell's equations (3) and all the above conditions and exponentially decaying at $x \to \pm\infty$.

Substituting (2) into Maxwell's equations (3), one obtains

$$\begin{cases} i\gamma\mathrm{H}_x(x) - \mathrm{H}_z'(x) = -i\omega\varepsilon_0\varepsilon\mathrm{E}_y(x), \\ -i\gamma\mathrm{E}_y(x) = i\omega\mu\mathrm{H}_x(x), \\ \mathrm{E}_y'(x) = i\omega\mu\mathrm{H}_z(x). \end{cases} \tag{8}$$

The third equation in the obtained system provides the relationship between the tangential components of the electric and magnetic fields expressed by the formula

$$\mathrm{H}_z(x) = -\frac{i}{\omega\mu}\mathrm{E}'_y(x). \tag{9}$$

Expressing H_x and H'_z from the second and the third equations of system (8), respectively, and substituting them into the first equation, one obtains

$$\gamma^2\mathrm{E}_y(x) - \mathrm{E}''_y(x) = \omega^2\mu\varepsilon_0\varepsilon\mathrm{E}_y(x).$$

Passing to dimensionless variables in the obtained equation by virtue of the formulas

$$\widetilde{x} = k_0 x, \quad \widetilde{h} = k_0 h, \quad \widetilde{\gamma} = k_0^{-1}\gamma,$$

where $k_0^2 = \omega^2\mu\varepsilon_0$, and, omitting the tilde, we obtain the equation

$$\mathrm{E}''_y(x) = -(\varepsilon - \gamma^2)\mathrm{E}_y(x). \tag{10}$$

Solving (10) in half-spaces $x < 0$, $x > h$ and using the condition at infinity, one finds

$$\mathrm{E}_y(x) = \begin{cases} C_1 e^{x\sqrt{\gamma^2-\varepsilon_1}}, & x < 0, \\ C_2 e^{-(x-h)\sqrt{\gamma^2-\varepsilon_3}}, & x > h, \end{cases} \tag{11}$$

where C_1 and C_2 are constants. Note that E_y must be real-valued; from here, taking into account inequality $\varepsilon_3 \geqslant \varepsilon_1$, it follows that γ must satisfy

$$\gamma^2 > \varepsilon_3. \tag{12}$$

From the above conditions imposed on field (1) and Formula (9), expressing the relationship between the tangential components of the magnetic and electric fields, it follows that E_y must satisfy the conditions

$$\begin{aligned} &\mathrm{E}_y|_{0-0} - \mathrm{E}_y|_{0+0} = 0, \\ &\mathrm{E}'_y|_{0-0} - \mathrm{E}'_y|_{0+0} = 0 \end{aligned} \tag{13}$$

and

$$\begin{aligned} &\mathrm{E}_y|_{h-0} - \mathrm{E}_y|_{h+0} = 0, \\ &\mathrm{E}'_y|_{h-0} - \mathrm{E}'_y|_{h+0} = -i\cdot 120\pi\sigma_g\mathrm{E}_y|_{x=h}, \end{aligned} \tag{14}$$

where

$$\sigma_g = \sigma^{(1)} + \sigma^{(3)}\mathrm{E}_y^2\Big|_{x=h}, \tag{15}$$

which results from (5), taking into account the form of field (2) and the realness of parameter γ.

Then, one can reformulate the problem $\mathcal{P}'$ in the following way. Problem $\mathcal{P}'$ is to find such values $\gamma = \gamma'$ satisfying inequality (12), where there exists the solution $\mathrm{E}_y \equiv \mathrm{E}_y(x;\gamma')$ to Equation (10), satisfying the boundary conditions (13) and (14).

3.2. An Eigenvalue Problem

Introducing the notation $u(x) := \mathrm{E}_y(x)$, Equation (10) in the layer can be written in the form

$$u''(x) = -(\varepsilon_2 - \gamma^2)u(x). \tag{16}$$

Using conditions (13) and (14), the solutions in half-spaces (11), and Formula (15), one can write the boundary conditions for function $u(x)$ as

$$u'(0) - \sqrt{\gamma^2 - \varepsilon_1}\cdot u(0) = 0, \tag{17}$$

$$u'(h) + \sqrt{\gamma^2 - \varepsilon_3}\cdot u(h) = -120\pi i(\sigma^{(1)} + \sigma^{(3)}u^2(h))u(h). \tag{18}$$

Besides this, we need one more condition in order to obtain a discrete set of solutions to the problem. It is in accordance with the physical process of the electromagnetic wave propagation in waveguiding structures. We use the following form of the additional condition

$$u(0) = A, \tag{19}$$

where $A \neq 0$ is a constant.

It is easy to determine that when a couple (γ, u) satisfies Equation (16), $(-\gamma, u)$ satisfies it as well. For this reason, we consider positive γ.

So, problem $\mathcal{P}'$ is equivalent to the boundary value problem; we call it problem $\mathcal{P}$, which is to find the positive $\gamma = \widehat{\gamma}$ satisfying inequality (12), such that there exists a twice continuously differentiable function $u \equiv u(x; \widehat{\gamma})$ that is a solution to Equation (16) satisfying conditions (17)–(19). A number $\widehat{\gamma}$ is called an *eigenvalue* of problem $\mathcal{P}$, and function $u(x; \widehat{\gamma})$ is called an *eigenfunction* of problem $\mathcal{P}$.

Note that condition (18) is nonlinear with respect to the unknown function $u(x)$; so, in fact problem $\mathcal{P}$ is nonlinear. If one sets $\sigma^{(3)} = 0$ in (18), then problem $\mathcal{P}$ degenerates into a linear problem, which we call problem $\mathcal{P}_0$.

Although the statement of problem $\mathcal{P}$ is given above, and problem $\mathcal{P}_0$ is its special case, we present the statement of problem $\mathcal{P}_0$ here as well. So, problem $\mathcal{P}_0$ is to find the positive $\gamma = \widetilde{\gamma}$ satisfying inequality (12), such that there exists a function $v \equiv v(x; \gamma)$, which is a solution to the equation

$$v''(x) = -(\varepsilon_2 - \gamma^2) v(x)$$

satisfying the following boundary conditions

$$\begin{aligned} v'(0) - \sqrt{\gamma^2 - \varepsilon_1} \cdot v(0) &= 0, \\ v'(h) + \sqrt{\gamma^2 - \varepsilon_3} \cdot v(h) &= -i \cdot 120\pi\sigma^{(1)} v(h); \end{aligned}$$

the number $\gamma = \widetilde{\gamma}$ is called an eigenvalue of problem $\mathcal{P}_0$, and function $v(x; \widetilde{\gamma})$ is called an eigenfunction of problem $\mathcal{P}_0$.

We stress that problem $\mathcal{P}_0$ does not require (19).

Problem $\mathcal{P}_0$ and problem $\mathcal{P}$ serve as mathematical models of the monochromatic TE-wave propagation in a plane waveguide having a graphene layer on one of its boundaries, but in the case of problem $\mathcal{P}_0$ graphene is characterized by $\sigma_g = \sigma_1$, i.e., the nonlinear response of graphene is neglected. Note that such a model allows one to obtain results close to the experimental data only at intensities of electromagnetic radiation.

3.3. *Dispersion Equation of Problem $\mathcal{P}$*

The solution to Equation (16) has the form

$$u = C_3 \cos \sqrt{\varepsilon_2 - \gamma^2} x + C_4 \sin \sqrt{\varepsilon_2 - \gamma^2} x,$$

where C_3 and C_4 are constants. Using condition (17), one can express C_4 by C_3 and then, using condition (19), determine C_3; doing this, one obtains

$$u = \frac{1}{\sqrt{\varepsilon_2 - \gamma^2}} A \left(\sqrt{\gamma^2 - \varepsilon_1} \sin \sqrt{\varepsilon_2 - \gamma^2} x + \sqrt{\varepsilon_2 - \gamma^2} \cos \sqrt{\varepsilon_2 - \gamma^2} x \right).$$

Using condition (18), one obtains the following equation

$$\left(\sqrt{\gamma^2-\varepsilon_1}\sqrt{\gamma^2-\varepsilon_3}+120\pi i\sigma^{(1)}\sqrt{\gamma^2-\varepsilon_1}-(\varepsilon_2-\gamma^2)\right)\sin h\sqrt{\varepsilon_2-\gamma^2}+$$
$$+\sqrt{\varepsilon_2-\gamma^2}\left(\sqrt{\gamma^2-\varepsilon_1}+\sqrt{\gamma^2-\varepsilon_3}+120\pi i\sigma^{(1)}\right)\cos h\sqrt{\varepsilon_2-\gamma^2}=$$
$$=-120\pi i\frac{A^2}{\varepsilon_2-\gamma^2}\left(\sqrt{\gamma^2-\varepsilon_1}\sin h\sqrt{\varepsilon_2-\gamma^2}+\sqrt{\varepsilon_2-\gamma^2}\cos h\sqrt{\varepsilon_2-\gamma^2}\right)^3. \quad (20)$$

Let us introduce notation $\lambda^2=\varepsilon_2-\gamma^2$, $\epsilon_1^2=\varepsilon_2-\varepsilon_1$, $\epsilon_3^2=\varepsilon_2-\varepsilon_3$, $\bar{\sigma}_1=i\cdot 120\pi\sigma^{(1)}$, $\bar{\sigma}_3=i\cdot 120\pi\sigma^{(3)}$; taking into account Formulas (6) and (7), it is clear that $\bar{\sigma}_1$ and $\bar{\sigma}_3$ are real, $\bar{\sigma}_1<0$, and $\bar{\sigma}_3>0$. Using this notation, Equation (20) can be written in the form

$$\left(\sqrt{\epsilon_1^2-\lambda^2}\sqrt{\epsilon_3^2-\lambda^2}-|\bar{\sigma}_1|\sqrt{\epsilon_1^2-\lambda^2}-\lambda^2\right)\sin\lambda h+$$
$$+\left(\sqrt{\epsilon_3^2-\lambda^2}+\sqrt{\epsilon_1^2-\lambda^2}-|\bar{\sigma}_1|\right)\lambda\cos\lambda h=$$
$$=-|\bar{\sigma}_3|\lambda^{-2}A^2\left(\sqrt{\epsilon_1^2-\lambda^2}\sin\lambda h+\lambda\cos\lambda h\right)^3. \quad (21)$$

Expression (21) can be considered as the characteristic equation of problem $\mathcal{P}$. This means that any solution $\lambda=\widehat{\lambda}$ to Equation (21) corresponds to an eigenvalue $\widehat{\gamma}=\sqrt{\varepsilon_2-\widehat{\lambda}^2}$ of problem $\mathcal{P}$, and any eigenvalue $\widehat{\gamma}$ of problem $\mathcal{P}$ corresponds to the solution $\widehat{\lambda}=\sqrt{\varepsilon_2-\widehat{\gamma}^2}$ of Equation (21).

From a physical point of view, relation (21) is the so-called dispersion equation, as it provides the relationship between the thickness of the waveguide and its propagation constants.

Setting $\bar{\sigma}_3=0$ in (21), one obtains the dispersion equation for problem $\mathcal{P}_0$ in the form

$$\operatorname{tg}\lambda h=\frac{\lambda\left(\sqrt{\epsilon_3^2-\lambda^2}+\sqrt{\epsilon_1^2-\lambda^2}-|\bar{\sigma}_1|\right)}{\lambda^2-\sqrt{\epsilon_1^2-\lambda^2}\cdot\sqrt{\epsilon_3^2-\lambda^2}+|\bar{\sigma}_1|\sqrt{\epsilon_1^2-\lambda^2}}. \quad (22)$$

Further, setting $\bar{\sigma}_1=0$ in the previous formula, one obtains the classical dispersion equation of the form

$$\operatorname{tg}\lambda h=\frac{\lambda\left(\sqrt{\epsilon_3^2-\lambda^2}+\sqrt{\epsilon_1^2-\lambda^2}\right)}{\lambda^2-\sqrt{\epsilon_1^2-\lambda^2}\cdot\sqrt{\epsilon_3^2-\lambda^2}} \quad (23)$$

for the problem of the electromagnetic TE-wave propagation in a plane dielectric layer sandwiched between two half-spaces; the above equation is given in [25] and in [26] (only for the case $\epsilon_1=\epsilon_3$).

3.4. Solvability of Problem $\mathcal{P}$

Problem $\mathcal{P}$ might have eigenvalues in the interval $\gamma\in(\sqrt{\varepsilon_3},\sqrt{\varepsilon_2})$, as well as in the unbounded domain $\gamma>\sqrt{\varepsilon_2}$ (unless the contrary is proved). The former kind of eigenvalues corresponds to solutions of Equation (21) belonging to the interval $\lambda\in\Lambda$, where $\Lambda=(0,\epsilon_3)$; the latter one corresponds to solutions of Equation (21) having the form $\lambda=i\lambda'$ with $\lambda'>0$.

Statement 1. *Equation* (21) *does not have solutions of the form* $\lambda=i\lambda'$, $\lambda'>0$ *if* $\lambda'>\lambda'_*$, *where*

$$\lambda'_*=\max\left\{\frac{\sqrt{|\bar{\sigma}_1|^2+|\bar{\sigma}_1|\sqrt{|\bar{\sigma}_1|^2+16\epsilon_1^2}}}{2\sqrt{2}},\sqrt{\frac{|\bar{\sigma}_1|}{2}}\right\}. \quad (24)$$

Proof. So, let $\lambda = i\lambda'$, where $\lambda' > 0$. Substituting $\lambda = i\lambda'$ into (21), one obtains the equation, with respect to λ', of the form

$$\begin{aligned}\Big(\sqrt{\epsilon_1^2+\lambda'^2}\sqrt{\epsilon_3^2+\lambda'^2}-|\bar{\sigma}_1|\sqrt{\epsilon_1^2+\lambda'^2}+\lambda'^2\Big)\operatorname{sh}\lambda' h+\\ +\Big(\sqrt{\epsilon_3^2+\lambda'^2}+\sqrt{\epsilon_1^2+\lambda'^2}-|\bar{\sigma}_1|\Big)\lambda'\operatorname{ch}\lambda' h=\\ =-|\bar{\sigma}_3|\lambda'^{-2}A^2\Big(\sqrt{\epsilon_1^2+\lambda'^2}\operatorname{sh}\lambda' h+\lambda'\operatorname{ch}\lambda' h\Big)^3.\end{aligned} \quad (25)$$

It is clear that the right-hand side of (25) is negative for all $\lambda' > 0$, whereas the left-hand side is positive starting with some $\lambda' = \lambda'_*$. This means that Equation (25) does not have solutions in the domain $\lambda' > \lambda'_*$.

Let us obtain the estimate (24) for λ'_*. It is clear that the left-hand side is positive as soon as both terms in brackets are positive. Taking into account the inequality

$$\sqrt{\epsilon_1^2+\lambda'^2}\sqrt{\epsilon_3^2+\lambda'^2}-|\bar{\sigma}_1|\sqrt{\epsilon_1^2+\lambda'^2}+\lambda'^2 \geqslant 2\lambda'^2-|\bar{\sigma}_1|\sqrt{\epsilon_1^2+\lambda'^2}$$

and the inequality $2\lambda'^2-|\bar{\sigma}_1|\sqrt{\epsilon_1^2+\lambda'^2} \geqslant 0$, which holds for all $\lambda' \geqslant \lambda'_1$, where

$$\lambda'_1 = \frac{\sqrt{|\bar{\sigma}_1|^2+|\bar{\sigma}_1|\sqrt{|\bar{\sigma}_1|^2+16\epsilon_1^2}}}{2\sqrt{2}},$$

one finds that the expression in the first bracket is positive for all $\lambda' > \lambda'_1$. Further, the inequality

$$\sqrt{\epsilon_3^2+\lambda'^2}+\sqrt{\epsilon_1^2+\lambda'^2}-|\bar{\sigma}_1| \geqslant 2\lambda'-|\bar{\sigma}_1|$$

together with $2\lambda'-|\bar{\sigma}_1| \geqslant 0$, which holds for all $\lambda' \geqslant \lambda'_2$, where $\lambda'_2 = \sqrt{\frac{|\bar{\sigma}_1|}{2}}$, imply that the expression in the second bracket is positive for all $\lambda' > \lambda'_2$. Combining the obtained results, one obtains Formula (24). □

Now, let us pass to the case $\lambda \in \Lambda$. For the further analysis, it is convenient to rewrite Equation (21) in the following way

$$\operatorname{tg}\lambda h = -\frac{\lambda\cdot\theta_1(\lambda)}{\theta_2(\lambda)}, \quad (26)$$

where

$$\begin{aligned}\theta_1(\lambda) &= \sqrt{\epsilon_1^2-\lambda^2}+\sqrt{\epsilon_3^2-\lambda^2}-|\bar{\sigma}_1|+|\bar{\sigma}_3|A^2\Big(1+\frac{3\epsilon_1^2-4\lambda^2}{\lambda^2}\sin^2\lambda h\Big),\\ \theta_2(\lambda) &= \sqrt{\epsilon_1^2-\lambda^2}\Big[\sqrt{\epsilon_3^2-\lambda^2}-|\bar{\sigma}_1|+|\bar{\sigma}_3|A^2\Big(3\cos^2\lambda h+\frac{\epsilon_1^2-\lambda^2}{\lambda^2}\sin^2\lambda h\Big)\Big]-\lambda^2.\end{aligned}$$

Since functions θ_1 and θ_2 in the right-hand side of (26) do not have accumulation points of zeros, then (26) can have only a finite number of solutions $\lambda \in \Lambda$.

The following result provides a sufficient condition for the existence of at least one solution to Equation (26).

Statement 2. *If inequalities*

$$|\sigma_3|A^2\min\Big\{3;\epsilon_1^2\epsilon_3^{-2}\Big\} > |\sigma_1| \quad (27)$$

and

$$h > \frac{3\pi\sqrt{\epsilon_1^2 + \epsilon_3^2}}{2\epsilon_1\epsilon_3} \tag{28}$$

are fulfilled, then Equation (26) *has at least one solution* $\widehat{\lambda} \in (0, \lambda_*)$, *where*

$$\lambda_* = \frac{\epsilon_1\epsilon_3}{\sqrt{\epsilon_1^2 + \epsilon_3^2}} (< \epsilon_3). \tag{29}$$

Proof. Let us consider the right-hand side of (26). It is clear that function $\theta_1(\lambda)$ is continuous for all $\lambda \in \Lambda$. Function $\theta_2(\lambda)$ is continuous for all $\lambda \in \Lambda$, but generally, it can change the sign. Nevertheless, one can check that under condition (27), function θ_2 preserves the sign and, to be more precise, is positive for $\lambda < \lambda_*$, where λ_* is defined in (29). Indeed, taking into account the chain of simple inequalities

$$3\cos^2\lambda h + \frac{\epsilon_1^2 - \lambda^2}{\lambda^2}\sin^2\lambda h \geqslant 3\cos^2\lambda h + \frac{\epsilon_1^2 - \lambda_*^2}{\lambda_*^2}\sin^2\lambda h \geqslant \min\left\{3; \epsilon_1^2\epsilon_3{}^2\right\}$$

and condition (27), one can see that

$$|\bar{\sigma}_3|A^2\left(3\cos^2\lambda h + \frac{\epsilon_1^2 - \lambda^2}{\lambda^2}\sin^2\lambda h\right) > |\bar{\sigma}_1|.$$

Combining this result with the inequality

$$\sqrt{\epsilon_1^2 - \lambda^2}\sqrt{\epsilon_3^2 - \lambda^2} \geqslant \lambda^2$$

taking place for $\lambda < \lambda_*$, one comes to conclusion that $\theta_2(\lambda) > 0$ for such λ. Thus, the right-hand side in (26) is continuous for $\lambda < \lambda^*$.

Condition (28) implies that, firstly, λh takes all values from zero to $3\pi/2$, and secondly, inequality $\lambda^* h > 3\pi/2$ is valid. Equation (26) is defined for $\lambda h \in (0, 3\pi/2)$, and the right-hand side of (26) is continuous for such λh. This means that a graph of the function in the right-hand side of (26) has at least one intersection with $\operatorname{tg}\lambda h$ in the left-hand side of (26), for $\lambda \in (h^{-1}\pi/2, h^{-1}3\pi/2)$. □

The sufficient condition for the existence of one solution to Equation (26) given in Statement 2 can easily be generalized to the case of the existence of $n > 1$ solutions. Indeed, the following result takes place.

Statement 3. *If inequalities* (27) *and*

$$h > \frac{(1+2n)\pi\sqrt{\epsilon_1^2 + \epsilon_3^2}}{2\epsilon_1\epsilon_3} \tag{30}$$

are fulfilled, then Equation (26) *has at least n solutions* $\widehat{\lambda}_k \in (0, \lambda_*)$, $k = \overline{1, n}$, *where* λ_* *is defined in* (29).

Proof. The proof of this statement repeats the proof of Statement 2. We have $\lambda_* h > (1 + 2n)\pi/2$, where λ_* is defined in (29). Further, λh takes all values from zero to $(1 + 2n)\pi/2$. Thus, Equation (26) is defined for $\lambda h \in (0, (1 + 2n)\pi/2)$, and the right-hand side of (26) is continuous in this interval. This means that a graph of the function in the right-hand side of (26) has at least n intersections with $\operatorname{tg}\lambda h$ in the left-hand side of (26) in interval $\lambda \in \left(\frac{\pi}{2h}, \frac{(1+2n)\pi}{2h}\right)$. □

Taking into account Statements 1–3 and the equivalence between Equation (21) and problem $\mathcal{P}$, one obtains the following result.

Statement 4. *If conditions* (27) *and* (30) *are fulfilled, then problem $\mathcal{P}$ has at least n eigenvalues $\widehat{\gamma}_k \in (\gamma_*, \gamma^*)$, $k = \overline{1, n}$, where*

$$\gamma_* = \sqrt{\frac{\varepsilon_2^2 - \varepsilon_1\varepsilon_3}{2\varepsilon_2 - \varepsilon_1 - \varepsilon_3}}, \quad \gamma^* = \sqrt{\varepsilon_2 + \lambda_*'^2};$$

here, λ_*' is defined in (24).

In fact, Statement 3 gives a sufficient condition for the existence of eigenmodes supported by the waveguide under consideration.

3.5. Numerical Results

In the calculations below, we used the following parameters: $\varepsilon_1 = \varepsilon_3 = 1$, $\varepsilon_2 = 11.7$, $\sigma^{(1)} = i \cdot 4.67 \cdot 10^{-4}$ S, $\sigma^{(3)} = -i \cdot 2.15 \cdot 10^{-17}$ S $\cdot$ m$^2 \cdot$ V^{-2}. The value of amplitude A of the electric field is given in figures' captions.

It is worth giving a comment about the chosen values of the parameters. The linear part of the electric conductivity of graphene was calculated by virtue of the formula given in [22]. In this formula, there are three optional parameters: the wave frequency ω, the chemical potential μ_c, and the absolute temperature T; for calculating σ_1, we used $\omega = 2\pi \cdot 4 \cdot 10^{12}$ Hz, $\mu_c = 0.1$ eV, and $T = 300$ K; we stress that we neglected the real part of σ_1. Quantity $\sigma^{(3)}$ was calculated by virtue of the formula presented in [7] using the same parameters as for calculating σ_1.

In Figure 2, the dispersion curves of problems $\mathcal{P}$ (blue curves) and $\mathcal{P}_0$ (red curves) are presented. The dispersion curves were plotted as the dependence of a wave number (a propagation constant) on either the wave frequency ω or thickness h of the waveguide. Since the statement of the problem does not involve ω explicitly due to the normalization by $k_0^2 = \omega^2\mu_0\varepsilon_0$, we plotted the dispersion curves as γ vs. h.

The vertical line $h = 3$ in Figure 2 corresponds to the waveguide of thickness $3k_0^{-1}$. The intersection points of the dispersion curves with this line denoted by diamonds are the eigenvalues of the corresponding problems, and these eigenvalues, in turn, correspond to the propagation constants γk_0 of the waveguide in the problems of electromagnetic wave propagation.

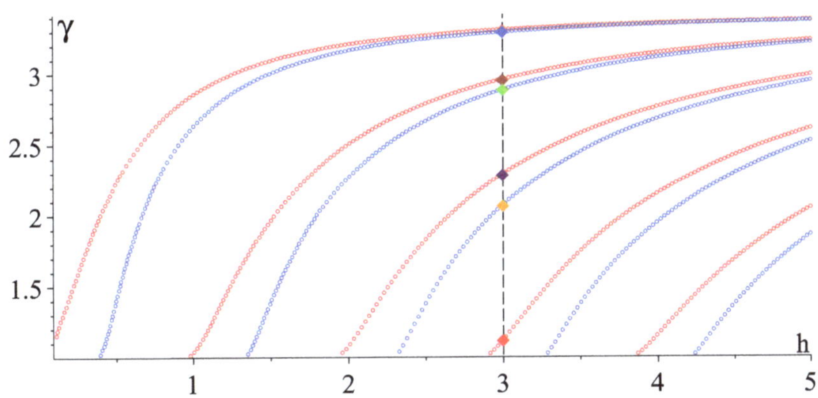

Figure 2. The first five dispersion curves of problems $\mathcal{P}$ (blue curves) and $\mathcal{P}_0$ (red curves); the value of the electric field at the boundary $x = 0$ is $A = 10^8$ V $\cdot$ m^{-1}. The diamonds denote the eigenvalues $\widehat{\gamma}_1 \approx 3.29$ (blue), $\widehat{\gamma}_2 \approx 2.88$ (green), and $\widehat{\gamma}_3 \approx 2.07$ (orange) of problem $\mathcal{P}$ and the eigenvalues $\widetilde{\gamma}_1 \approx 3.3$ (red), $\widetilde{\gamma}_2 \approx 2.95$ (brown), $\widetilde{\gamma}_3 \approx 2.29$ (purple), and $\widetilde{\gamma}_4 \approx 1.11$ (red) of problem $\mathcal{P}_0$.

In Figure 3, we plotted the eigenfunction $u(x; \widehat{\gamma}_3)$ of problem $\mathcal{P}$ and the eigenfunction $v(x; \widetilde{\gamma}_3)$ of problem $\mathcal{P}_0$. The eigenvalue $\widehat{\gamma}_3$ of problem $\mathcal{P}$ is a perturbation of eigenvalue $\widetilde{\gamma}_3$ of problem $\mathcal{P}_0$, and it can be shown that $\lim_{\sigma_3 \to +0} \widehat{\gamma}_3 = \widetilde{\gamma}_3$. Due to the closeness of the eigenvalues $\widehat{\gamma}_3$ and $\widetilde{\gamma}_3$, it is natural to expect the closeness of the corresponding eigenfunctions $u(x;\widehat{\gamma}_3)$ and $v(x;\widetilde{\gamma}_3)$; Figure 3 demonstrates this clearly; in addition, it can be shown that $\lim_{\sigma_3 \to +0} u(x;\widehat{\gamma}_3) = v(x;\widetilde{\gamma}_3)$. In Figure 3, one can see that in the nonlinear case, the absolute value of the eigenfunction (tangential component of the electric field) at the boundary $x = h$ was significantly smaller than in the linear case. This means that the nonlinearity arising in graphene led to a greater localization of the electromagnetic field inside the waveguide. At the same time, the maximum and minimum values of the eigenfunction (tangential component of the electric field) in the nonlinear case were smaller in absolute value than in the linear case, and the extremum points shifted to the left relative to their positions in the linear case.

In Figure 4, we plotted the eigenfunction $u(x;\widehat{\gamma}_2)$ of problem $\mathcal{P}$ and the eigenfunction $v(x;\widetilde{\gamma}_2)$ of problem $\mathcal{P}_0$. The eigenvalue $\widehat{\gamma}_2$ of problem $\mathcal{P}$ is a perturbation of eigenvalue $\widetilde{\gamma}_2$ of problem $\mathcal{P}_0$, and it can be shown that $\lim_{\sigma_3 \to +0} \widehat{\gamma}_2 = \widetilde{\gamma}_2$. Figure 4 shows that the eigenmode corresponding to the nonlinear case was more localized than its linear counterpart. In addition, the maximum and minimum values of the nonlinear eigenmode were smaller in absolute value than the maximum and minimum values of the linear eigenmode, and the extremum points shifted to the left relative to their positions in the linear case.

It also seems interesting to learn at which conditions the discussed nonlinear effect in graphene became significant. In Figure 5, as well as in Figure 2, we plotted the dispersion curves of problems $\mathcal{P}$ (blue curves) and $\mathcal{P}_0$ (red curves); however, in the calculations, we used a smaller value for the amplitude of the electric field, namely $A = 2 \cdot 10^7$ V $\cdot$ m^{-1}. In this case, the dispersion curves of the (nonlinear) problem $\mathcal{P}$ were no longer strongly different from the dispersion curves of (linear) problem $\mathcal{P}_0$. In Figures 6 and 7, the eigenfunctions of the problems $\mathcal{P}$ and $\mathcal{P}_0$, corresponding to the eigenvalues denoted in Figure 5, are presented. It can be seen that the eigenmode corresponding to the nonlinear case was more localized within the waveguide than its linear counterpart; however, this effect was much weaker than the one demonstrated in Figures 3 and 4, and the reason is that the amplitude of the electric field was smaller. It is worth noting that the strength of the nonlinearity in graphene depends on the value of the nonlinearity coefficient σ_3 in the formula as well as on the amplitude of the incident wave; in accordance with the formula given in [7], the coefficient σ_3 was proportional to ω^{-3}. Thus, in order to make the considered nonlinear effect more significant, one can either increase the amplitude of the incident wave or decrease its frequency.

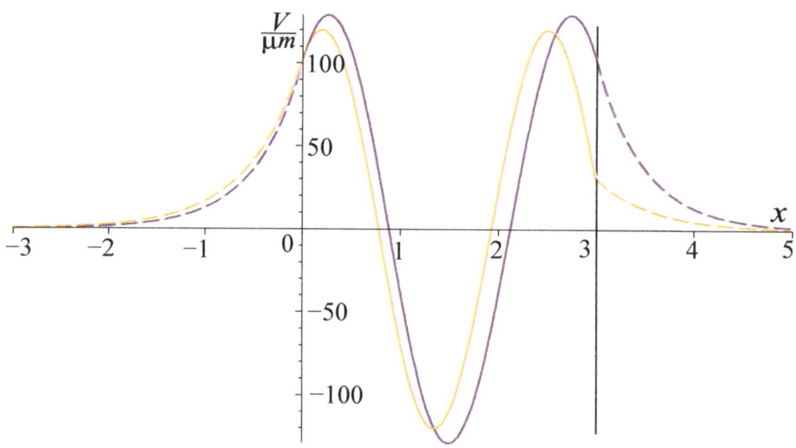

Figure 3. The orange curve presents the eigenfunction $u(x;\widehat{\gamma}_3)$ of problem $\mathcal{P}$, corresponding to the eigenvalue $\widehat{\gamma}_3 \approx 2.07$ denoted in Figure 2 by the orange diamond. The purple curve presents the eigenfunction $v(x;\widetilde{\gamma}_3)$ of problem $\mathcal{P}_0$, corresponding to the eigenvalue $\widetilde{\gamma}_3 \approx 2.29$ denoted in Figure 2 by the purple diamond.

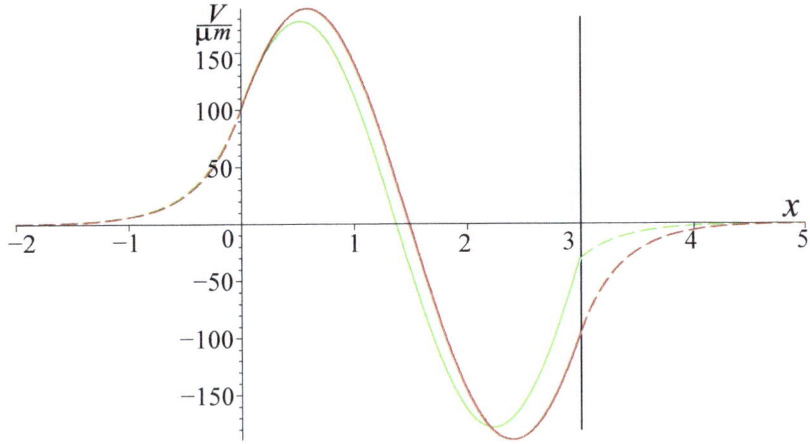

Figure 4. The green curve presents the eigenfunction $u(x;\widehat{\gamma}_2)$ of problem $\mathcal{P}$ corresponding to the eigenvalue $\widehat{\gamma}_2 \approx 2.88$ denoted in Figure 2 by the green diamond. The brown curve presents the eigenfunction $v(x;\widetilde{\gamma}_2)$ of problem $\mathcal{P}_0$ corresponding to the eigenvalue $\widetilde{\gamma}_2 \approx 2.95$, denoted in Figure 2 by the brown diamond.

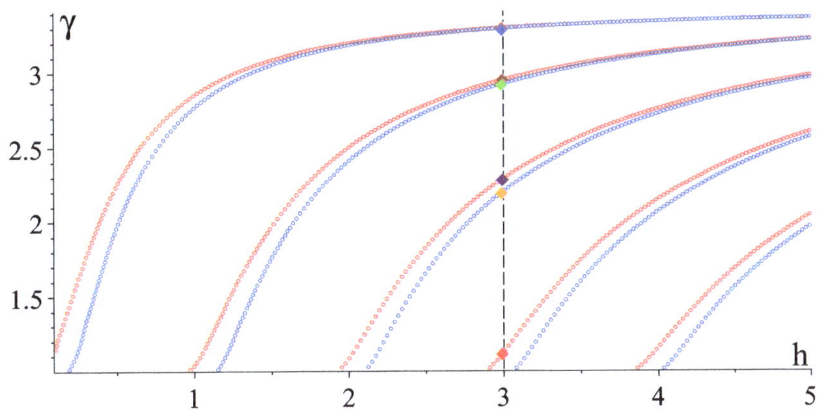

Figure 5. The first five dispersion curves of problems $\mathcal{P}$ (blue curves) and $\mathcal{P}_0$ (red curves); the value of the tangential component of the electric field at the boundary $x = 0$ is $A = 2 \cdot 10^7\ \mathrm{V} \cdot \mathrm{m}^{-1}$. The diamonds denote the eigenvalues $\widehat{\gamma}_1 \approx 3.3$ (blue), $\widehat{\gamma}_2 \approx 2.95$ (green), and $\widehat{\gamma}_3 \approx 2.19$ (orange) of problem $\mathcal{P}$ and the eigenvalues $\widetilde{\gamma}_1 \approx 3.3$ (red), $\widetilde{\gamma}_2 \approx 2.95$ (brown), $\widetilde{\gamma}_3 \approx 2.29$ (purple), and $\widetilde{\gamma}_4 \approx 1.11$ (red) of problem $\mathcal{P}_0$.

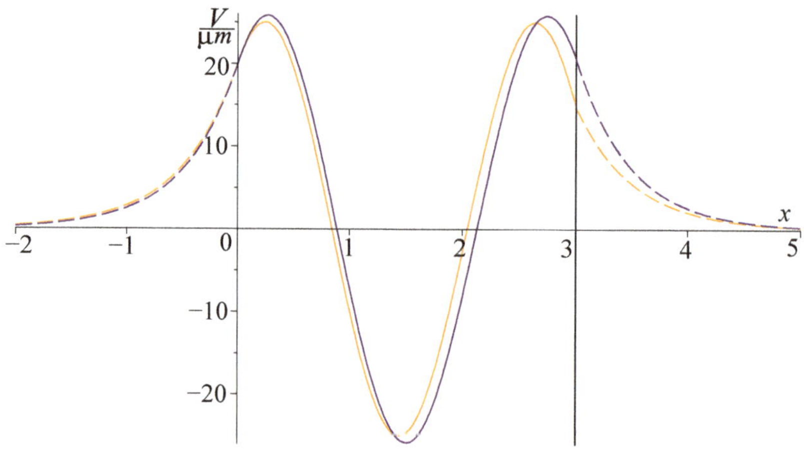

Figure 6. The orange curve presents the eigenfunction $u(x;\widehat{\gamma}_3)$ of problem $\mathcal{P}$, corresponding to the eigenvalue $\widehat{\gamma}_3 \approx 2.19$ denoted in Figure 5 by the orange diamond. The purple curve presents the eigenfunction $v(x;\widetilde{\gamma}_3)$ of problem $\mathcal{P}_0$, corresponding to the eigenvalue $\widetilde{\gamma}_3 \approx 2.29$ denoted in Figure 5 by the purple diamond.

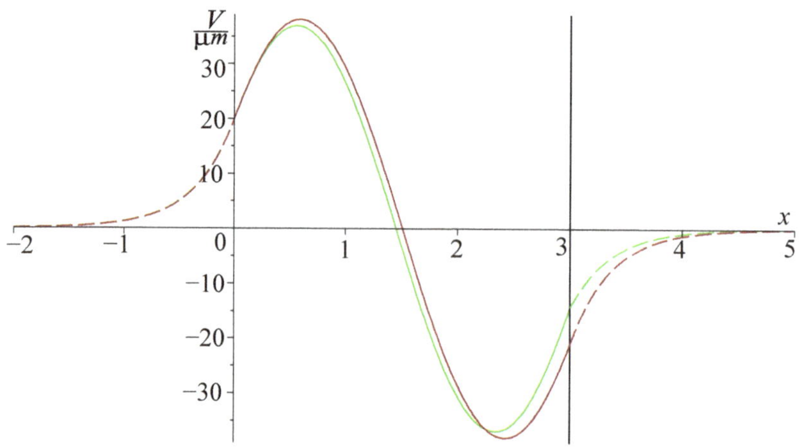

Figure 7. The green curve presents the eigenfunction $u(x;\widehat{\gamma}_2)$ of problem $\mathcal{P}$ corresponding to the eigenvalue $\widehat{\gamma}_2 \approx 2.92$ denoted in Figure 5 by the green diamond. The brown curve presents the eigenfunction $v(x;\widetilde{\gamma}_2)$ of problem $\mathcal{P}_0$ corresponding to the eigenvalue $\widetilde{\gamma}_2 \approx 2.95$, denoted in Figure 5 by the brown diamond.

4. Discussion

In the paper, the problem of the propagation of a monochromatic TE-polarized electromagnetic wave in a plane dielectric waveguide, covered with a layer of graphene on one side, is studied. The novelty of this study is that we take into account the nonlinearity of graphene, which becomes significant if the electromagnetic radiation is strong enough. The main result of this paper is the dispersion equation, which allows one to find, for a waveguide of a given thickness, a set of its propagation constants that determine the configuration of the so-called eigenmodes of the waveguide. We stress that whereas in most papers studying the problems similar to the one considered in this paper, the preference is given to numerical simulations, in this study we focused rather on obtaining qualitative results related to the solvability of the problem. Studying the dispersion equation, we found sufficient conditions for the existence of several propagation constants (and their corresponding eigenwaves) of the considered waveguiding structure, see Statements 2 and 3. In other words, we proved the ability to propagate the TE waves localized in the plane dielectric waveguide covered with a graphene layer, taking into account the third-order nonlinearity of graphene affecting the wave propagation.

The numerical results, see Section 3.5, clearly demonstrate that the nonlinear effects in graphene have a significant affect on the electromagnetic wave propagation process, if the radiation is strong enough. For example, Figures 3 and 4 allow one to compare the component of the electric field in the linear (for $\sigma_3 = 0$) and nonlinear cases: one can see that in the nonlinear case, it was more localized within the waveguide, and its extremum points were shifted to the left relative to their position in the linear case.

In this section, we also want to briefly clarify why we were forced to neglect losses in our study. Let us look at the dispersion Equation (20). It is easy to notice that if σ_1 and σ_3 are complex numbers with a nonzero real part, then Equation (20) does not have real solutions γ, and we need to assume parameter γ to be a complex number with a nonzero imaginary part. However, that leads to the following problem. In Formulas (2), the functions E_y, H_x, and H_z depend only on one spatial variable: the transverse coordinate x. This choice of the fields is possible only in the case of real γ. Indeed, substituting the fields $\mathbf{E}$ and $\mathbf{H}$ with components (2) into Formula (4), with σ_g defined in (5), we obtain the equation that depends on x and does not depend on z. To be more precise, the multiplier $e^{i\gamma z}$ is reduced on both sides of the equations, and the term $|\mathbf{E}|$ does not depend on z as $|e^{i\gamma z}| = 1$ for a real γ. At the same time, if γ is not real, then $|e^{i\gamma z}|$ is a function with respect to z. This means that components E_y, H_x, and H_z depend not only on x but also on z. This contradicts the choice of these components as functions depending only on x. All this, however, does not mean that it is not possible to consider a complex γ in a waveguiding problem with a nonlinear graphene layer. We just say that for a complex γ, the problem cannot be solved correctly mathematically for fields (1) and (2). To overcome the described problem, we need σ_1 and σ_3 to be purely imaginary, since in this case, the dispersion Equation (20) might have real solutions.

In this paper, we focus only on TE-waves and do not study the ability to propagate TM-polarized waves in the considered waveguiding structure. However, this does not mean that the TM case is less important and less interesting than the TE one. The reason why we do not study TM waves here is that the governing equations as well as the boundary conditions for the TE and TM-waves are different, and the analytical approach, developed in this paper to deal with the TE-waves, needs to be strongly modified for TM-polarized waves. So, we strongly believe that it is more convenient to study these two types of waves separately.

An essential continuation of this research will be the study of waveguiding structures of a more complex configuration. For example, it is interesting to study the problem of the propagation of a monochromatic TE-polarized wave in a plane waveguide covered on both sides by layers of graphene and to take into account the arising nonlinear effects at both boundaries of the waveguide. We stress that such a problem is very important from a practical point of view, as well as rich in mathematical content, because it involves two boundary conditions, which are nonlinear with respect to the sought-for function; in other words, there arises a new class of boundary value problems with nonclassical (nonlinear) boundary conditions. Finally, we stress that circular cylindrical waveguides combined with graphene are of great importance.

Author Contributions: Methodology, Y.S.; software, S.T.; investigation, S.T. and Y.S.; writing—original draft preparation, S.T.; writing—review and editing, Y.S.; visualization, S.T.; supervision, Y.S. All authors have read and agreed to the published version of the manuscript.

Funding: The research was funded by the Russian Science Foundation under the project 20-11-20087.

Institutional Review Board Statement: Not applicable.

Informed Consent Statement: Not applicable.

Data Availability Statement: Not applicable.

Conflicts of Interest: The authors declare no conflict of interest.

References

1. Geim, A.K.; Novoselov, K.S. The rise of graphene. *Nat. Mater.* **2007**, *6*, 183–191. [CrossRef] [PubMed]
2. Heydari, M.B.; Samiei, M.H.V. Plasmonic Graphene Waveguides: A Literature Review. *arXiv* **2018**, arXiv:1809.09937. https://doi.org/10.48550/ARXIV.1809.09937.
3. Shangguan, Q.; Chen, Z.; Yang, H.; Cheng, S.; Yang, W.; Yi, Z.; Wu, X.; Wang, S.; Yi, Y.; Wu, P. Design of Ultra-Narrow Band Graphene Refractive Index Sensor. *Sensors* **2022**, *22*, 6483. [CrossRef] [PubMed]
4. Shangguan, Q.; Zhao, Y.; Song, Z.; Wang, J.; Yang, H.; Chen, J.; Liu, C.; Cheng, S.; Yang, W.; Yi, Z. High sensitivity active adjustable graphene absorber for refractive index sensing applications. *Diam. Relat. Mater.* **2022**, *128*, 109273. [CrossRef]
5. Ye, Z.; Wu, P.; Wang, H.; Jiang, S.; Huang, M.; Lei, D.; Wu, F. Multimode tunable terahertz absorber based on a quarter graphene disk structure. *Results Phys.* **2023**, *48*, 106420. [CrossRef]
6. Cheng, Z.; Liao, J.; He, B.; Zhang, F.; Zhang, F.; Huang, X.; Zhou, L. One-Step Fabrication of Graphene Oxide Enhanced Magnetic Composite Gel for Highly Efficient Dye Adsorption and Catalysis. *ACS Sustain. Chem. Eng.* **2015**, *3*, 1677–1685. [CrossRef]
7. Mikhailov, S.A. Non-linear electromagnetic response of graphene. *Europhys. Lett.* **2007**, *79*, 27002. [CrossRef]
8. Mikhailov, S.A.; Ziegler, K. Nonlinear electromagnetic response of graphene: Frequency multiplication and the self-consistent-field effects. *J. Phys. Condens. Matter* **2008**, *20*, 384204. [CrossRef]
9. Hendry, E.; Hale, P.J.; Moger, J.; Savchenko, A.K.; Mikhailov, S.A. Coherent Nonlinear Optical Response of Graphene. *Phys. Rev. Lett.* **2010**, *105*, 097401. [CrossRef]
10. Ooi, K.J.; Tan, D.T. Nonlinear graphene plasmonics. *Proc. R. Soc. Math. Phys. Eng. Sci.* **2017**, *473*, 20170433. [CrossRef]
11. Mikhailov, S.A. Quantum theory of the third-order nonlinear electrodynamic effects of graphene. *Phys. Rev. B* **2016**, *93*, 085403. [CrossRef]
12. Liang, H.; Ruan, S.; Zhang, M.; Su, H.; Li, I.L. Graphene surface plasmon polaritons with opposite in-plane electron oscillations along its two surfaces. *Appl. Phys. Lett.* **2015**, *107*, 091602. [CrossRef]
13. Gan, C.H.; Chu, H.S.; Li, E.P. Synthesis of highly confined surface plasmon modes with doped graphene sheets in the midinfrared and terahertz frequencies. *Phys. Rev. B* **2012**, *85*, 125431. [CrossRef]
14. Zhu, B.; Ren, G.; Zheng, S.; Lin, Z.; Jian, S. Nanoscale dielectric-graphene-dielectric tunable infrared waveguide with ultrahigh refractive indices. *Opt. Express* **2013**, *21*, 17089–17096. [CrossRef]
15. Svintsov, D.; Vyurkov, V.; Ryzhii, V.; Otsuji, T. Voltage-controlled surface plasmon-polaritons in double graphene layer structures. *J. Appl. Phys.* **2013**, *113*, 053701. [CrossRef]
16. Belonenko, M.B.; Lebedev, N.G.; Yanyushkina, N.N. Solitons in a system of coupled graphene waveguides. *Phys. Solid State* **2012**, *54*, 174–177. [CrossRef]
17. Buslaev, P.I.; Iorsh, I.V.; Shadrivov, I.V.; Belov, P.A.; Kivshar, Y.S. Plasmons in waveguide structures formed by two graphene layers. *JETP Lett.* **2013**, *97*, 535–539. [CrossRef]
18. Evseev, D.A.; Eliseeva, S.V.; Sementsov, D.I. Waves in a plane graphene-dielectric waveguide structure. *Eur. Phys. J. Appl. Phys.* **2017**, *80*, 10501. [CrossRef]
19. Wright, A.R.; Xu, X.G.; Cao, J.C.; Zhang, C. Strong nonlinear optical response of graphene in the terahertz regime. *Appl. Phys. Lett.* **2009**, *95*, 072101. [CrossRef]
20. Cheng, J.L.; Vermeulen, N.; Sipe, J.E. Third order optical nonlinearity of graphene. *New J. Phys.* **2014**, *16*, 053014. [CrossRef]
21. Hanson, G.W. Dyadic Green's functions and guided surface waves for a surface conductivity model of graphene. *J. Appl. Phys.* **2008**, *103*, 064302. [CrossRef]
22. Falkovsky, L.A. Optical properties of graphene. *J. Phys. Conf. Ser.* **2008**, *129*, 012004. [CrossRef]
23. Bludov, Y.V.; Smirnova, D.A.; Kivshar, Y.S.; Peres, N.M.R.; Vasilevskiy, M.I. Nonlinear TE-polarized surface polaritons on graphene. *Phys. Rev. B* **2014**, *89*, 035406. [CrossRef]
24. Hajian, H.; Rukhlenko, I.D.; Leung, P.T.; Caglayan, H.; Ozbay, E. Guided plasmon modes of a graphene-coated Kerr slab. *Plasmonics* **2016**, *11*, 735–741. [CrossRef]
25. Valovik, D.V.; Smirnov, Y.G. Calculation of the Propagation Constants of TM Electromagnetic Waves in a Nonlinear Layer. *J. Commun. Technol. Electron.* **2008**, *53*, 883–889. [CrossRef]
26. Snyder, A.; Love, J. *Optical Waveguide Theory*; Chapman and Hall: London, UK, 1983.

photonics

MDPI

Article

Some Optical Properties of $LiNbO_3:Gd^{3+}(0.003):Mg^{2+}(0.65$ wt %) Single Crystal: A Promising Material for Laser Radiation Conversion

Nikolay Sidorov [1], Mikhail Palatnikov [1,*], Natalya Teplyakova [1], Alexander Pyatyshev [2] and Alexander Skrabatun [2]

[1] Tananaev Institute of Chemistry—Subdivision of the Federal Research Centre "Kola Science Centre of the Russian Academy of Sciences", 184206 Apatity, Russia; n.sidorov@ksc.ru (N.S.); n.tepliakova@ksc.ru (N.T.)
[2] P. N. Lebedev Physical Institute of the Russian Academy of Sciences, 119991 Moscow, Russia; jb_valensia@mail.ru (A.P.); skrabatunav@lebedev.ru (A.S.)
* Correspondence: m.palatnikov@ksc.ru

Abstract: A nonlinear optical double-doped single-crystal $LiNbO_3$:Gd:Mg (Gd concentration is 0.003, Mg—0.65 wt % in the crystal) has been researched by several optical methods: laser conoscopy, photoinduced light scattering (PILS), optical spectroscopy, and Raman scattering. The crystal has been shown to have no photorefraction effect and a high optical uniformity. Fine features of the crystal structure have been studied via Raman spectra. Spectra have been registered in the first and second orders, they have been excited by visible (532 nm) and near-IR (785 nm) lasers. Registered Raman spectra have the fundamental vibrations of the crystal lattice of the A_1(TO,LO)- and E(TO,LO)-type symmetry located in the range of 150–900 cm^{-1}. A number of low-intensity Raman bands in the 900–2000 cm^{-1} region have been determined to correspond to the second-order Raman spectrum. These bands are polarized and appear only in certain polarization-scattering geometries. They appear in the spectrum excited by visible radiation, but their number and intensity are much lower than those excited by near-IR lasers. Oxygen-octahedral MeO_6 clusters in our case can contain Li, Nb, Gd, or Mg in the Me site. The clusters in the $LiNbO_3$:Gd:Mg crystal structure are slightly distorted compared with similar clusters in the nominally pure $LiNbO_3$ crystal. It has been established that the value R = [Li]/[Nb] in the studied crystal is ≈ 1. Such a ratio usually characterizes a nominally pure stoichiometric crystal.

Keywords: lithium niobate; single crystal; double doping; photorefraction; laser conoscopy; photoinduced light scattering; Raman scattering; second order spectra

1. Introduction

Modern physical materials science is highly interested in the development of optical materials with a low photorefraction effect (also known as optical damage). Such materials can be based on a nonlinear optical single crystal of lithium niobate (LN, $LiNbO_3$). LN in a crystal state has high nonlinear optical coefficients, which is crucial for the development of functional materials. Laser radiation can be generated, converted, and modulated using such materials [1–6]. In this case, an important requirement in the conversion of both laser and broadband radiation is minimization of various distortions that are inevitably present in the crystalline matrix. There are a number of factors that significantly limit the possibilities of creating highly advanced optical materials for various purposes [6–11] based on $LiNbO_3$. The optical damage phenomenon is an effect of a photoinduced change in the refractive indices. It is the most important factor, since it causes strong distortions of photoluminescence and destroys the laser radiation wavefront passing through the $LiNbO_3$ crystal. We should also expect such compositional non-uniformities as microstructures (clusters) in the structure of the $LiNbO_3$ crystal; they have various sizes and their structure differs from that of the crystalline matrix.

$LiNbO_3$:Mg(5.0 mol % or 0.85 wt %) is $LiNbO_3$ single-crystal-based optically highly perfect and compositionally uniform material. It has a low photorefraction effect and it is currently used for optical frequency conversion in the phase quasi-phase-matching mode on regular domain structures (PPLN—periodically-poled $LiNbO_3$) [12–15]. However, such a strong doping decreases compositional and optical uniformity of a single-doped crystal compared with a nominally pure congruent crystal. The decrease is caused by the closeness of the magnesium concentration in the $LiNbO_3$:Mg (0.85 wt %) crystal to the threshold value of the magnesium concentration in LN (0.935 wt % or 5.5 mol %). Dopants tend to incorporate into certain positions of LN structure. The threshold changes abruptly the incorporation mechanism of the dopant [16]. Thus, it is interesting to investigate the possibilities of double doping of the LN crystal. Double and triple doping can create highly perfect LN single crystals with an improved compositional uniformity [17]; the magnesium concentration is expected to be significantly lower than the threshold concentration. Double doping provides finer control than single doping. Doping controls many properties of an LN crystal. The most important ones are the alteration of the cations along the LN crystal polar axis and the polarizability of MeO_6 clusters. Me sites in general can be occupied with Li, Nb, vacancy, impurity, or doping metal. The alteration of cations determines the magnitude of spontaneous polarization, and the MeO_6 cluster polarizability determines the nonlinear optical properties of the crystal. Doping (including double and multiple doping) also influences the type and concentration of defects in a crystal. Defects can be both point and complex (extended). These defects produce deep and shallow energy levels in the LN band gap, which significantly affects the magnitude of the photorefraction effect and the optical damage resistance of the crystal [6]. Nonlinear optical coefficients are higher in perfect double-doped $LiNbO_3$ crystals, including those co-doped with magnesium, compared with single-doped crystals [18,19].

Important information about the structural perfection, compositional uniformity, and optical damage resistance of LN crystals can be obtained using a set of methods: laser conoscopy, photoinduced light scattering (PILS), and Raman spectroscopy of the first and second orders. The Raman spectra are sensitive to the slightest rearrangements in the structural units of the LN crystal [7,8,10,20–22]. Second-order spectra are especially sensitive and informative because they are absent from highly perfect crystals. Second-order Raman spectra exist only because of two factors: the disordering of the crystal structure and the anharmonic interaction between the structural units of the crystal [10,23–25]. Thus, the more perfect the crystal structure is, the less intense and less "rich" the second-order Raman spectrum should be. The Raman spectra of various substances obtained upon excitation in the visible and near IR regions can noticeably differ from each other [26,27]. This may be caused by several reasons: different sensitivities of their secondary structures, structural disorder, and anharmonicity, which inevitably exist in different crystals. By the secondary structure we mean micro- and macrostructures, clusters, and point defects. Point defects can be caused by irregularly located intrinsic, doping, and impurity atoms.

In this work, the features of the defect structure, compositional uniformity, and photorefractive properties are investigated in a double-doped $LiNbO_3$:Gd:Mg (Gd concentration is 0.003, Mg—0.65 wt % in the crystal) single crystal. We investigated Raman spectroscopy in the frequency range 50–2000 cm^{-1}, and spectra were excited in the visible and near-IR region. Optical spectroscopy, laser conoscopy, and PILS were also applied in the study. The compositional uniformity and photorefractive properties of the $LiNbO_3$:Gd:Mg crystal were controlled with those for nominally pure (undoped) crystals of congruent (CLN, R = [Li]/[Nb] = 0.946) and stoichiometric (SLN, R = 1) compositions. The Raman spectra of a series of doubly doped $LiNbO_3$:Gd:Mg crystals of various compositions have been previously studied in [20]. The crystal structure of samples with relatively low concentrations of Gd^{3+} and Mg^{2+} are characterized by better ordering in the cation arrangement along the polar axis. However, in [20], only the Raman spectra of fundamental vibrations of the crystal lattice have been studied in the frequency range 50–1000 cm^{-1}. Second-order Raman spectra (above 1000 cm^{-1}) have not been recorded in [20] due to the insufficient sensitivity of the spectrometers of that time. In addition, the $LiNbO_3$:Gd:Mg crystal has not been previously studied by optical spectroscopy, laser conoscopy, and PILS. Doping Gd^{3+} and Mg^{2+} cations have different valencies and ionic radii. Both are non-photorefractive dopants. Doping affects the crystal defectivity, spontaneous polarization, the geometry of NbO_6 clusters, and the effect of photorefraction. The two different dopants have different influences on these factors. At the same time, additional Raman bands appear upon doping with magnesium > 3.0 wt %. The bands correspond to pseudoscalar fundamental lattice vibrations of the A_2 symmetry type, which are forbidden by the selection rules for the R3c symmetry space group (SPGR) [28]. An increase in the magnesium concentration in single- and double-doped LN crystals increases the intensity of E-symmetry-type Raman vibrations without a noticeable frequency shift [29–31].

2. Materials and Methods

The creation of optically and compositionally uniform LN crystals with double doping by the Czochralski method is quite a difficult task. It becomes even more complex when the concentration of one of the dopants is significantly higher than the concentration of the other one. Mg and Gd have strongly different dopant distribution coefficients in LN: $K_{DMg} \approx 1.1$, $K_{DGd} \approx 0.25$ [20]. Consequently, the composition of the melt layer near the crystallization front during crystal growth can simultaneously have an excess of one dopant and a lack of the other. Thus, the composition of a doped crystal cone almost always fails to coincide with the composition of the bottom of the crystal; such a difference causes compositional and optical non-uniformity. Thus, different parts of the crystal will have substantially different characteristics. To minimize such shortcomings, it is necessary to experimentally select the optimal technological parameters of crystal growth and their combinations: temperature gradients in the melt and growth zones, and the speed of rotation and pulling of the crystal. The shortcomings are also eliminated by the effective technological methods: melt pre-treatment before crystal growth, a special design of the thermal unit, the use of low crystallization rates, long after-growth annealing, and appropriate conditions for electro-thermal treatment of the crystal [6,20].

The charge was synthesized using Nb_2O_5 ($<3 \times 10^{-4}$ wt % of each impurity, Solikamsk Magnesium Works, Solikamsk, Russia) prepared due to Russian technical standard 1763–025-00545484–2000 and Li_2CO_3 ($<3 \times 10^{-4}$ wt % of each impurity, Solikamsk Magnesium Works, Solikamsk, Russia). These components were mixed under certain conditions in order to prepare granulated congruent charge ($[Li_2O]$ = 48.6 mol %). Ready charge of any composition had a high bulk density (~3.4 g/cm^3). The preparation of the charge is described in detail in [6,32].

The congruent melt ([Li]/[Nb] = 0.946) was directly doped with Gd_2O_3 (not more than 10^{-3} of each impurity, Neva Reactiv, Russia) and MgO (not more than 10^{-4} of each impurity, Himkraft, Russia) oxides. Table 1 shows the concentration of basic impurities in the granular charge and the grown $LiNbO_3$:Gd:Mg single crystal.

Table 1. Impurity concentrations (C, wt %) in the granular charge, in the cone and bottom parts of the $LiNbO_3$:Gd:Mg crystal.

	Impurity	Mn	Ni	Al	Fe	Cr, Cu, V	Pb, Sn	Bi	Mg	Si, Ti, Mo, Ca, Co	Sb	Zr
$C \cdot 10^{-3}$, wt %	Charge	<0.2	<0.3	<0.3	<0.3	0.3	<0.5	0.5	0.5	1	2.1	<10
	Crystal cone	<0.2	<0.3	<0.3	0.32	0.3	<0.5	0.5	0.53	1	1.7	10
	Crystal bottom	<0.2	<0.3	1	0.38	0.3	<0.5	0.5	0.58	1	2	10

The melt was overheated for 8–11 h by 180–200 °C relative to the regular LN melting temperature (T_{melt} = 1263 °C) before the seeding. This procedure homogenizes the impurities in the melt.

An $LiNbO_3$:Gd:Mg single crystal was grown in air from a melt by the Czochralski method. Platinum crucibles, 75 mm in diameter, were used during growth. Crystals were grown at a low axis gradient (~2.5 deg/cm). According to the Czochralski method, pulling was carried out along the polar axis (Z-cut, (001)). The rotation speed was 16–18 rpm and the speed of crystal pulling was 0.7 mm/h. The crystal increment rate was 1.02–1.04 mm/h. Growth was carried out at the Kristall-2 facility (Voroshilovgradsky zavod electronnogo mashinostroeniya, Voroshilovgrad, USSR) equipped with automatic crystal diameter control. The $LiNbO_3$:Gd:Mg growth parameters of the crystal were selected based on the need to obtain a flat crystallization front, which should ensure a sufficiently high structural perfection of the crystal. The growth was considered finished when the $LiNbO_3$:Mg:Gd crystal achieved ≤275 g; a growth unit weight sensor determined the crystal weight. An exact weight of the $LiNbO_3$:Gd:Mg crystal was determined after extraction of the crystal seed. Approximately 25% of the total melt weight transferred into the crystal. The grown crystal had a flat crystallization front and the following geometric dimensions: diameter (∅) ≈ 38 mm, length of the cylindrical part L_c ≈ 40 mm. To relieve thermoelastic stresses, the grown crystal was subjected to heat treatment in an air atmosphere at 1500 K for 15 h in a high-temperature annealing furnace «Lantan» (Voroshilovgradsky zavod electronnogo mashinostroeniya, Voroshilovgrad, USSR). The design of the thermal unit used had a double "insulation", it is described in [33]. The thermal unit creates an isothermal zone in the bulk of the platinum screen. The zone serves for postgrowth annealing of the crystal. It also provides a small temperature gradient at the crystallization front during crystal growth.

After growth, the $LiNbO_3$:Gd:Mg crystal was not cooled immediately; it was annealed at 1200 °C in a growth facility for 10 h and only then cooled at a controlled rate of ~50 °C/h. The prolonged post-growth annealing was carried out in the isothermal zone under the platinum screen. This step homogenizes the composition of the doped crystal and relieves possible thermal and mechanical stresses.

Control CLN and SLN single crystals were grown using a similar sequencing: CLN from a congruent melt (R = 0.946) and SLN from a melt with 58.5 mol % Li_2O.

As-grown CLN, SLN, and $LiNbO_3$:Gd:Mg crystals are not strictly single domains. All LN crystals are supposed to be turned to a single domain state; the procedure is called high-temperature electrodiffusion annealing: a constant electric voltage is applied to the polar cuts of the LN crystal during its cooling at a slow rate of 20 °C/h in the temperature range ~1230–870 °C.

The distribution of a dopant concentration in the charge and $LiNbO_3$:Gd:Mg crystal was determined by an analysis of plates cut from the top (closest to the cone) and bottom parts of the boule. The concentrations of different dopants in the milled samples were determined by different methods. Inductively coupled plasma atomic emission spectroscopy was used to study Mg concentration (ICP AES) on an Optima 8300 ICP-OES spectrometer (PerkinElmer, Hopkinton, MA, USA). Atomic absorption spectrometry (AAS) was used to study Gd concentration on a Kvant-FA instrument (GRANAT, Russia, Moscow).

The samples for investigations by Raman spectroscopy, optical spectroscopy, laser conoscopy, and PILS were parallelepipeds. Pieces were cut from the boule bulk. Their dimensions were 4.93 × 5.98 × 8.41 mm^3, and the ribs of the samples coincided with the directions of the principal LN crystallophysical axes. Careful polishing was applied to the faces of the parallelepipeds.

Experimental setups for recording conoscopic patterns by laser conoscopy and PILS patterns are described in the book [32]. Conoscopic patterns were excited by a laser radiation with a wavelength of 532 nm and powers of 1 and 90 mW. The optical transmission spectra of the single crystal were recorded on a UNICO 2800 UV/VIS spectrophotometer (United Products & Instruments, Dayton, NJ, USA). The range of detection was

190–1100 nm with a resolution of 0.2 nm. A BWS465-532S spectrometer (B&W Tek, Plainsboro Township, NJ, USA) was used while recording the Raman spectra in the visible region. This setup is able to fix spectra in the range of 50–4000 cm^{-1} with a resolution of 4.5 cm^{-1}. A BWS465-785H spectrometer (B&W Tek, Plainsboro Township, NJ, USA) recorded the Raman spectra in the near-IR region in the range of 50–2850 cm^{-1} with a resolution of 3.5 cm^{-1}. To excite the Raman spectrum in the visible and near-IR regions, cw lasers with a wavelength of 532 nm and 785 nm were used. The power of green laser radiation (λ_0 = 532 nm) during registration of Raman spectra was 30 mW, and the power of radiation in the near-IR region (λ_0 = 785 nm) was 340 mW. In all cases, the numerical aperture of the focusing system was 0.22. The size of the laser spot at the focus was 85 μm. All Raman spectra were recorded at room temperature. The geometry in all cases was backscattering. The use of long-wavelength (785 nm) exciting laser radiation ensured suppression of the photorefraction effect observed in LN crystals when illuminated with shorter-wavelength radiation.

3. Results and Discussion

3.1. Transmission Spectra of the $LiNbO_3$:Gd:Mg Crystal

Figure 1 shows the transmission spectrum of a $LiNbO_3$:Gd:Mg crystal.

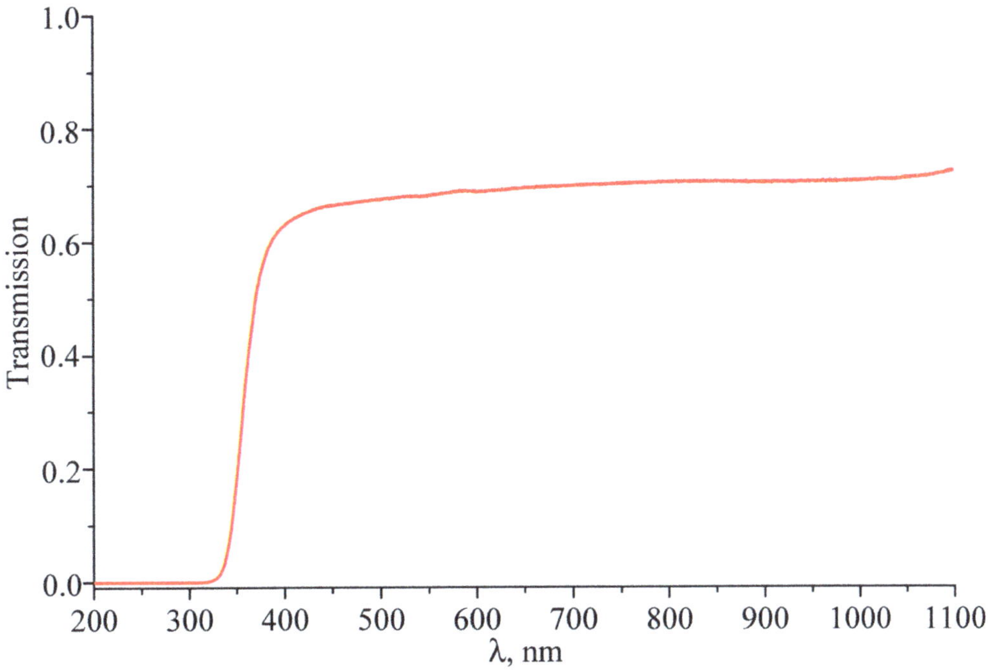

Figure 1. Transmission spectrum of a $LiNbO_3$:Gd:Mg crystal.

Figure 1 demonstrates that the crystal under study has a fairly high and stable transmittance in the spectral range of 450–1100 nm.

The bandgap width was determined according to the formula $\Delta E_g = hc/\lambda$, where λ is the wavelength corresponding to the fundamental absorption edge, h is Planck's constant, and c is the speed of light in vacuum. The accuracy of determining the boundary of the fundamental absorption edge was ±1.0 nm. The band gap in the $LiNbO_3$:Gd:Mg crystal is ~3.62 eV. The band gap has an intermediate value between the values of CLN~3.72 eV and SLN~3.48 eV [32]. The electrical conductivity of a CLN crystal is ~10^{-16}–10^{-15} $(\Omega\ cm)^{-1}$ [34]. The smaller value of the band gap in the SLN crystal indicates a higher density of small energy levels ("sticking levels") in the band gap, which increases the photorefractive effect and the electrical conductivity of the crystal.

3.2. *Laser Conoscopy of the $LiNbO_3$:Gd:Mg Crystal*

Figure 2 illustrates conoscopic patterns of the $LiNbO_3$:Gd:Mg crystal. The patterns were obtained upon excitation by a laser radiation with a wavelength of 532 nm and powers of 1 and 90 mW. For comparison, conoscopic patterns of nominally pure CLN, SLN crystals are also given in Figure 2.

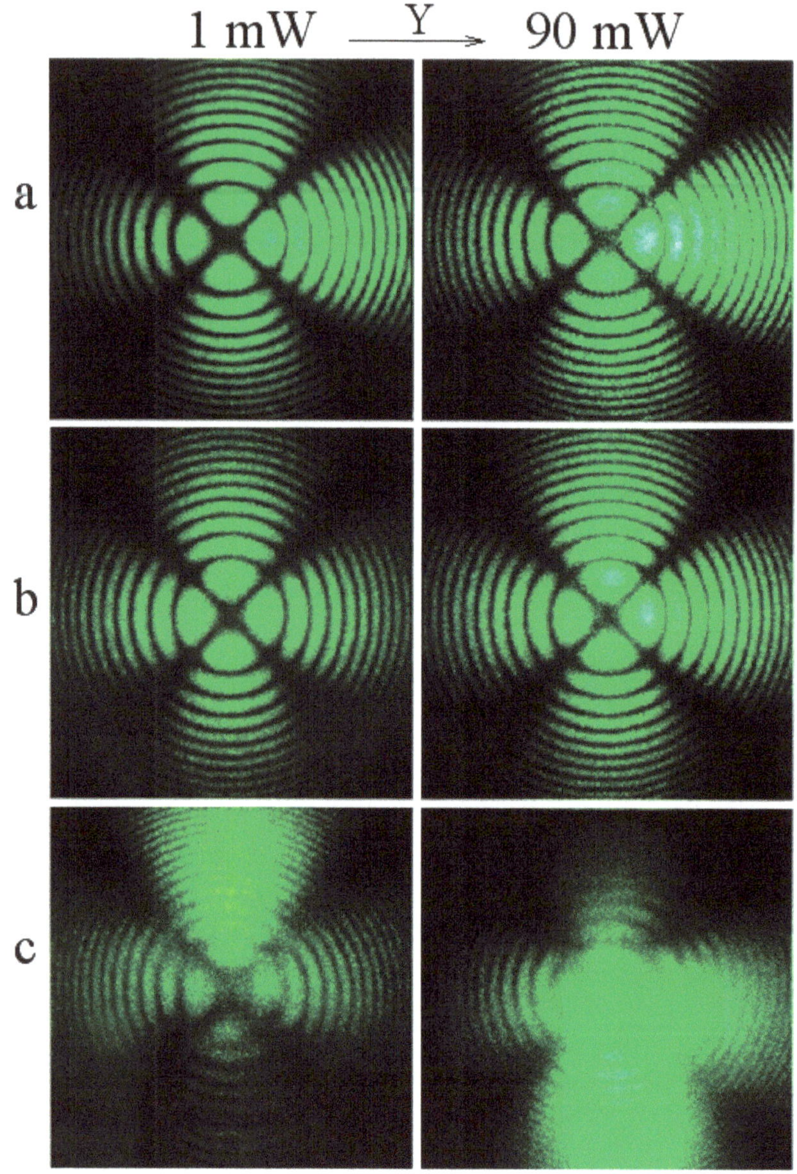

Figure 2. Conoscopic patterns of the crystals: (**a**)—$LiNbO_3$:Gd:Mg; (**b**)—CLN; (**c**)—SLN, λ_0 = 532 nm, P = 1 and 90 mW.

Conoscopic patterns obtained at a laser power of 1 mW for $LiNbO_3$:Gd:Mg and CLN crystals correspond to uniaxial crystals. Moreover, such a low power of laser radiation (1 mW) fails to excite the photorefractive effect; thus conoscopic patterns characterize only the structural defectiveness of crystals. The conoscopic patterns of $LiNbO_3$:Gd:Mg and CLN crystals are ideal patterns for a uniaxial crystal. The conoscopic patterns have the background of concentric isochrome rings. The branches of the Maltese cross are formed by two isogyres of minimum intensity. Branches intersect in the center of the field of view; they are perpendicular to each other, and coincide with the transmission axes of the polarizer and analyzer. This type of conoscopic pattern indicates the high optical uniformity of the CLN and $LiNbO_3$:Gd:Mg crystals and their high optical quality. Scanning with a laser beam along the input face of these crystals also confirms that the optical uniformity over the volume of the $LiNbO_3$:Gd:Mg crystal is high, coincides with that for the CLN crystal, and is much higher than the optical uniformity of the SLN crystal. It is a known fact that congruent LN crystals are distinguished by the maximum optical homogeneity among all nominally pure (with different Li/Nb ratios) and doped with different metals LN single crystals [32].

Figure 2 reveals that the conoscopic patterns of the SLN crystal differ significantly from those of the $LiNbO_3$:Gd:Mg and CLN crystals. SLN conoscopic patterns contain signs of anomalous optical biaxiality: a noticeable deformation of the optical indicatrix of the crystal in the direction corresponding to the direction of the shift of the parts of the Maltese cross. The isochromes are stretched in the direction of displacement of the cross fragments and take the form of ellipses. The conoscopic pattern itself is significantly blurred (Figure 2). The detected anomalies in the conoscopic patterns mean that the SLN crystals are optically non-uniform.

With an increase in the laser radiation power to 90 mW, both the intrinsic defectiveness of the crystal (determined by the composition and conditions of crystal growth) and the imperfections induced by laser radiation appear in the conoscopic patterns. To study the effect of PILS on conoscopic patterns, the crystals were preliminarily irradiated with ~160 mW laser radiation. Increasing the laser power to 90 mW leads to a certain enhancement of the contrast and clarity of the image; however, no additional distortions appear in the conoscopic patterns of $LiNbO_3$:Gd:Mg and CLN crystals (Figure 2a,b). The absence of a photorefractive response of these crystals might be the reason for such behavior (Figure 3a,b, see Section 3.3). The anomalies of the conoscopic patterns of the SLN crystal increase with an increase in the laser radiation power; the contrast and sharpness of the image generally decrease; and the branches of the Maltese cross noticeably blur (Figure 2c). A significant distortion of the conoscopic patterns for the SLN crystal with an increase in the laser radiation power to ~90 mW is obviously associated with the presence of a significant photorefractive response. The response is confirmed by the data of PILS (Figure 3c). In this case, the effect of photorefraction is due to the capture of photoelectrons by deep traps in the bandgap; photoelectrons drift in the electric field when the crystal is illuminated due to the manifestation of the photovoltaic effect.

3.3. PILS of the $LiNbO_3$:Gd:Mg Crystal

Figure 3 demonstrates the PILS patterns of $LiNbO_3$:Gd:Mg, CLN, and SLN crystals, obtained upon excitation by a laser radiation with a wavelength of 532 nm and a power of 160 mW. The PILS speckle structure in LN crystals of different compositions predominantly opens along the polar axis of the crystal [32]. In our case, the exciting laser radiation is directed perpendicularly to the polar axis. Figure 3 shows that the PILS patterns of $LiNbO_3$:Gd:Mg and CLN crystals stay the same regardless of the time of laser radiation exposure to the crystal. Moreover, for these crystals, even after exposure to laser radiation with wavelength λ_0 = 532 nm and power of 160 mW for 360 s, the indicatrix of the PILS speckle structure is not revealed, and only circular scattering by static structural defects is observed (Figure 3a,b). This fact indicates the absence of photoinduced defects in the

$LiNbO_3$:Gd:Mg and CLN crystals exposed to laser radiation with a wavelength of 532 nm and a power of 160 mW.

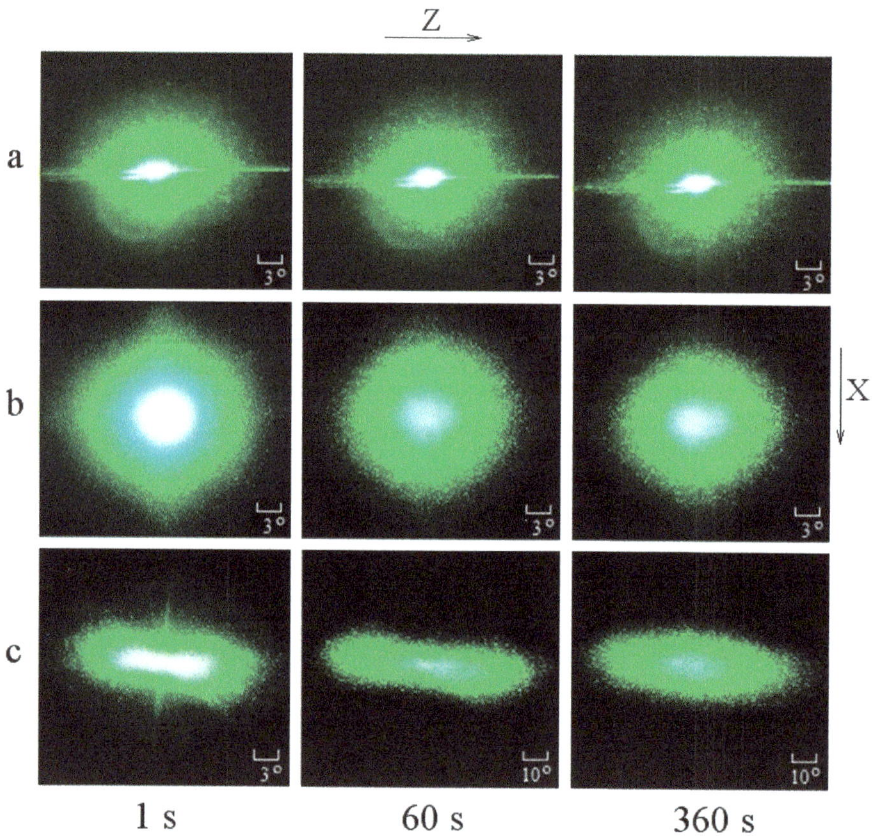

Figure 3. PILS patterns of the crystals: (**a**)—$LiNbO_3$:Gd:Mg; (**b**)—CLN; (**c**)—SLN, λ_0 = 532 nm, P = 160 mW.

For the SLN crystal, the speckle structure of the PILS indicatrix is completely opened. Moreover, the dynamics of its disclosure over time is unusual. The PILS indicatrix opens up very quickly, during the first second of exposure of the crystal to laser radiation. Subsequently, the central layer of the speckle structure, corresponding to the cross section of the laser beam at the exit from the crystal, almost completely disappears during the first 60 s of crystal irradiation. Then the central layer appears again, but it has a much lower brightness than at the first seconds of irradiation of the crystal. This indicates that the energy of the laser beam after the first seconds of impact begins to dissipate significantly on structural defects in the bulk of the crystal. The energy of the laser beam also dissipates on structural defects of CLN crystal, but much less, and the dissipation is not accompanied by the opening of the speckle structure of the PILS indicatrix (Figure 3). No energy dissipation is observed for the $LiNbO_3$:Gd:Mg crystal, which also indicates a small number of structural defects (Figure 3a).

The differences found in the features of PILS in $LiNbO_3$:Gd:Mg, SLN, and CLN crystals indicate the following: the band gap of the SLN crystal has a high density of shallow energy levels (sticking levels [9]) near the bottom of the band gap. Shallow levels are added to deep energy traps. Shallow levels are caused by shallow electron traps in the form of defects V_{Li},

V_{Nb}, Cu_V, etc., which significantly increase the diffusion photoelectric fields that determine the features of the dynamics of the opening of the speckle structure of the PILS indicatrix in time. At the same time, the CLN crystal is distinguished by a high concentration (≈5 wt %) of deep electron traps in the form of Nb_{Li} point defects [6,20]. However, CLN has much fewer of such shallow electron traps, and, therefore, the fraction of the photovoltaic mechanism in the formation of the photorefractive effect is much larger. Indeed, the values of the photoelectric fields (diffusion E_d and photovoltaic E_{pv}) excited in certain conditions (λ_0 = 532 nm, P = 160 mW (6.29 W/cm^2)) were: E_{pv} = 3178 V/cm, E_d = 1360 V/cm for SLN crystal; E_{pv} = 5620 V/cm, E_d = 104 V/cm for CLN crystal; E_{pv} = 3700 V/cm, E_d = 735 V/cm for $LiNbO_3$:Gd^{3+}(0.003):Mg^{2+}(0.65 wt %) crystal. The values of the band gap ΔE_g were 3.48, 3.72, and 3.62 eV, respectively, for SLN, CLN, and $LiNbO_3$:Gd^{3+}(0.003):Mg^{2+}(0.65 wt %) crystals [35].

Thus, the data show an increase in the diffusion field (104, 745, and 1360 V/cm) in the series of CLN, $LiNbO_3$:Gd:Mg, SLN crystals. At the same time, the band gap decreases. The predominant mechanism of photorefraction in all studied crystals is photovoltaic, since the value of the photovoltaic field is much greater than that of the diffusion field. The number of deep electron traps caused by multiply charged transition metal impurity cations (Fe, Cu, etc.), which affect the photorefraction effect and PILS, is the same in all the crystals under study since single crystals were grown using the same granular charge with the same composition and concentration of trace metal impurities (Table 1).

Deep electron traps in LN are usually Nb_{Li} point defects and impurities of multiply charged transition metal cations Fe, Cu, etc.; shallow electron traps are formed by point defects V_{Li}, V_{Nb}, Cu_V, etc. The presence of both can also lead to a high dissipation of the absorbed energy on lattice phonons, which should show up in the vibrational spectrum of the crystal. The magnitude of the photorefraction effect depends on the wavelength of the laser radiation. Thus, there should be noticeable differences in the spectra when excited by the visible laser radiation (λ_0 = 532 nm when the photorefraction effect is present) and the near-IR (λ_0 = 785 nm when there is no photorefraction effect).

3.4. Raman Spectra of the $LiNbO_3$:Gd:Mg Crystal

Figure 4 shows the Raman spectra of the $LiNbO_3$:Gd:Mg crystal in scattering geometries: $x(zz,zy)\bar{x}$ and $z(xx,yy,xy)\bar{z}$, recorded in the range 50–1000 cm^{-1}. The spectra in Figure 4 were excited by lasers with wavelengths of 532 and 785 nm.

Figure 4 shows that the spectra contain intense bands with frequencies 253–256 and 627–629 cm^{-1}, attributed to fundamental phonons of A_1(TO) symmetry type along the polar Z axis; bands with frequencies 295–296, 615, and 875–876 cm^{-1}, attributed to fundamental phonons of A_1(LO) symmetry type; and all 9 bands with frequencies 152, 237, 262, 328, 363, 429, and 580 cm^{-1}, attributed to fundamental doubly degenerate phonons of E(TO) symmetry type; these phonons vibrate perpendicularly to the polar Z axis.

A weak band with a frequency 186 cm^{-1} was detected in $x(zz,zy)\bar{x}$ scattering geometry and 188 cm^{-1} in $z(xx,yy,xy)\bar{z}$ geometry upon excitation of Raman spectra with the 785 nm laser. According to [36–39], these bands are attributed to a fundamental line 1E(LO). The $z(xx,yy,xy)\bar{z}$ geometry contains a wide Raman band with a frequency of 732 cm^{-1}. The attribution of the band is still discussed today. In [40], this band is attributed to the ilmenite phase. In article [41], the authors believe that it corresponds to the second-order Raman line. Finally, in a number of works [42,43], this band is assigned to the fundamental mode E(TO).

The Raman spectrum of a LN single crystal of any composition (both nominally pure with different [Li]/[Nb] ratios and doped) is distinguished by an abundance of low-intensity bands not provided by the selection rules for fundamental phonons for the SPGR R3c (C_{3V}^6). This SPGR describes the unit cell symmetry of a LN crystal with the exact composition $LiNbO_3$ and two formula units in it. Such bands are called "extra bands". According to numerous studies in the literature, the number of extra low-intensity bands in the Raman spectrum only in the range of 50–1000 cm^{-1} is approximately equal to the number of bands (26) corresponding to fundamental phonons [6,8,10,20,24–26,28].

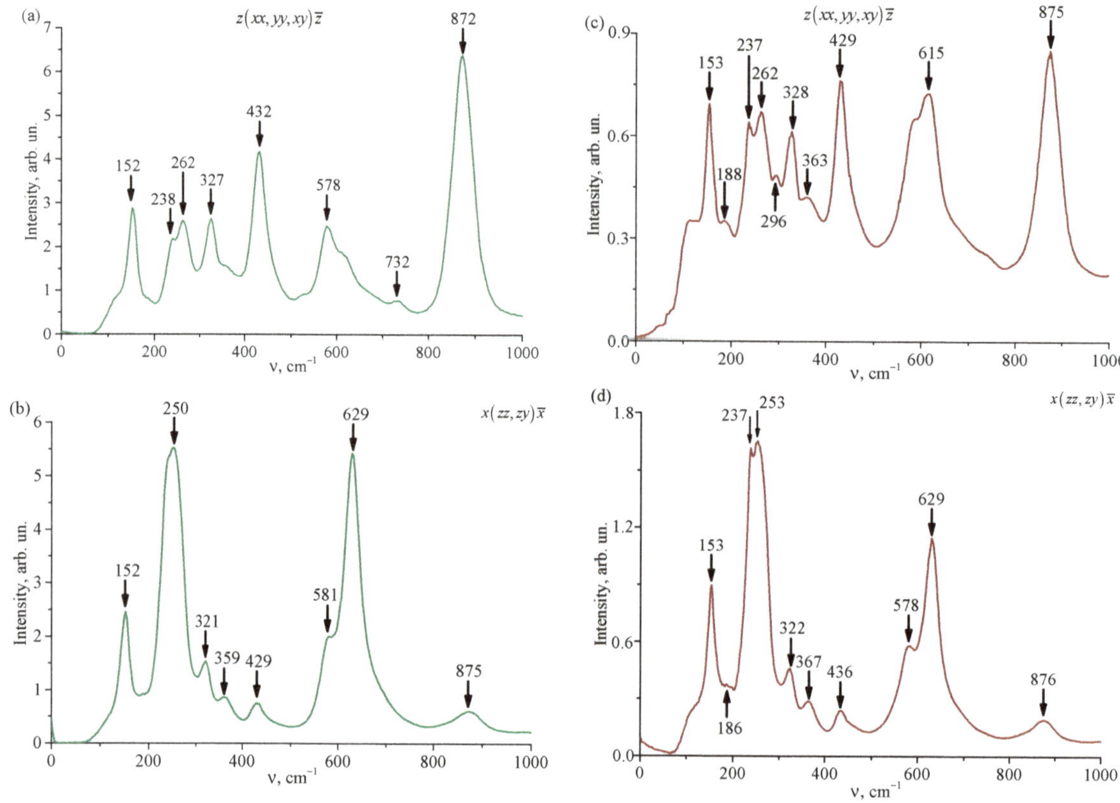

Figure 4. Raman spectra of a $LiNbO_3$:Gd:Mg crystal. Spectra were recorded in scattering geometries $x(zz, zy)\bar{x}$ and $z(xx, yy, xy)\bar{z}$ and excited by a laser line with a wavelength of (**a**,**b**) 532 nm and (**c**), (**d**) 785 nm in the range 0–1000 cm^{-1}.

The Raman spectra of the $LiNbO_3$:Gd:Mg and SLN crystals did not contain a low-intensity band with a frequency of 120 cm^{-1} of the A_1(TO) symmetry type. The band corresponds to two-particle states of acoustic phonons with a total wave vector equal to zero. It is usually observed in the involved scattering geometries in the spectra of all nonstoichiometric LN crystals [6,8,10,20,24,25]. It was experimentally shown in [6,8,10,20,24,25] that the Raman spectrum of a highly perfect stoichiometric $LiNbO_3$ crystal has no bands in the range of 110–130 cm^{-1}. The absence of Raman bands in the range of 110–130 cm^{-1} of a highly perfect SLN crystal was first noted in a theoretical work [44]. It was explained by the fact that acoustic phonons with frequencies of 110–130 cm^{-1} are not active in the Raman spectrum since their wave vector is not equal to zero. Thus, if a 120 cm^{-1} A_1(TO) band is absent from the Raman spectrum, then the R = [Li]/[Nb] value in the $LiNbO_3$:Gd:Mg crystal approaches unity. Such an R value is usually characteristic of a nominally pure stoichiometric $LiNbO_3$ crystal [6,20,24,25,44]. The 110–130 cm^{-1} Raman bands in the spectrum of nonstoichiometric LN crystals of various compositions indicates a strong anharmonic interaction of the lowest-frequency fundamental phonons of the A_1(TO) symmetry type (quasi-soft mode) with the acoustic continuum [20,24,25]. In this case, if two acoustic phonons with oppositely directed wave vectors interact with each other in such a way that they form a quasiparticle (biphonon) whose wave vector is equal to zero, then such an acoustic biphonon will be active in the Raman scattering of light. Moreover, the intensity of the band in the Raman spectrum corresponding to this acoustic biphonon will be proportional to the interaction energy due to the Fermi resonance of the biphonon with low-frequency optical fundamental phonons of the A_1(TO) symmetry type.

In the Raman spectrum of the $LiNbO_3$:Gd:Mg crystal, we also did not find extra bands found in the spectrum of LN crystals heavily doped with magnesium (>4.0 mol %) [28]. The bands can be explained in the Raman spectrum by a decrease in the local crystal symmetry SPGR from C_{3V} to C_3 due to the distortion of MeO_6 clusters by doping cations [8,20]. For the point symmetry group C_3, the optical representation has the form 9A(z) + 9E(x,y) [20]. Thus, pseudoscalar phonons of the A_2 symmetry type transfer into phonons of the A(z) symmetry type. Phonons of the A_2 symmetry type are forbidden in the Raman spectrum for the C_{3V} symmetry group by the selection rules, but phonons of the A(z) symmetry type are allowed. Thus, if phonons of the A_2 symmetry type are absent, than co-doping with Gd^{3+} and Mg^{2+} cations has little effect on the geometry of oxygen-octahedral MeO_6 clusters in the $LiNbO_3$:Gd:Mg.

Figure 5 shows fragments of the second-order Raman spectra of the $LiNbO_3$:Gd:Mg crystal in the scattering geometries $x(zz,zy)\overline{x}$, $y(zz,zx)\overline{y}$, and $z(xx,yy,xy)\overline{z}$, recorded in the range 1000–2000 cm^{-1}, excited by both laser lines.

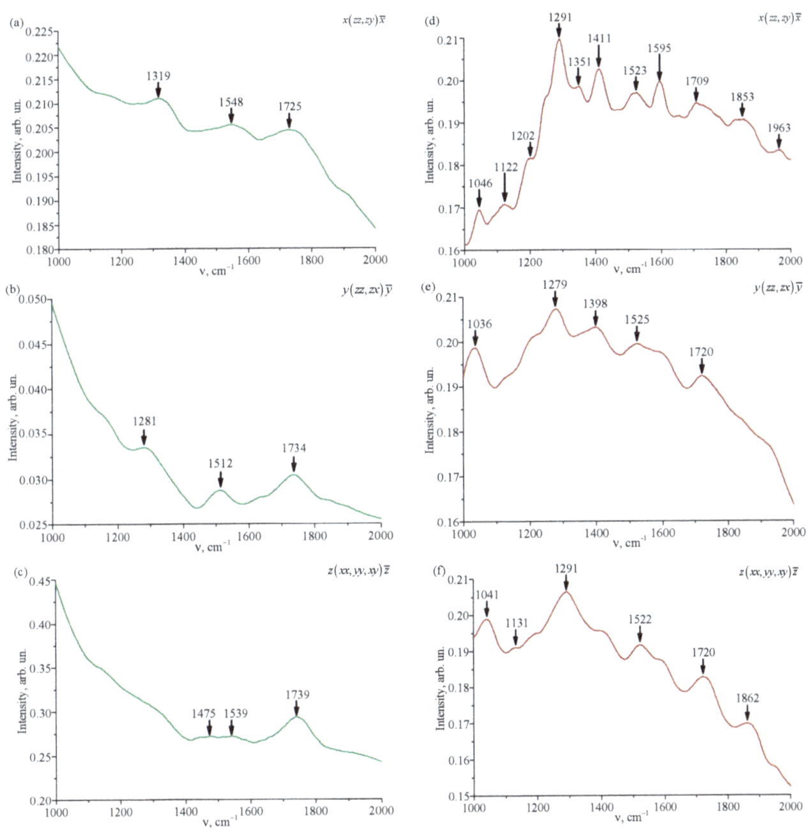

Figure 5. Fragments of the Raman spectra (on an enlarged scale) of the $LiNbO_3$:Gd:Mg crystal, obtained in scattering geometries $x(zz,zy)\overline{x}$, $y(zz,zx)\overline{y}$, and $z(xx,yy,xy)\overline{z}$, excited by laser radiation (**a–c**) 532 nm and (**d–f**) 785 nm in the range 1000–2000 cm^{-1}.

As is known, LN is a uniaxial crystal. Despite this, we have discovered that the second-order Raman spectra are polarized; for two orientations perpendicular to the polar axis, they differ noticeably. A number of broad bands appear in the range of 1000–2000 cm^{-1}, the frequencies of which do not correspond to the fundamental phonons of the crystal lattice (Figures 4 and 5). The intensity of these extra low-intence bands is almost two orders of magnitude less than the intensity of the strongest bands corresponding to the fundamental phonons of the lattice. Figure 5 also shows that, upon excitation by the visible laser line (532 nm), three second-order Raman lines bands were recorded; the frequencies of the bands do not exceed the exact value of the doubled frequency of the overtone of the fully symmetric fundamental mode $4A_1$(z)LO (875 cm^{-1}, Figure 4b). When excited by near-IR laser radiation, significantly more second-order Raman bands were recorded (Figure 5a–c). The frequencies of two second-order Raman bands significantly exceed the exact value of the doubled frequency of the overtone of the fully symmetric fundamental mode $4A_1$(z)LO (876 cm^{-1}, Figure 4d).

In the Raman spectra of nominally pure $LiNbO_3$ and $LiTaO_3$ single crystals, according to the data of numerous studies by different authors carried out over decades, no bands were found at all in the range of 900–2000 cm^{-1}. Quite intense bands were observed in the range of 900–2500 cm^{-1} [23,45–47] in the spectra of single crystals and ceramics of single and double doping $LiNbO_3$:Tb^{3+}(2.24 wt %), $LiTaO_3$:Cr^{3+}(0.005 wt %), $LiTaO_3$:Cr(0.2):Nd(0.45 wt %), and $LiNb_xTa_{1-x}O_3$. The bands are attributed to the second-order Raman spectrum. These samples have strongly disordered cationic sublattices. Their MeO_6 clusters are noticeably distorted compared with the MeO_6 clusters of a congruent and stoichiometric crystal. The reasons for the appearance in the range of 900–2000 cm^{-1} of low-intensity bands in the Raman spectrum of $LiNbO_3$ and $LiTaO_3$ single crystals and ceramics of various compositions remain unclear at present. The most probable is the attribution of these extra bands to the second-order spectrum, which arises in disordered and highly anharmonic crystals and ceramics. It is also unclear why the intensity of the extra bands is higher when the Raman spectrum is excited by laser radiation in the near-IR region (785 nm) than when it is excited in the visible region (532 nm). The currently existing methods for calculating spectra and theories do not yet allow us to explain this fact.

4. Conclusions

Optical transmission, conoscopy, PILS, optical spectroscopy, and Raman spectroscopy were used to study nonlinear optical single crystals of double doping $LiNbO_3$:Gd:Mg (Gd concentration is 0.003, Mg—0.65 wt % in the crystal).

The absence of distortion of the conoscopic patterns of $LiNbO_3$:Gd:Mg was established, which demonstrates the high optical uniformity of this crystal. When the crystal under study was irradiated for 360 s with laser radiation with a wavelength of 532 nm and a power of 160 mW, the PILS speckle structure did not open. During this time, only circular scattering by static structural defects was observed. These facts indicate the absence of photoinduced defects in the $LiNbO_3$:Gd:Mg crystal.

According to the Raman spectra of the first and second orders obtained in the range of 50–2000 cm^{-1} upon excitation by lasers with wavelengths of 532 and 785 nm, it was shown that the structures of the $LiNbO_3$:Gd:Mg MeO_6 clusters are slightly distorted compared with those of the nominally pure crystal and, in addition, the value of R = [Li]/[Nb] is increased compared with that for a congruent crystal. This indicates an increase in the stoichiometry of $LiNbO_3$ with its double doping with magnesium and gadolinium. The increase in stoichiometry is also confirmed by the fact that the Raman spectrum of the $LiNbO_3$:Gd:Mg crystal lacks a low-intensity band with a frequency of 120 cm^{-1} (A_1(TO)). This band corresponds to two-particle states of acoustic phonons with a total wave vector equal to zero. Its absence in the spectrum also indicates a small anharmonic interaction of the lowest-frequency fundamental vibration A_1(TO) of the symmetry type (quasi-soft mode) with the acoustic continuum. A low-intensity band with a frequency of 120 cm^{-1} (A_1(TO)) is definitely observed in the spectra of non-stoichiometric $LiNbO_3$ crystals, nominally pure and doped, and it is absent from the Raman spectrum of a stoichiometric crystal [6,20,24,25,35]. The results obtained indicate a high structural perfection of the crystal. The results allow us to claim that the $LiNbO_3$:Gd:Mg crystal is close in some of its properties to the stoichiometric $LiNbO_3$ crystal. One such property of stoichiometric and magnesium-doped $LiNbO_3$ crystals, which are important for creating materials for laser radiation conversion on periodically polarized submicron-sized domains with flat boundaries [14], is a low value of the coercive field ($\approx$2.3 kV/cm). In a congruent $LiNbO_3$ crystal, the coercive field is much higher, $\approx$23.0 kV/cm. At the same time, the $LiNbO_3$:Gd:Mg crystal has a much higher optical uniformity than the stoichiometric crystal and is also distinguished by the absence of the photorefraction effect.

It was found that, upon excitation of the Raman spectrum of $LiNbO_3$:Gd:Mg by a laser line with a wavelength of 785 nm, the spectrum manifests itself as a number of bands corresponding to the fundamental vibrations of the crystal lattice A_1(TO,LO)- and E(TO,LO)-type symmetry, as well as bands corresponding to the second-order Raman spectrum with frequencies of 1046, 1122, 1202, 1291, 1291, 1351, 1411, 1523, 1595, 1709, 1853, and 1963 cm^{-1}. Simultaneously, when the Raman spectrum of the $LiNbO_3$:Gd:Mg crystal is excited by a laser line with a wavelength of 532 nm, the intensity and number of bands corresponding to the second-order spectrum are significantly (by an order of magnitude or more) reduced. At the same time, in the Raman spectrum in the region below 900 cm^{-1}, no bands were found corresponding to the second-order spectrum, as well as bands with frequencies of 209, 230, 298, and 880 cm^{-1} of disordered $LiNbO_3$:Mg crystals [28].

Author Contributions: Conceptualization, N.S. and A.P.; methodology, A.P.; formal analysis, A.S.; investigation, A.S., A.P. and N.T.; writing—original draft preparation, N.S., A.P. and N.T.; writing—review and editing, N.S., A.P. and M.P.; visualization, A.S., A.P. and N.T.; supervision, M.P. All authors have read and agreed to the published version of the manuscript.

Funding: This research was funded by the Ministry of Science and Higher Education Russian Federation scientific topic 0186-2022-0002 (registration # FMEZ-2022-0016) and by RFBR and BRFBR (Grant No. 20-52-04001 Bel_mol_a).

Institutional Review Board Statement: Not applicable.

Informed Consent Statement: Not applicable.

Data Availability Statement: The raw data required to reproduce these findings are available from author A.P. on reasonable request.

Conflicts of Interest: The authors declare no conflict of interest. The funders had no role in the design of the study; in the collection, analyses, or interpretation of data; in the writing of the manuscript; or in the decision to publish the results.

References

1. Arizmendi, L. Photonic applications of lithium niobate crystals. *Phys. Stat. Sol.* **2004**, *201*, 253–283. [CrossRef]
2. Günter, P.; Huighard, J.P. *Photorefractive Materials and Their Applications. 1. Basis Effects*; Springer: New York, NY, USA, 2006.
3. Wang, Y.; Wang, R.; Yuan, J.; Wang, Y. Terahertz generation from Cu ion implantation into lithium niobate. *J. Lumin.* **2014**, *147*, 242–244. [CrossRef]
4. Warrier, A.M.; Lin, J.; Pask, H.M.; Lee, A.J.; Spence, D.J. Multiwavelength ultrafast $LiNbO_3$ Raman laser. *Opt. Express* **2015**, *23*, 25582–25587. [CrossRef]
5. Yu, M.; Okawachi, Y.; Cheng, R.; Wang, C.; Zhang, M.; Gaeta, A.L.; Lončar, M. Raman lasing and soliton mode-locking in lithium niobate microresonators. *Light Sci. Appl.* **2020**, *9*, 9. [CrossRef]
6. Lengyel, K.; Péter, Á.; Kovács, L.; Corradi, G.; Pálfalvi, L.; Hebling, J.; Unferdorben, M.; Dravecz, G.; Hajdara, I.; Szaller, Z.; et al. Growth, defect structure, and THz application of stoichiometric lithium niobate. *Appl. Phys. Rev.* **2015**, *2*, 040601. [CrossRef]
7. Sidorov, N.V.; Palatnikov, M.N.; Teplyakova, N.A.; Birukova, I.V.; Titov, R.A.; Makarova, O.V.; Masloboeva, S.M. *Single Crystals of Lithium Niobate and Tantalate of Different Composition and Genesis*; RAS: Moscow, Russia, 2022. (In Russian)
8. Fontana, M.D.; Bourson, P. Microstructure and defects probed by Raman spectroscopy in lithium niobate crystals and devices. *Appl. Phys. Rev.* **2015**, *2*, 040602. [CrossRef]
9. Blistanov, A.A.; Lyubchenko, V.M.; Goryunov, A.N. Recombination processes in $LiNbO_3$ crystals. *Crystallogr. Rep.* **1998**, *43*, 78–82.

10. Xing, L.; Yang, W.; Lin, J. Impact of crystal defect on upconverted white-light emission in lanthanide-doped $LiNbO_3$ single crystal. *Opt. Mater.* **2018**, *84*, 215–220. [CrossRef]
11. Grachev, V.G.; Malovichko, G.I. Structures of impurity defects in lithium niobate and tantalate derived from electron paramagnetic and electron nuclear double resonance data. *Crystals* **2021**, *11*, 339. [CrossRef]
12. Kemlin, V.; Jegouso, D.; Debray, J.; Boursier, E.; Segonds, P.; Boulanger, B.; Ishizuki, H.; Taira, T.; Mennerat, G.; Melkonian, J.-M.; et al. Dual-wavelength source from 5% MgO:PPLN cylinders for the characterization of nonlinear infrared crystals. *Opt. Express* **2013**, *21*, 28886–28891. [CrossRef]
13. Murray, R.T.; Runcorn, T.H.; Guha, S.; Taylor, J.R. High average power parametric wavelength conversion at 3.31–3.48 μm in MgO:PPLN. *Opt. Express* **2017**, *25*, 6421–6430. [CrossRef]
14. Shur, V.Y.; Akhmatkhanov, A.R.; Baturin, I.S. Micro- and nano-domain engineering in lithium niobate. *Appl. Phys. Rev.* **2015**, *2*, 040604. [CrossRef]
15. Choge, D.K.; Chen, H.; Guo, L.; Li, G.; Liang, W. Simultaneous second-harmonic, sum-frequency generation and stimulated Raman scattering in MgO:PPLN. *Materials* **2018**, *11*, 2266. [CrossRef]
16. Sidorov, N.V.; Bobreva, L.A.; Teplyakova, N.A.; Palatnikov, M.N.; Makarova, O.V. A comparative study of the structure and chemical homogeneity of $LiNbO_3$:Mg (~5.3 mol%) crystals grown from charges of different origins. *Inorg. Mater.* **2019**, *55*, 1132–1137. [CrossRef]
17. Kong, T.; Luo, Y.; Wang, W.; Kong, H.; Fan, Z.; Liu, H. Enhanced ultraviolet damage resistance in magnesium doped lithium niobate crystals through zirconium co-doping. *Materials* **2021**, *14*, 1017. [CrossRef]
18. Liu, J.; Liu, A.; Chen, Y.; Tu, X.; Zheng, Y. Growth and optical properties of Pr-Mg co-doped $LiNbO_3$ crystal using Bridgman method. *Phys. B Condens. Matter* **2022**, *224*, 413419. [CrossRef]
19. Zhang, P.; Yin, J.; Zhang, L.; Liu, Y.; Hong, J.; Ning, K.; Chen, Z.; Wang, X.; Shi, C.; Hang, Y. Efficient enhanced 1. *54 μm emission in Er/Yb:$LiNbO_3$ crystal codoped with Mg^{2+} ions, Opt. Mater.* **2014**, *36*, 1986–1990. [CrossRef]
20. Sidorov, N.V.; Volk, T.R.; Mavrin, B.N.; Kalinnikov, V.T. *Lithium Niobate: Defects, Photorefraction, Vibrational Spectra, Polaritons;* Nauka: Moscow, Russia, 2003. (In Russian)
21. Irzhak, A.; Irzhak, D.; Pundikov, K.; Sergeev, A. Changing of $LiNbO_3$ Raman spectra under external electric field applying. *Opt. Mater.* **2022**, *133*, 112988. [CrossRef]
22. Irzhak, A.; Irzhak, D.; Khvostikov, V.; Pundikov, K.; Roshchupkin, D.; Fahrtdinov, R. Effect of local changes in the composition of the $LiNb_{1-x}Ta_xO_3$ single crystal on the Raman spectra. *J. Raman Spectrosc.* **2022**, *53*, 969–976. [CrossRef]
23. Gorelik, V.S.; Abdurakhmonov, S.D. Overtone Raman scattering in lithium niobate single crystal doped with terbium. *Crystallogr. Rep.* **2022**, *67*, 252–255. [CrossRef]
24. Anikiev, A.A.; Umarov, M.F.; Scott, J.F. Processing and characterization of improved congruent lithium niobate. *AIP Adv.* **2018**, *8*, 115016. [CrossRef]
25. Anikiev, A.A.; Umarov, M.F. Quasi-elastic light scattering in congruent lithium niobate crystals. *Opt. Spectrosc.* **2018**, *125*, 22–27. [CrossRef]
26. Rygula, A.; Majzner, K.; Marzec, K.M.; Kaczor, A.; Pilarczyk, M.; Baranska, M. Raman spectroscopy of proteins: A review. *J. Raman Spectrosc.* **2013**, *44*, 1061–1076. [CrossRef]
27. Antunes, E.F.; Lobo, A.O.; Corat, E.J.; Trava-Airoldi, V.J.; Martin, A.A.; Veríssimo, C. Comparative study of first- and second-order Raman spectra of MWCNT at visible and infrared laser excitation. *Carbon* **2006**, *44*, 2202–2211. [CrossRef]
28. Singh, E.; Pionteck, M.N.; Reitzig, S.; Lange, M.; Rusing, M.; Eng, L.M.; Sanna, S. Vibrational properties of $LiNbO_3$ and $LiTaO_3$ under uniaxial stress. *Phys. Rev. Mater.* **2023**, *7*, 024420. [CrossRef]
29. Quispe-Siccha, R.; Mejía-Uriarte, E.V.; Villagrán-Muniz, M.; Jaque, D.; García Solé, J.; Jaque, F.; Sato-Berrú, R.Y.; Camarillo, E.; Hernández, J.; Murrieta, H. The effect of Nd and Mg doping on the micro-Raman spectra of $LiNbO_3$ single-crystals. *J. Phys. Condens. Matter* **2009**, *21*, 145401. [CrossRef] [PubMed]
30. Rahman, M.K.R.; Riscob, B.; Bhatt, R.; Bhaumik, I.; Ganesamoorthy, S.; Vijayan, N.; Bhagavannarayana, G.; Karnal, A.K.; Nair, L. Investigations on crystalline perfection, Raman spectra and optical characteristics of transition metal (Ru) co-doped Mg:$LiNbO_3$ single crystals. *ACS Omega* **2021**, *6*, 10807–10815. [CrossRef] [PubMed]
31. Tejerina, M.R.; Pereira da Silva, K.; Goñi, A.R.; Torchia, G.A. Hydrostatic-pressure dependence of Raman-active optical phonons in Nd:Mg:$LiNbO_3$. *Opt. Mater.* **2013**, *36*, 581–583. [CrossRef]
32. Sidorov, N.V.; Pikoul, O.Y.; Teplyakova, N.A.; Palatnikov, M.N. *Laser Conoscopy and Photoinduced Light Scattering in Research of Properties of Optically Nonlinear Lithium Niobate Crystal;* RAS: Moscow, Russia, 2019. (In Russian)
33. Palatnikov, M.N.; Sidorov, N.V.; Kadetova, A.V.; Teplyakova, N.A.; Makarova, O.V.; Manukovskaya, D.V. Concentration threshold in optically nonlinear $LiNbO_3$:Tb crystals. *Opt. Laser Technol.* **2021**, *137*, 106821. [CrossRef]
34. Akhmadullin, I.S.; Golenishchev-Kutuzov, V.A.; Migachev, S.A.; Mironov, S.P. Low-temperature electrical conductivity of congruent lithium niobate. *Phys. Solid State* **1998**, *40*, 1190–1192. [CrossRef]
35. Goulkov, M.; Imlau, M.; Woike, T. Photorefractive parameters of lithium niobate crystals from photoindused light scattering. *Phys. Rev. B* **2008**, *77*, 235110. [CrossRef]
36. Ridah, A.; Bourson, P.; Fontana, M.D.; Malovichko, G. The composition dependence of the Raman spectrum and new assignment of the phonons in $LiNbO_3$. *J. Phys. Cond. Matter* **1997**, *9*, 9687–9693. [CrossRef]
37. Parlinsky, K.; Li, Z.Q.; Kawazoe, Y. Ab initio calculations of phonons in $LiNbO_3$. *Phys. Rev. B* **2000**, *61*, 272–278. [CrossRef]

38. Li, R. Vibrational spectra and electro-optic effect of ferroelectric $LiNbO_3$ crystal. *Phys. Stat. Sol.* **2023**, *260*, 2300051. [CrossRef]
39. Friedrich, M.; Riefer, A.; Sanna, S.; Schmidt, W.G.; Schindlmayr, A. Phonon dispersion and zero-point renormalization of $LiNbO_3$ from density-functional perturbation theory. *J. Phys. Cond. Matter* **2015**, *27*, 385402. [CrossRef]
40. Kong, Y.; Xu, J.; Chen, X.; Zhang, C.; Zhang, W.; Zhang, G. Ilmenite-like stacking defect in nonstoichiometric lithium niobate crystals investigated by Raman scattering spectra. *J. Appl. Phys.* **2000**, *87*, 4410–4414. [CrossRef]
41. Yang, X.; Lan, G.; Li, B.; Wang, H. Raman spectra and directional dispersion in $LiNbO_3$ and $LiTaO_3$. *Phys. Stat. Sol.* **1987**, *142*, 287–300. [CrossRef]
42. Veithen, M.; Ghosez, P. First-principles study of the dielectric and dynamical properties of lithium niobate. *Phys. Rev. B* **2002**, *65*, 214302. [CrossRef]
43. Nogueira, B.A.; Milani, A.; Castiglioni, C.; Fausto, R. The correlation between experimental polarized Raman spectra and their density functional theory prediction in the LCAO framework: The R3c $LiNbO_3$ crystal as a test case. *J. Raman Spectrosc.* **2021**, *52*, 995–1010. [CrossRef]
44. Okamoto, Y.; Wang, P.-C.; Scott, J.F. Analysis of quasielastic light scattering in $LiNbO_3$ near T_C. *Phys. Rev. B* **1985**, *32*, 6787–6792. [CrossRef]
45. Sidorov, N.; Palatnikov, M.; Pyatyshev, A. Raman scattering in double-doped single crystal $LiTaO_3$:Cr(0.2):Nd(0.45wt.%). *Photonics* **2022**, *9*, 712. [CrossRef]
46. Sidorov, N.; Palatnikov, M.; Pyatyshev, A.; Sverbil, P. Second-order Raman scattering in ferroelectric ceramic solid solutions $LiNb_xTa_{1-x}O_3$. *Crystals* **2022**, *12*, 456. [CrossRef]
47. Palatnikov, M.; Sidorov, N.; Pyatyshev, A.; Skrabatun, A. Comparison of Raman spectra of optically nonlinear $LiTaO_3:Cr^{3+}$ (0.005 wt%) crystal laser excited in visible (532 nm) and near-IR (785 nm) areas. *Photonics* **2023**, *10*, 439. [CrossRef]

photonics

MDPI

Article

Second Harmonic Generation Versus Linear Magneto-Optical Response Studies of Laser-Induced Switching of Pinning Effects in Antiferromagnetic/Ferromagnetic Films

Irina A. Kolmychek [1], Vladimir B. Novikov [1], Nikita S. Gusev [2], Igor Yu. Pashen'kin [2], Evgeny A. Karashtin [2] and Tatiana V. Murzina [1,*]

[1] Physics Department, M. V. Lomonosov Moscow State University, 119991 Moscow, Russia; irisha@shg.ru (I.A.K.)
[2] Institute for Physics of Microstructures, RAS, 603087 Nizhny Novgorod, Russia
* Correspondence: mur@shg.ru

Abstract: Composite magnetic nanostructures are a subject of high research interest, as they provide a number of exciting effects absent in live nature. Among others, much attention has been paid to the studies of exchange coupling in antiferromagnetic/ferromagnetic (AFM/FM) films, which leads to the pinning effect. It manifests itself as a widening and shift of the magnetic hysteresis loop with respect to zero value of the external magnetic field oriented along the pinning direction. In this work, we report on comparative studies of linear and nonlinear magneto-optical effects under the laser-induced switching of the pinning effect in IrMn/CoFe films of various thickness of the ferromagnetic CoFe layer. We found that the magneto-optical response of the pinned AFM/FM nanofilms appears with different hysteresis loop parameters in the transverse magneto-optical Kerr effect (MOKE) and interface-sensitive magnetization-induced second harmonic generation (SHG), indicating the diversity of the magnetic effects at interfaces compared to the bulk of the films.

Keywords: exchange bias; pinning; magneto-optics; second harmonic generation

1. Introduction

Modern technologies enable the production of high-quality metal magnetic films with nanometer thickness offering new possibilities for the development of novel magnetic and transport phenomena in artificial nanomaterials [1]. Fascinating effects appear when different types of magnetic materials are composed in a single structure. Indeed, great research endeavors are aimed towards harnessing the magnetic interaction between adjacent antiferromagnetic (AFM) and ferromagnetic (FM) layers, where a proper composition results in the exchange coupling of spins of the two constituent materials. This leads to the appearance of the unidirectional magnetic anisotropy, i.e., to the pinning effect [2,3].

Interface exchange coupling can arise after cooling a thin AFM/FM bilayer in the presence of a static magnetic field **H** from the temperature exceeding the Neel temperature T_N of an antiferromagnetic film, which in turn should be lower than the Curie temperature T_C of the FM layer. After this, at the temperature $T < T_N$ the AFM/FM system exhibits the magnetic hysteresis loop shifted in the direction opposite to the magnetic field applied during the cooling process [2–5]. The following intuitive physical picture can be exploited to explain this effect. For the AFM/FM layered system subjected to the procedure described above, the magnetization reversal of the FM layer requires higher **H** than in the absence of the pinning effect as it has to overcome the microscopic torque exerted by ordered spins at the interface with the AFM film. Thus, the unidirectional magnetic anisotropy is formed, the main attribute of which is the magnetic hysteresis loop asymmetry with respect to the zero value of the external magnetic field applied parallel to the pinning direction. The accompanying hysteresis loop shift is known as the exchange bias field, H_{EB}.

It has also been recognized that the treatment of an AFM layer using the described thermal-assisted approach results in the magnetic anisotropy even in the absence of an adjacent ferromagnetic film, while its value can be lower as compared to that for the AFM/FM bilayer made under similar conditions [6]. In the case of AFM/FM layers with low AFM anisotropy, the field cooling results in broadening the magnetic hysteresis loop, i.e., it leads to an increase in coercivity H_C as compared to that of the same structure prior to the field cooling. The mechanism of this effect is the rotation of spins of both the FM and AFM layers under the reversion of the external magnetic field that requires more energy and thus broadens the magnetic hysteresis loop [3].

Importantly, the magnetic interface coupling is rather sensitive to the temperature of an AFM/FM structure [7,8]. Typically, the exchange bias in the AFM/FM films breaks down at temperatures exceeding T_N, while in nanostructured systems this coupling has been shown to disappear at temperatures much lower than T_N due to size effects. The temperature at which the hysteresis loop shift vanishes is denoted as blocking temperature T_B [9,10].

A promising method for optical control of the pinning effect is laser-induced heating of the AFM/FM interface; namely, as the structure is heated above the blocking temperature, the pinning is destroyed. This provides a possibility to switch this effect and thus to control the magnetic interaction between the FM and AFM layers [3,11]. A complete reversal of the exchange bias at the AFM/FM interface can be induced as well by a single femtosecond pulse of sufficiently high energy [12–17]. Spin precession in magnetic structures subjected by ultrashort excitation provides extensive information on magnetic interactions that was harnessed for the studies of the exchange bias in pinned films by ultrafast pump-probe [18] and terahertz emission spectroscopy [11]. Laser pulse perturbation is found to modify the exchange bias as evidenced by the shift of the hysteresis loop with respect to the applied DC magnetic field as observed by the magneto-optical Kerr effect [11]. The modification of the exchange bias by means of intense laser beam irradiation of the magnetic AFM/FM structure looks attractive as the spatial region of this impact is comparable to the laser beam spot, which can be used for optical data writing and storage [13,19–23].

The optical second harmonic generation (SHG) technique is known as a sensitive tool for studies of interfaces and nanostructures made of materials with inversion symmetry. Moreover, when being applied to studies of magnetic structures, magneto-optical effects at the SHG wavelength appear with the typical values being one–two orders of magnitude higher compared to their linear analogues such as magneto-optical Kerr effect (MOKE) [24,25]. This nonlinear optical method allows one to distinguish various types of nontrivial magnetic states, such as magnetic vortices and magnetic toroidal moment, gradients of magnetization, etc. [26–29]. At the same time, to the best of our knowledge, no detailed studies of the SHG effect in the exchange-biased AFM/FM systems have been performed up to now. In this paper, we present the results of the experimental studies of linear and nonlinear optical response of a set of AFM/FM (IrMn/CoFe) structures with thickness of the FM layer from 2 nm up to 50 nm under the laser-induced switching of the pinning effect. The difference in linear and nonlinear optical responses of the structures is unveiled.

2. Materials and Methods

The experimental structures under study are planar bilayered IrMn/CoFe films with a thickness of the antiferromagnetic IrMn layer of 10 nm, and the thickness of the ferromagnetic CoFe layer (d_{CoFe}) ranging from 2 nm to 50 nm. The films were covered by a 2 nm Si protective film. These structures were deposited on a glass substrate by magnetron sputtering in an argon atmosphere at a pressure of 2×10^{-3} Torr with a static in-plane magnetic field of about 130 Oe applied during the deposition of metals. All the structures were annealed at 220 °C for 90 min at low pressure (10^{-5} Torr) when applying the DC magnetic field of 1500 Oe in the same direction as during the IrMn sputtering. As the blocking temperature of thin IrMn films is about T_B = 200 °C [9], the Curie temperature of CoFe is T_C = 977 °C [30]; the CoFe layer becomes pinned to the adjacent IrMn layer at room temperature after this treatment.

Experimental studies were performed using the setup shown schematically in Figure 1a. It was specially assembled to provide the magnetic hysteresis loop measurements by means of both MOKE and SHG methods. The former was measured using the halogen lamp as a light source, while the SHG studies were performed when using the radiation of a Ti-sapphire laser. Both p-polarized light beams were focused by a lens to the same area on the sample, allowing us to directly compare the linear and nonlinear response of the films. The angle of incidence was equal to 45 °. Laser-induced switching of the pinning of the AFM/FM film was performed by extra optical beam using Ti-sapphire (Ti-Sa) laser.

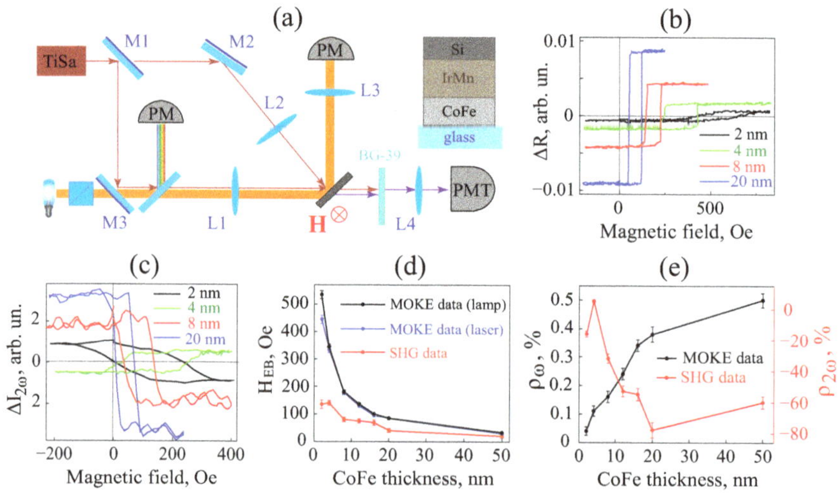

Figure 1. (**a**) Experimental setup and scheme of the structures under study; (**b**) transverse MOKE hysteresis loops; (**c**) SHG hysteresis loops; (**d**) dependence of the values of the hysteresis loops' shifts obtained from the MOKE and SHG data on CoFe thickness; (**e**) dependence of the magnetic contrast for the linear reflection coefficient (black points, left axis) and transmitted SHG intensity (red points, right axis).

In more detail, the setup operated as follows. For the linear MOKE measurements, the mirror M3 was flipped down, whereas the lens L1 focused the broadband light on the studied film into a spot with a diameter of 70–100 μm. Two photodiode power-meters (PMs) Thorlabs S121C (Newton, NJ, USA) were used as signal and reference detectors, so that the normalized power of light reflected from the sample was measured. The DC magnetic field of up to 1 kOe formed by a programmatically controlled electromagnet was applied to the films in the transversal geometry.

In the nonlinear optical experiments, a Ti-Sa laser was used as a source of fundamental radiation with a wavelength of 800 nm, the repetition rate of 80 MHz, and pulse duration of 30–50 fs. In this case, mirrors M1 and M3 were flipped up and the lens L1 focused the laser beam on the sample into a spot of about 20 μm in diameter. The second harmonic power generated in transmission through the sample was detected by a photomultiplier tube (PMT, Hamamatsu R4220, Shizuoka, Japan), while the transmitted light at the fundamental frequency was blocked by a color filter BG-39. The magnetization-induced SHG hysteresis loops were obtained for the average power of the incident laser beam of about 12–13 mW, which corresponds to the peak intensity of the laser pulse of 1 GW/cm^2. Applied light power ensured that the structure did not get hotter than the T_B under the laser beam irradiation.

For the laser-induced switching of the pinning effect in the AFM/FM structure, the mirror M1 was flipped down and the heating Ti-Sa beam was focused at near-normal incidence by the lens L2 with the focal distance of 70 mm. This lens was placed on a two-coordinate motorized translator, the application of which allowed us to perform the laser heating for a square area of 100 μm × 100 μm. The beam had an average power of 55–60 mW and was focused into a spot of 30 μm in diameter, which corresponds to the mean light intensity of about 8.5 kW/cm^2 (peak intensity of about 3.5 GW/cm^2). It is crucially important that (i) the laser-treated region of the film is larger than the focal spot of the lamp and of the probe Ti-Sa light sources and (ii) all three beams (one from the halogen lamp and the two laser beams used for heating the sample and for the SHG measurements) were aligned in the same spot of the structure using a CMOS camera.

3. Results

3.1. Moke vs. SHG Magnetic Measurements

Firstly, we address the results of measurements of the transverse MOKE and SHG hysteresis loops for as-prepared IrMn/CoFe structures. For this, we acquired the magnetic field dependencies of the magnetization-induced change of (a) the reflection coefficient (ΔR) and (b) the SHG intensity ($\Delta I_{2\omega}$) relative to the averaged SHG intensity value. It is necessary to outline that the nonlinear optical experiments were performed at low input laser intensity (12–13 mW) in order to prevent the influence of the laser radiation on the magnetic state of the film (see Section 3.2). The obtained hysteresis loops are shown in Figure 1b,c for MOKE and SHG, respectively, for various thicknesses of the CoFe layer, when light is incident on the sample from the IrMn side. We found that all dependencies $\Delta R(H)$ and $\Delta I_{2\omega}(H)$ are shifted towards the positive direction of the magnetic field H whereas the values of the exchange bias and the width of the hysteresis loops vary with the ferromagnet thickness (Figure 1d). Namely, a monotonous decrease in H_{EB} is observed with the increase in d_{CoFe}.

To check that the low-intensity laser beam used in the SHG measurements did not change the magnetic state of the studied structure, we performed additional MOKE measurements when irradiating the samples by the fundamental laser beam with the power of 12–13 mW. We used the mirrors M1 and M3 to direct the femtosecond laser beam towards the film and the two identical power-meters (PM) as detectors (Figure 1a). The values of H_{EB}^{MOKE} obtained for this low-intensity laser irradiation are shown in Figure 1d, blue points. One can see that for all CoFe thicknesses the value of H_{EB}^{MOKE} coincides with that obtained when using the halogen lamp as the probe source, except a small difference observed for the thinnest film with d_{CoFe} = 2 nm, confirming the preservation of the magnetic properties of the samples during the SHG measurements.

We found that for all the structures in the series, the shift of the MOKE hysteresis loops is larger compared to that obtained in the SHG measurements, $H_{EB}^{SHG} < H_{EB}^{MOKE}$ (Figure 1d). This is probably attributed to different localization of the linear and nonlinear optical source terms within the films as is discussed below.

As a measure of the transversal magnetization-induced effects in the linear and nonlinear optical response, we used the magnetic contrast defined as:

$$\rho_{\omega,2\omega} = \frac{I_{\omega,2\omega}(+H) - I_{\omega,2\omega}(-H)}{I_{\omega,2\omega}(+H) + I_{\omega,2\omega}(-H)}, \quad (1)$$

where $I_{\omega,2\omega}(\pm H)$ are the reflected or transmitted light intensity at the fundamental or the SHG frequency obtained for the opposite directions of the saturated external magnetic field, $\pm H$. The obtained ρ_ω and $\rho_{2\omega}$ values (Figure 1e) are tenths and dozens of percent, respectively, which is typical for ferromagnetic metals [26]. The magnitudes of the magnetization-induced effects in linear and nonlinear optical response increase with the CoFe thickness (Figure 1e), indicating the growth of the volume of the magnetic material and of the relevant specific magnetization. Interestingly, $\rho_{2\omega}$ changes its sign with the thickness of the FM layer; namely, it is positive for $d_{CoFe} = 4$ nm and negative for other structures from the series.

To investigate whether different interfaces play a role in the linear and nonlinear magneto-optical effect, similar experiments were performed when irradiating the structures from the substrate side. In this case, the optical beam first reached the glass side and the CoFe layer, and then passed through the IrMn/CoFe interface. It was found that the parameters of the hysteresis loops in both MOKE and SHG response coincide with those presented in Figure 1d within the experimental accuracy. When irradiating the structure from the substrate side, the values and sign of the magnetic contrast ρ_{ω} coincide with those shown by black points in Figure 1e, whereas the values of $\rho_{2\omega}$ are of the opposite sign to the data presented in Figure 1e by red points.

3.2. Laser-Induced Pinning Switching

The next stage of the study was devoted to the laser-induced switching of the pinning effect in the AFM/FM structures. Below, we present the experimental data for the IrMn/CoFe film with the CoFe thickness of 4 nm, while it was checked that the samples with all the other CoFe thicknesses from the considered series demonstrated qualitatively similar effects. The laser-induced switching experiment involved illuminating a 100 μm × 100 μm area of the AFM/FM film with an intense laser beam (see Section 2), resulting in the IrMn/CoFe bilayer being heated above the blocking temperature T_B. Subsequently, the illumination was turned off and the sample was cooled down to room temperature while exposed to the saturating magnetic field of 800 Oe. To illustrate different steps of this process, we sketched them in Figure 2a,b, where the alignment of spins of the AFM material for the two opposite directions of **H** are shown schematically. In such a scheme, the direction of the external magnetic field applied during the films' cooling defines the pinning direction.

Firstly, we found the critical power of Ti-Sa heating beam that leads to the pinning switching. The sample was positioned in such a way that the external magnetic field was applied along the pinning direction and perpendicular to the plane of the light incidence. The MOKE loop was obtained in the same film region before (Figure 2c, wine-colored points) and after the laser beam heating process performed for different average laser power (15–60 mW), with positive-in-sign magnetic field being applied during the consequent cooling process. It can be seen that irradiating the sample by the laser beam of the power 15 mW or less (red points) does not change the magnetic properties of the film. The power range of 15–45 mW (orange and green points) can be considered as transitional in the sense that the MOKE hysteresis loop of the films is modified as compared to that of the as-prepared sample. This gentle change instead of an abrupt one can be explained by the inhomogeneity of the magnetic properties of the sample caused by the laser exposure. Finally, using the laser beam power exceeding 45 mW and until thermal destroying of the structure, the reversal of the pinning effect was realized. It appears as the MOKE hysteresis loop undergoing a transfer that is nearly mirror-symmetric with respect to $H = 0$ due to the described laser impact (blue points, Figure 2c). It is worth noting that the same change behavior was observed for both the femtosecond and continuous wave regimes of the Ti-sapphire laser used for the pinning switching which indicates the thermal mechanism of this effect. Further, the switching of the pinning effect was realized using a light power of 60 mW.

Then, we demonstrate the multiple switching of the pinning effect for IrMn/CoFe films harnessing the MOKE and SHG response. Figure 2d–f shows consequently the MOKE hysteresis loops of the non-perturbed IrM/CoFe(4nm) film and after the laser-induced switching performed by 60 mW laser beam in the presence of the positive (Figure 2a) and negative (Figure 2b) magnetic field. Analogous SHG measurements performed under the same conditions were carried out and the corresponding results are shown in Figure 2g–i. Obtained SHG hysteresis loops were acquired using the fundamental beam with the power of 12–13 mW, which excluded its influence on the magnetic state of the samples still keeping measurable values of the transmitted second harmonic intensity. We found that high-intensity laser irradiation of the film region followed by cooling in the saturated positive magnetic field induces the pinning switching (Figure 2a). It is sustained by the shift of the MOKE and SHG hysteresis loops towards the negative magnetic field (Figure 2e,h). The assumed spin alignments in the AFM/FM system in saturated magnetic fields are schematically illustrated at the top of Figure 2e. Importantly, the absolute value of the shift and the width of the magnetic hysteresis loops coincide with those obtained for cooling in negative **H** within the experimental accuracy (Figure 2f,i). These results are a direct manifestation of switching the pinning effect.

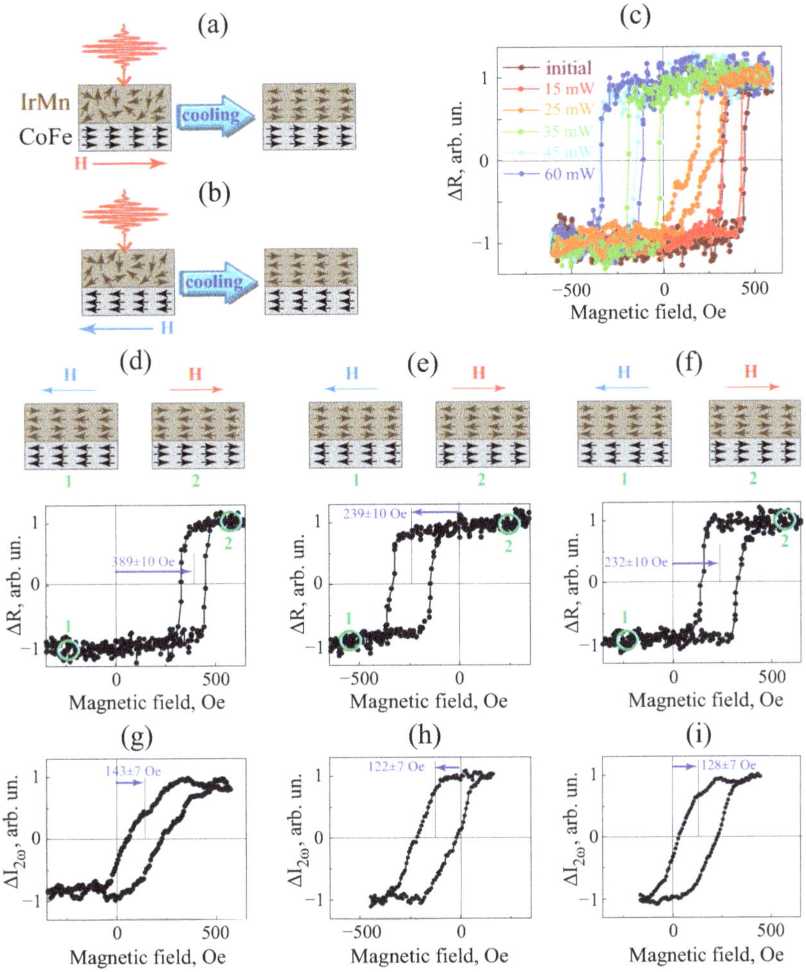

Figure 2. (**a**,**b**) Schemes of the pinning switching process; (**c**) MOKE hysteresis loops obtained in an as-prepared IrMn/CoFe film (wine-colored points) and after the laser heating with different light intensities, indicated in the legend. MOKE hysteresis loops obtained in an (**d**) as-prepared IrMn/CoFe film, (**e**) after cooling in the positive magnetic field. and (**f**) after cooling in the negative field. (**g**–**i**) SHG intensity hysteresis loops obtained in the same conditions as in panels (**d**–**f**). Magnetic field was applied parallel to the initial pinning direction and perpendicular to the plane of light incidence, $d_{CoFe} = 4$ nm. All the data were normalized to maximum values.

We pay special attention to the parameters of the hysteresis loops. We found that, for as-prepared films, (i) initial shifts of the centers of the MOKE and SHG hysteresis loops, i.e., H_{EB}^{MOKE} and H_{EB}^{SHG}, differ from each other and are 389 ± 10 Oe and 143 ± 7 Oe, respectively, and (ii) the widths of the loops are also different and are 120 ± 10 Oe and 180 ± 10 Oe in the linear and nonlinear response, correspondingly. This difference of hysteresis loops takes place as well in the 100 × 100 μm^2 film region subjected to heating by the laser beam (Figure 2e,h). Indeed, the H_{EB}^{MOKE} and H_{EB}^{SHG} are −239 ± 10 Oe and −122 ± 7 Oe, respectively, whereas the corresponding loop widths are 180 Oe and 200 Oe.

In order to check whether the switching effect depends on the magnetic anisotropy of the film induced during the films' deposition and annealing, we performed similar experiments as the sample was oriented in such a way that the initial pinning direction was orthogonal to the external magnetic field (Figure 3). In this case, MOKE and SHG hysteresis loops have near-zero width and almost symmetric shape relative to $H = 0$ (Figure 3a,d). Schematic distributions of the spin alignment in two hysteresis points near the saturation are shown in Figure 3a. We revealed that a clear pinning effect appears as a negative shift of both MOKE and SHG hysteresis loops in the film region irradiated by intense laser beam and cooled in the presence of positive magnetic field (Figure 3b,e). In turn, irradiation of the structure in the negative magnetic field is found to induce the pinning effect in the opposite direction (Figure 3c,f). In other words, a laser-induced heating procedure with subsequent cooling in the DC magnetic field allows us to introduce and control the pinning effect independently on the initial orientation of the pinning direction.

Importantly, all the results described above are perfectly reproducible; moreover, laser-induced switching in a single region of the AFM/FM film can be repeated many times back and forth (see Figures S1 and S2 in Supplementary). We note that after such a subjection, the parameters of the magnetic linear and nonlinear hysteresis loops, H_{EB}^{MOKE} and H_{EB}^{SHG}, remain stable and the considered structures do not exhibit the training effect, which consists of a decrease in the exchange bias during the magnetic field cycling at hysteresis loop measurements for some of the AFM/FM heterostructures [31] (see Figure S1d in Supplementary).

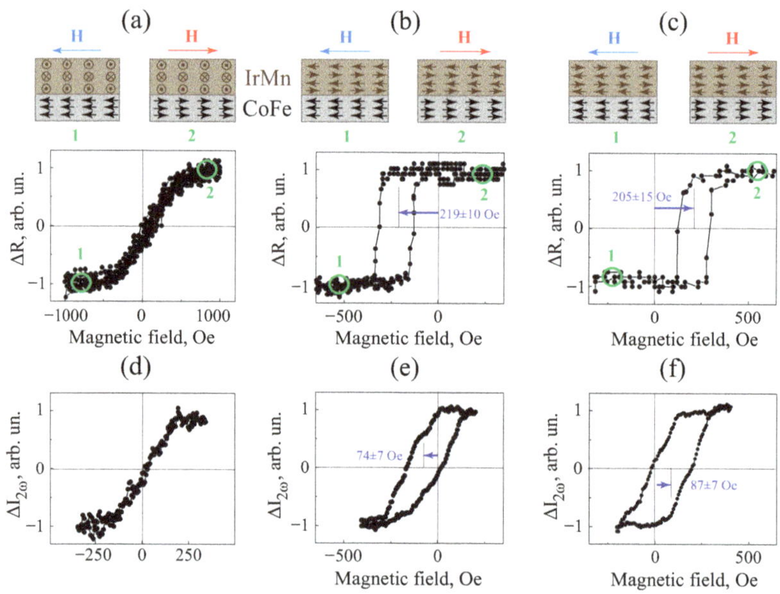

Figure 3. MOKE hysteresis loops obtained for (**a**) as-prepared film, (**b**) area of the sample after cooling in the positive field, and (**c**) area of the sample after cooling in the negative field, respectively. (**d**–**f**) SHG intensity hysteresis loops obtained in the same conditions. Magnetic field was oriented perpendicular to the plane of incidence, while the pinning direction lies in the plane of incidence parallel to the film's surface, $d_{CoFe} = 4$ nm. All the data were normalized to maximum (MOKE and SHG) values.

4. Discussion

When analysing the MOKE data for the set of samples with different CoFe thickness (Figure 1b), one can see that H_{EB} is nearly inversely proportional to d_{CoFe}, as shown in Figure 1d. Such a behavior is expected and consistent with the interfacial nature of the exchange interaction responsible for the pinning effect [32].

The data shown in Figures 2 and 3 clearly illustrate the possibility of switching the pinning effect in AFM/FM films through laser-induced heating of the samples; the role of heating is confirmed by the dependence of the switching parameters on the average intensity of the laser beam and not on the peak intensity. We also show that the suggested procedure of the laser treatment of the AFM/FM films allows one to reorient the pinning direction even to 90 degrees. At the same time, there is a difference between the original values of $|H_{EB}|$ and coercivity (Figure 2d) for as-prepared films and after their laser-induced switching (Figure 2e). This may be attributed to different conditions (temperature, magnetic field) during laser- induced switching compared to those used during the film preparation. Furthermore, laser-induced heating is local and therefore temperature gradients can arise in the illuminated (switching) areas of the film. This may lead to slightly different conditions for switching of the IrMn layer at different points on the sample, which in turn results in broadening the hysteresis loop. Nevertheless, the magnetic hysteresis loops' parameters remain consistent during multiple rounds of laser-induced switching.

Based on the comparison of the linear and nonlinear optical data, we conclude that the shift of the MOKE hysteresis loop relative to zero magnetic field exceeds that of the SHG loop and the strongest difference is observed in the structure with the thinnest CoFe layer of 2 nm (Figure 1d). We note that the linear MOKE characterizes the magnetic properties of the "bulk" of the structure at the penetration depth of light, while the electric dipole SHG sources are localized exclusively at the surface and interfaces of a nanofilm made of a centrosymmetric material. We suppose that just this diversity is responsible for the observed difference of the hysteresis loop parameters of linear and nonlinear magneto-optical response of IrMn/CoFe films. As in the experiment, we detect the averaged SHG response from two CoFe interfaces; the difference in the magnetic properties of these may be important for the observed shift of the SHG hysteresis loop relative to $H = 0$.

The extra contribution to the nonlinear response can be associated with the gradients of the magnetization, $\nabla_z M$, along the normal to the AFM/FM interface, z, which results in the difference of the SHG and MOKE hystereses. Indeed, it was found earlier that a nonuniform magnetization can give contributions to the magnetization-induced SHG response [28,29], whereas linear MOKE resulting from the average value of the magnetization of the structure is much less sensitive to it. Obviously, the largest values of $\nabla_z M$ occur in the thinnest sample which sustains our observation of the pronounced difference between H_{EB}^{SHG} and H_{EB}^{MOKE}.

It should be noted that the width of the SHG hysteresis loop usually exceeds that of the MOKE loop, while the slope is smaller (see Figure 2d,g). We associate this with roughness of the film interfaces and high sensitivity of the SHG response to the defects of the surface. Analogous effects were observed in polycrystalline FeMn/NiFe bilayers, where areas with different exchange bias fields were introduced by the He ion irradiation [33].

5. Conclusions

In conclusion, we studied experimentally the appearance of the switchable pinning effect in IrMn/CoFe films in their linear and nonlinear magneto-optical response. Switching is established as the shift of the magnetic hysteresis loops of the MOKE and SHG response relative to the zero value of the magnetic field applied parallel to the pinning direction. It was found that the hysteresis loop shift for AFM/FM films decreases as the thickness of the ferromagnetic layer increases, as expected for the interfacial exchange interaction in these structures. We revealed a pronounced difference in the shape and parameters of the magnetic hysteresis loops in the linear and nonlinear magneto-optical response of the IrMn/CoFe films; namely, the shift of the SHG hysteresis loop is notably smaller than that of the MOKE. This distinction increases with decreasing CoFe thickness which can be associated with gradients of magnetization at the IrMn/CoFe interface and the roughness of the surfaces.

Supplementary Materials: The following supporting information can be downloaded at: https://www.mdpi.com/article/10.3390/photonics10121303/s1, Figure S1: MOKE hysteresis loops for as-prepared and multiple switched pinned films; Figure S2: SHG hysteresis loops for as-prepared and multiple switched pinned films.

Author Contributions: Conceptualization, T.V.M. and I.A.K.; methodology, V.B.N.; formal analysis, E.A.K.; sample preparation, N.S.G. and I.Y.P.; investigation, I.A.K. and V.B.N.; writing—original draft preparation, I.A.K.; project administration, T.V.M. All authors have read and agreed to the published version of the manuscript.

Funding: This work is supported by the Russian Science Foundation, grant No. 19-72-20103.

Data Availability Statement: The data presented in this study are available on request from the corresponding author.

Conflicts of Interest: The authors declare no conflict of interest.

Abbreviations

The following abbreviations are used in this manuscript:

SHG	second harmonic generation
MOKE	magneto-optical Kerr effect
PM	power-meter
PMT	photomultiplier tube
AFM	antiferromagnet
FM	ferromagnet

References

1. Zhang, H.; Zhang, Y.; Hou, Z.; Qin, M.; Gao, X.; Liu, J. Magnetic skyrmions: Materials, manipulation, detection, and applications in spintronic devices. *Mater. Futur.* **2023**, *2*, 032201. [CrossRef]
2. Nogués, J.; Schuller, I.K. Exchange bias. *J. Magn. Magn. Mater.* **1999**, *192*, 203–232. [CrossRef]
3. Nogués, J.; Sort, J.; Langlais, V.; Skumryev, V.; Suriñach, S.; Muñoz, J.; Baró, M. Exchange bias in nanostructures. *Phys. Rep.* **2005**, *422*, 65–117. [CrossRef]
4. Kravets, A.F.; Timoshevskii, A.N.; Yanchitsky, B.Z.; Bergmann, M.A.; Buhler, J.; Andersson, S.; Korenivski, V. Temperature-controlled interlayer exchange coupling in strong/weak ferromagnetic multilayers: A thermomagnetic Curie switch. *Phys. Rev. B* **2012**, *86*, 214413. [CrossRef]
5. Vdovichev, S.N.; Polushkin, N.I.; Rodionov, I.D.; Prudnikov, V.N.; Chang, J.; Fraerman, A.A. High magnetocaloric efficiency of a NiFe/NiCu/CoFe/MnIr multilayer in a small magnetic field. *Phys. Rev. B* **2018**, *98*, 014428. [CrossRef]
6. Berkowitz, A.; Takano, K. Exchange anisotropy—A review. *J. Magn. Magn. Mater.* **1999**, *200*, 552–570. [CrossRef]
7. Peng, S.; Zhu, D.; Li, W.; Wu, H.; Grutter, A.J.; Gilbert, D.A.; Lu, J.; Xiong, D.; Cai, W.; P. Shafer, K.L.W.; et al. Exchange bias switching in an antiferromagnet/ferromagnet bilayer driven by spin–Orbit torque. *Nat. Electron.* **2020**, *3*, 757–764. [CrossRef]
8. Fang, B.; Sánchez-Tejerina San José, L.; Chen, A.; Li, Y.; Zheng, D.; Ma, Y.; Algaidi, H.; Liu, K.; Finocchio, G.; Zhang, X. Electrical Manipulation of Exchange Bias in an Antiferromagnet/Ferromagnet-Based Device via Spin–Orbit Torque. *Adv. Funct. Mater.* **2022**, *32*, 2112406. [CrossRef]
9. Devasahayam, A.J.; Kryder, M.H. The dependence of the antiferromagnet/ferromagnet blocking temperature on antiferromagnet thickness and deposition conditions. *J. Appl. Phys.* **1999**, *85*, 5519–5521. [CrossRef]
10. Gritsenko, C.; Dzhun, I.; Volochaev, M.; Gorshenkov, M.; Babaytsev, G.; Chechenin, N.; Sokolov, A.; Tretiakov, O.A.; Rodionova, V. Temperature-dependent magnetization reversal in exchange bias NiFe/IrMn/NiFe structures. *J. Magn. Magn. Mater.* **2019**, *482*, 370–375. [CrossRef]
11. Saito, Y.; Kholid, F.N.; Karashtin, E.; Pashenkin, I.; Mikhaylovskiy, R.V. Terahertz Emission Spectroscopy of Exchange-Biased Spintronic Heterostructures: Single- and Double-Pump Techniques. *Phys. Rev. Appl.* **2023**, *19*, 064040. [CrossRef]
12. Guo, Z.; Wang, J.; Malinowski, G.; Zhang, B.; Zhang, W.; Wang, H.; Liu, C.; Peng, Y.; Vallobra, P.; Xu, Y.; et al. Single-shot laser-induced switching of an exchange biased antiferromagnet. *arXiv* **2023**, arXiv:2302.04510.
13. Porat, A.; Bar-Ad, S.; Schuller, I.K. Novel laser-induced dynamics in exchange-biased systems. *Europhys. Lett.* **2009**, *87*, 67001. [CrossRef]

14. Seu, K.A.; Reilly, A.C. Ultrafast laser excitation of spin waves and the permanent modification of the exchange bias interaction in IrMn/Co. *J. Appl. Phys.* **2008**, *103*, 07C104. [CrossRef]
15. Dalla Longa, F.; Kohlhepp, J.T.; de Jonge, W.J.M.; Koopmans, B. Resolving the genuine laser-induced ultrafast dynamics of exchange interaction in ferromagnet/antiferromagnet bilayers. *Phys. Rev. B* **2010**, *81*, 094435. [CrossRef]
16. Ju, G.; Nurmikko, A.V.; Farrow, R.F.C.; Marks, R.F.; Carey, M.J.; Gurney, B.A. Ultrafast Time Resolved Photoinduced Magnetization Rotation in a Ferromagnetic/Antiferromagnetic Exchange Coupled System. *Phys. Rev. Lett.* **1999**, *82*, 3705–3708. [CrossRef]
17. Weber, M.C.; Nembach, H.; Fassbender, J. Picosecond optical control of the magnetization in exchange biased NiFe/FeMn bilayers. *J. Appl. Phys.* **2004**, *95*, 6613–6615. [CrossRef]
18. Dalla Longa, F.; Kohlhepp, J.T.; de Jonge, W.J.M.; Koopmans, B. Laser-induced magnetization dynamics in Co/IrMn exchange coupled bilayers. *J. Appl. Phys.* **2008**, *103*, 07B101. [CrossRef]
19. Almeida, M.; Matthes, P.; Ueberschär, O.; Müller, M.; Ecke, R.; Exner, H.; Albrecht, M.; Schulz, S. Optimum Laser Exposure for Setting Exchange Bias in Spin Valve Sensors. *Phys. Procedia* **2015**, *75*, 1192–1197. [CrossRef]
20. Guo, Z.; Malinowski, G.; Vallobra, P.; Peng, Y.; Xu, Y.; Mangin, S.; Zhao, W.; Hehn, M.; Zhang, B. Ultrafast antiferromagnet rearrangement in Co/IrMn/CoGd trilayers. *Chin. Phys. B* **2023**, *32*, 087507. [CrossRef]
21. Vallobra, P.; Fache, T.; Xu, Y.; Zhang, L.; Malinowski, G.; Hehn, M.; Rojas-Sánchez, J.C.; Fullerton, E.E.; Mangin, S. Manipulating exchange bias using all-optical helicity-dependent switching. *Phys. Rev. B* **2017**, *96*, 144403. [CrossRef]
22. Berthold, I.; Loschner, U.; Schille, J.; Ebert, R.; Exner, H. Exchange Bias Realignment Using a Laser-based Direct-write Technique. *Phys. Procedia* **2014**, *56*, 1136–1142. [CrossRef]
23. Sharma, A.; Hoffmann, M.; Matthes, P.; Busse, S.; Selyshchev, O.; Mack, P.; Exner, H.; Horn, A.; Schulz, S.; Zahn, D.; et al. Exchange bias and diffusion processes in laser annealed CoFeB/IrMn thin films. *J. Magn. Magn. Mater.* **2019**, *489*, 165390. [CrossRef]
24. Kirilyuk, A. Nonlinear optics in application to magnetic surface and thin films. *J. Magn. Mag. Mater.* **2002**, *35*, 189–207. [CrossRef]
25. Pan, R.P.; Wei, H.D.; Shen, Y.R. Optical second-harmonic generation from magnetized surfaces. *Phys. Rev. B* **1989**, *39*, 1229–1234. [CrossRef]
26. Kolmychek, I.A.; Krutyanskiy, V.L.; Murzina, T.V.; Sapozhnikov, M.V.; Karashtin, E.A.; Rogov, V.V.; Fraerman, A.A. First and second order in magnetization effects in optical second-harmonic generation from a trilayer magnetic structure. *J. Opt. Soc. Am. B* **2015**, *32*, 331–338. [CrossRef]
27. Krutyanskiy, V.L.; Kolmychek, I.A.; Gribkov, B.A.; Karashtin, E.A.; Skorohodov, E.V.; Murzina, T.V. Second harmonic generation in magnetic nanoparticles with vortex magnetic state. *Phys. Rev. B* **2013**, *88*, 094424. [CrossRef]
28. Kolmychek, I.; Radovskaya, V.; Mamonov, E.; Maydykovskiy, A.; Sadovnikov, A.; Sheshukova, S.; Nikitov, S.; Temiryazeva, M.; Gusev, N.; Fraerman, A.; et al. Interface-induced optical effects in magnetic two- and three-layer films. *J. Magn. Magn. Mater.* **2021**, *528*, 167780. [CrossRef]
29. Petukhov, A.V.; Lyubchanskii, I.L.; Rasing, T. Theory of nonlinear magneto-optical imaging of magnetic domains and domain walls. *Phys. Rev. B* **1997**, *56*, 2680–2687. [CrossRef]
30. MacLaren, J.M.; Schulthess, T.C.; Butler, W.H.; Sutton, R.; McHenry, M. Electronic structure, exchange interactions, and Curie temperature of FeCo. *J. Appl. Phys.* **1999**, *85*, 4833–4835. [CrossRef]
31. Binek, C. Training of the exchange-bias effect: A simple analytic approach. *Phys. Rev. B* **2004**, *70*, 014421. [CrossRef]
32. Xu, X.Y.; Gao, Y.J.; Wang, Y.L.; Hu, J.G. Thickness dependence of positive exchange bias in ferromagnetic/antiferromagnetic bilayers. *Solid State Commun.* **2011**, *151*, 952–955. [CrossRef]
33. Sampaio, L.C.; Mougin, A.; Ferré, J.; Georges, P.; Brun, A.; Bernas, H.; Poppe, S.; Mewes, T.; Fassbender, J.; Hillebrands, B. Probing interface magnetism in the FeMn/NiFe exchange bias system using magnetic second-harmonic generation. *Europhys. Lett.* **2003**, *63*, 819. [CrossRef]

photonics

MDPI

Article

Numerical Investigation of Localized Surface Plasmons in Gold Nano-Ridge Dimer-on-Mirror Structures

Mohamed El Ghafiani, Adnane Noual *, Madiha Amrani, Mohammed Moutaouekkil and El Houssaine El Boudouti

Laboratory of Physics of Matter and Radiation, University Mohammed First, Oujda 60000, Morocco; medghafiani@gmail.com (M.E.G.); m1.amrani@ump.ac.ma (M.A.); mohamed.moutaouekkil@gmail.com (M.M.); e.elboudouti@ump.ac.ma (E.H.E.B.)

* Correspondence: noualad@yahoo.fr

Abstract: The study of localized surface plasmons (LSPs) in nanoscale structures is an essential step towards identifying optimal plasmonic modes that can facilitate robust optomechanical coupling and deepen our understanding of light–matter interactions at the nanoscale. This paper investigates, numerically, using the finite element method, LSP modes in a design comprising two coupled nano-ridges deposited on a gold layer with an interposing polymer spacer layer. Such a structure, usually referred to as a particle-on-mirror structure, shows exquisite optical properties at the nanoscale. We first examine the LSP modes of a single nano-ridge through the analysis of its scattering cross-section in the visible and infrared ranges. To enhance the plasmonic response, a thin polymer layer is placed at the middle of the ridge, which introduces additional LSP modes confined within the former. Then, we extend the analysis to the dimer configuration, which exhibits more complex and enhanced plasmonic behavior compared to a single nano-ridge. In particular, the dimer configuration yields LSP resonances with a quality factor enhancement of approximately threefold relative to a single nano-ridge. Furthermore, the presence of the polymer layer within the ridges significantly improves plasmon field localization and the quality factor. These findings underscore the potential of nano-ridge-based structures in advancing optomechanical coupling and offering valuable insights for the development of high-performance acousto-plasmonic devices. In particular, the proposed device could help significantly improve the design of nano-acousto-optic modulators, operating in the visible or in the near-infrared ranges, that require an enhanced light–phonon coupling rate.

Keywords: localized surface plasmons; gold nano-ridges; nanoparticles on mirror; dimer; optomechanical coupling

1. Introduction

Plasmonics and its applications are based on the optical processes that occur in metallic nanostructures involving the coupling of light and collective oscillations of conduction band electrons (known as plasmons) at the metal–dielectric interface. The sizes, shapes, and compositions of the nanostructures can be engineered to manipulate the properties of the plasmons, such as their resonance frequency and localization, which, in turn, can enable a wide range of applications in various fields [1–5]. Notably, advanced concepts such as plasmonically induced transparency (PIT) have emerged lately; such resonance is shown to allow the precise on-chip control of light at the nanoscale [6]. Recent studies have explored the realization of PIT using graphene nanoribbons, providing robust methodologies to achieve transparency effects that are highly tunable and resilient to structural perturbations [7,8].

In addition, the application of plasmonics represents a good opportunity for nano-optomechanics to explore the interaction between light and mechanical motion at the nanoscale, enabling ultra-sensitive mass detection, the precision measurement of forces, and advancements in quantum information processing [9–12]. These systems typically utilize nanoscale resonators where light controls and measures mechanical vibrations. Achieving strong optomechanical coupling relies on the ability to localize and enhance electromagnetic fields at the nanoscale. Metallic nanostructures possess the ability to confine light below the diffraction limit via the excitation of localized surface plasmons (LSPs), the oscillations of free electrons at the surfaces of metallic nanostructures, offering a powerful means of confining light to sub-wavelength dimensions [13,14]. LSPs can interact with mechanical vibrations, making them ideal for studying nanoscale mechanical oscillations and enhancing optomechanical interactions [15–17]. Such strong acousto-plasmonic coupling could be harnessed in a multitude of advanced applications like ultra-sensitive accelerometers, force sensors, and optical signal processing [18], in the designing of multifunctional sensors with the possibility of integration into a wireless network [19], and in quantum computing [20]. Investigating systems with high LSP confinement is crucial for advancing nano-optomechanical applications.

Among the nanostructures supporting highly confined LSPs, one can cite nanoparticle-on-mirror (NPOM) structures (or film-coupled nanoparticles) [21,22], where a metal nanoparticle is separated from a metal film by a thin dielectric layer acting as a spacer; these structures have attracted significant attention due to their ease of fabrication [23] and the ability to trap plasmon modes in ultrasmall volumes [24–26]. The response of NPOM structures can be significantly influenced by the thickness of the metal film [27,28], as thinner metal films tend to result in stronger coupling between the nanoparticle and the film, leading to more pronounced plasmonic effects, while thicker films may reduce this interaction and broaden the resonance linewidths. Similarly, the interspacing between the gold film and the nanoparticles [29] and the shapes of the nanoparticles play crucial roles in determining the plasmonic properties [30]. Also, NPOM structures with dimers or trimers atop the metal film, as opposed to monomers, are expected to exhibit more interesting and/or improved features such as narrower plasmon linewidths and more pronounced scattered intensity [16,31].

This paper focuses on investigating, numerically, based on the finite element method (FEM), LSPs in a structure design in the category of NPOM devices, consisting of nano-ridges deposited on a gold layer, with a polymer layer acting as a spacer. It is worth mentioning that such a design could be readily fabricated experimentally given the state-of-the-art technology [32]. We first investigate the LSP modes of a single ridge by examining its scattering cross-section in the visible and near-infrared ranges. To introduce additional LSP modes, we insert a thin polymer layer at the middle of the ridge, which results in new LSPs confined within such a layer. Then, we explore the dimer configuration, which has been proven to show more interesting characteristics [16,31]. Our findings indicate that dimerization, achieved by positioning two ridges in close proximity, enhances the interaction between the modes, leading to LSP resonances with significantly improved quality factors—approximately a threefold enhancement compared to a single monomer. Additionally, the polymer layer at the middle of the ridges plays an important role in enhancing the localization of the plasmon field and the quality factor of the different modes as compared to previous reports [16]. In a recent paper [33], we studied localized mechanical vibrations in similar monomer and dimer structures. Here, we show, prospectively, based on the dimer LSP modes' effective volumes, how such modes would couple with the localized mechanical vibrations in the dimer, thereby enabling their classification in terms of optomechanical coupling intensity.

This paper is organized as follows. In Section 2, we present a detailed investigation of the LSPs of a single gold nano-ridge positioned atop a multilayer structure. In Section 3, we extend this analysis to coupled gold nano-ridges, examining how the dimer configuration influences the localization and enhancement of the plasmonic modes. In Section 4, we provide an analysis of the mode volume for each LSP mode, enabling the classification of the different LSP modes and discussing their potential applications in acousto-plasmonic interactions.

2. Localized Surface Plasmons of a Single Gold Nano-Ridge

The proposed design is shown in Figure 1a. It consists of a gold nano-ridge with a thin polymer layer lying in the middle, deposited above a polymer thin spacer layer made of the same material, that covers a gold thick film beneath. Experimentally speaking, the whole design could be grown on a silicon dioxide substrate (SiO_2), but in terms of simulations, the thick gold film absorbs and reflects all incident light before reaching SiO_2. Hence, the latter medium is excluded from simulations. Due to the system's symmetry along the *y*-axis, all calculations were performed in the x–y plane.

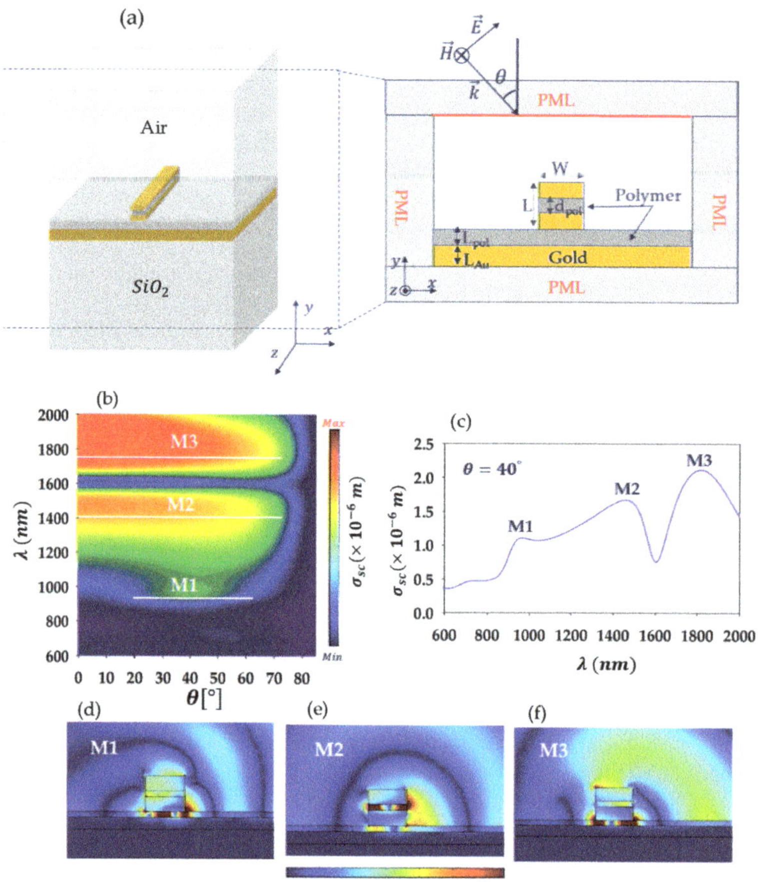

Figure 1. (**a**) Design of the 3D studied system composed of a gold nano-ridge with a thin polymer layer of thickness $d_{pol} = 50$ nm in the middle, placed above a gold layer with a polymer layer acting as a spacer in between. A 2D cross-section of the studied structure is shown on the right side. (**b**) Simulated scattering cross-section of the system in (**a**) as a function of wavelength λ and the angle of incidence θ. (**c**) Same as in (**b**), but for the fixed incidence angle $\theta = 40°$. (**d–f**) Field maps of the scattered electric field within and around the nano-ridge corresponding to modes M1, M2, and M3 shown in (**c**) at wavelengths ~975 nm, ~1493 nm, and ~1689 nm, respectively.

The modeled cross-section of the system is depicted in Figure 1a. The parameters of the studied system are as follows: *W* is the nano-ridge width while *L* is its height, d_{pol} is the thickness of the polymer layer inserted in the middle of the ridge, L_{pol} is the thickness of the polymer spacer layer, and L_{Au} is the thickness of the gold layer beneath. The simulation results presented in this work were obtained using electromagnetic frequency domain analysis based on the FEM using COMSOL Multiphysics software. In order to excite the system, a plane light-wave with TM polarization is launched from the incidence medium at the top at oblique incidence onto the latter, as shown in Figure 1a. To eliminate undesirable reflections at the edges of the system, perfectly matched layers (PMLs) have been placed on all boundaries. The materials used include a polymer with the refractive index $n_{pol} = 1.5$ while, for dispersive materials, namely the gold metal, a frequency-dependent complex-valued relative permittivity based on the Lorentz–Drude model is adopted [34]:

$$\varepsilon_r = \varepsilon_\infty + \sum_{m=1}^{M} \frac{f_m\, \omega_p^2}{\omega_m^2 - \omega^2 + i\Gamma_m \omega} \tag{1}$$

Here, the sum over m corresponds to different oscillators, ε_∞ is the high-frequency contribution to the relative permittivity, ω_p is the plasma frequency of free electrons, f_m is the oscillator strength, ω_m is the resonance frequency, and Γ_m is the damping coefficient. We use $M = 6$ damped harmonic oscillators in order to take the inter-band transitions of gold. The values of the different parameters have been taken from [35] and are listed in Table 1. Throughout the paper, unless stated otherwise, the geometrical parameters are taken as follows: $W = 320$ nm, $L = 300$ nm, $d_{pol} = 50$ nm, $L_{pol} = 50$ nm, and $L_{Au} = 100$ nm.

Table 1. Lorentz–Drude parameters values adopted for modeling the relative permittivity of gold.

Oscillator m	f_m	ω_p (rad/s)	ω_m (rad/s)	Γ_m (rad/s)
1	0.76	1.37188×10^{16}	0	8.05202×10^{13}
2	0.024	1.37188×10^{16}	6.30488×10^{14}	3.66139×10^{14}
3	0.01	1.37188×10^{16}	1.26098×10^{15}	5.24141×10^{14}
4	0.71	1.37188×10^{16}	4.51065×10^{15}	1.32175×10^{15}
5	0.601	1.37188×10^{16}	6.53885×10^{15}	3.78901×10^{15}
6	4.384	1.37188×10^{16}	2.02364×10^{16}	3.36362×10^{15}

To obtain the dispersion relation of the system, we have simulated the scattering cross-section σ_{sc} as a function of both the angle of incidence in the range [0°–85°] and the wavelength in the range of interest [600–2000 nm], which covers the visible and near-infrared domains. The corresponding results are shown in Figure 1b in the color scale. Let us indicate that the red color indicates the maximum scattering intensity while blue indicates the minimum (scattering) intensity. Three distinctive modes are observed, highlighted with horizontal white lines as a guide for the eye. We refer to these modes as M1 for the lower-wavelength (or higher-frequency) mode, M2 for the middle-wavelength mode, and M3 for the higher-wavelength mode. It is worth noting that modes M2 and M3 are excited for nearly all values of $\theta \in [0°, \sim 70°]$ while mode M1 is only excited in the range of θ between approximately ~20° and ~60°. To give an idea about the behavior of these modes in the scattering spectra, Figure 1c shows σ_{sc} as a function of wavelength at $\theta = 40°$, indicating three peaks corresponding to LSP resonances at wavelengths 975 nm (mode M1), 1493 nm (mode M2), and 1689 nm (mode M3). Figure 1d–f displays the maps of the scattered electric field at these wavelengths, respectively. The field localization shows the presence of two types of modes: one confined underneath the ridges (modes M1 and M3) and one at the middle of the ridge (mode M2). A characteristic that is common between the modes is the fact that they are trapped in between two metal materials within a dielectric, which suggests that they correspond to metal–insulator–metal (MIM)-type modes. To explore, in more detail, this aspect of the modes and their general behavior versus the key geometrical parameters, we analyze the effect of the nano-ridge width in Figure 2a, where σ_{sc} is plotted as a function of both the wavelength and the nano-ridge width, W. The remaining parameters are kept as in Figure 1 while the incidence angle is set to $\theta = 40°$, ensuring the excitation of the three modes. One notes that the resonance wavelengths of the three modes (corresponding to the scattering maximum/peaks) increase linearly with an increase in W. This trend is typical in an effective plasmonic cavity acting as a Fabry–Pérot cavity, where MIM-type resonances correspond to confined plasmonic standing waves [16,36]. Accordingly, the thin dielectric layers either underneath or at the middle of the nano-ridge, where the modes are localized, act as an effective plasmonic Fabry–Pérot cavity (PFPC) of nominal length W. In particular, the plasmonic MIM standing waves build up due to reflections off the PFPC edges on each side. The resonance wavelength of the (plasmonic) modes can be expressed as [36]:

$$\lambda_r = \frac{2n_{eff}W}{m} \tag{2}$$

Here, $n_{eff} = k_{spp}/k_0$ is the effective refractive index of the of the propagating MIM-like plasmon in the cavity, k_{spp} is its wave vector, k_0 is the wavevector in the vacuum, and $m = 1, 2, 3, \ldots$ is an integer representing the mode number. This equation is derived from the resonance condition corresponding to a round-trip of the wave in the cavity, namely, $2\,k_{spp}W + 2\phi_{ref} = 2\,\pi m$, where ϕ_{ref} is the phase shift due to the reflection off the edges of the nano-ridge, whose value can be approximated as $\phi_{ref} \approx k_{spp}\delta W$ [36], where δW is the apparent length extension of the mode off the nano-ridge edges. Therefore, W is actually replaced by the length, $W_{eff} = W + \delta W$. By fitting the LSP resonance wavelengths (shown with open white circles in Figure 2a), we estimate δW to be approximately $\delta W \approx 142$ nm for mode M1, $\delta W \approx 152$ nm for mode M2, and $\delta W \approx 48$ nm for mode M3. As for the values of m for each mode, these can be determined from the plot of the localized field inside the PFPC along the x-axis. From Figure 1d–f, we can deduce that m = 2 for mode M1 while for modes M2 and M3, m = 1. To provide an example of the calculation of the phase shift off the ridge edges, ϕ_{ref}, we use the relation $\phi_{ref} \approx k_{spp}\delta W$ such that $k_{spp} = n_{eff} \times 2\pi/\lambda_r$, where λ_r is the resonance wavelength of the mode, and n_{eff} (effective refractive index) is estimated numerically using a MIM waveguide model in Comsol. We calculate the values of ϕ_{ref} at the incidence angle $\theta = 40°$ and for the ridge width $W = 320$ nm (with the remaining parameters being set as in Figure 1). The phase shifts are approximately found to be $\phi_{ref}(M1) = 0.27\pi$, $\phi_{ref}(M2) \approx 0.22\pi$, and $\phi_{ref}(M3) \approx 0.05\pi$ for modes M1, M2, and M3, respectively. The solid white lines in Figure 2a represent the fits of the numerical data (white open circles) using Equation (3). We observe, using the PFPC model based on the latter equation, that we have successfully fitted the numerical results (Figure 2a), which confirms the mechanism behind the LSP modes (M1, M2, and M3) formation.

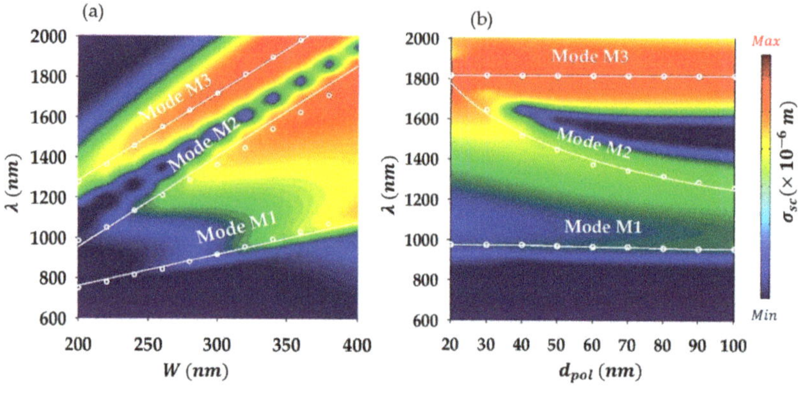

Figure 2. (**a**) Scattering cross-section as a function of wavelength and the width of the nano-ridge W. White open circles indicate the position of the LSP resonances for various values of W while white solid lines represent a fit of these points using the MIM-like waveguide model (Equation (2)). (**b**) Same as in (**a**), except that the scattering cross-section is plotted as a function of wavelength and the thickness of the polymer layer at the middle of nano-ridge.

In order to better understand the physics behind these modes, we examine, in Figure 2b, the effect of the polymer layer thickness d_{pol} in the middle of the nano-ridge for $\theta = 40°$. The rest of the parameters are set similar to Figure 1. Let us recall that the field is essentially confined inside the polymer layer for mode M2, whereas for modes M1 and M3, the field is localized underneath the ridge. This means that only mode M2 should be sensitive to d_{pol}, which is evident in Figure 2b as only mode M2 is affected by changes in d_{pol}. In particular, one remarks that the resonance wavelength of mode M2, λ_{r2}, drops nearly hyperbolically as d_{pol} increases. This trend reminds us of the effective refractive index (n_{eff}) of a metal–insulator–metal-type waveguide versus its width, whereby n_{eff} changes as the inverse of the guide width d, i.e., $n_{eff} \propto 1/d$ [37]. This relationship can be exploited to explain the behavior of λ_{r2} versus d_{pol} considering what we mentioned above, namely, mode M2 (or M1/M3) arises out of the trapping of a MIM-type mode in an effective plasmonic cavity (in the form of a MIM standing wave). Hence, the effective refractive index of such MIM plasmonic modes should change as $n_{eff} \propto 1/d_{pol}$, which enables one to estimate the resonance wavelength as follows:

$$\lambda_r = \frac{2W}{m}\frac{C}{d_{pol}} \tag{3}$$

Here, C is a fitting parameter. Applying Equation (3), we have fitted the numerically calculated (FEM) resonance wavelength of mode M2 in Figure 2b. Note that modes M1 or M3 have not been fitted given their insensitivity to d_{pol}. The curve based on Equation (3) is depicted in a white solid line while the FEM results are shown in white open circles; we can notice the very good agreement achieved. The fitting provided in Figure 2 using Equations (2) and (3) enables us to definitely validate the physical mechanism behind the formation of modes M1, M2, and M3 mentioned above. It is important to note that the plasmonic modes investigated in this paper are not to be confused with the localized surface plasmons associated with metallic nanoparticles. As is well known, the latter modes are optical eigen-modes of the particles that depend on their shapes, the surrounding medium, and the nature of metal. They correspond to coherent collective vibrations of free electrons in the metal constituting the nanoparticle and are localized near the latter surface boundary.

3. Localized Surface Plasmons of Coupled Gold Nano-Ridges (Dimer)

In this section, we investigate the effect of two identical coupled nano-ridges that form a dimer. This configuration is expected to enhance the localization of the plasmonic field and the quality factors of the modes, leading to stronger light–matter interaction [16]. The two nano-ridges are set to be separated by a distance d_m, as shown in the 2D (x–y) plane of the simulated system in Figure 3a. Note that d_m is a key parameter as it enables one to control the coupling magnitude in between the nano-ridges (forming the dimer). Figure 3b shows the simulated scattering cross-section as a function of wavelength and incidence angle, with $d_m = 50$ nm, while the other parameters are set as earlier (Figure 1). The plot reveals that for $\theta < 30°$, the modes are indistinguishable and not excited but rather correspond to large scattering bands. However, for $\theta > 30°$, four modes, labelled D1, D2, D3, and D4, show up. To examine these modes more closely, Figure 3c shows the scattering cross-section at $\theta = 55°$ as a function of wavelength (blue line), where four peaks are clearly identified at roughly 816 nm (D1), 975 nm (D2), 1483 nm (D3), and 1690 nm (D4). A significant observation is that the quality factor of the modes in the dimer case is improved by approximately a factor of 3 compared to the single monomer case (red dashed line). For instance, if we focus on modes D3 and D4, which are the most strongly confined within the dimer, their quality factors are estimated as $Q_{D3} \approx 14.5$ and $Q_{D4} \approx 18.2$, respectively. In comparison, mode M3 (of the monomer), which is the closest in terms of wavelength, has a quality factor of $Q_{M3} \approx 6.25$. This improvement is attributed to the coupling between the ridges, which enhances the localization of the electric field as is evident from the field maps of the scattered field shown in Figure 3d–g. This enhanced localization is particularly evident for modes D2 and D4, where the field is strongly confined underneath and between the ridges, as well as inside the polymer layers in the middle of the ridges. For mode D2, the field is primarily localized in the polymer layers at the middle, but also in between the ridges. In contrast, mode D1 exhibits weaker localization, having antenna-like components associated with multipolar localizations at the edges, that causes the field to radiate. Let us mention that the enhanced quality factor of the dimer modes comparing with the monomer case has been reported elsewhere [16,31].

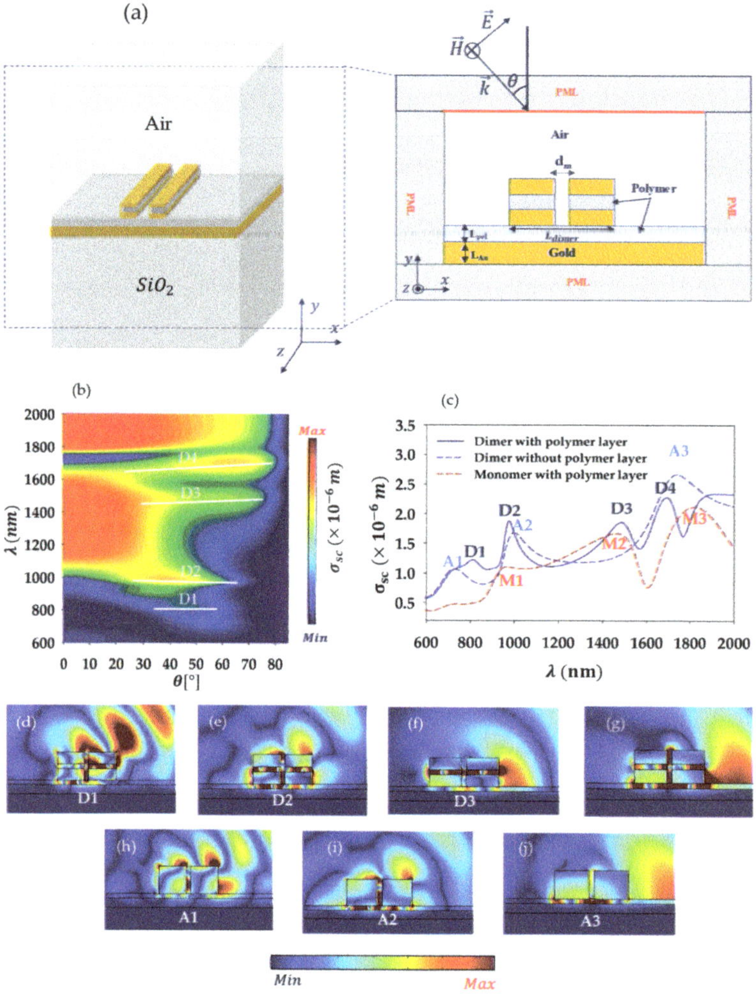

Figure 3. (**a**) Two-dimensional schematic representation of the simulated system consisting of two nano-ridges sitting atop a multilayer structure, separated by a distance d_m. (**b**) Scattering cross-section of the system in (**a**) as a function of wavelength λ and the incidence angle θ. (**c**) Same as in (**b**), but at $\theta = 55°$, for the monomer structure (red dashed lines), the dimer structure with the polymer layer is set at the middle of the ridges (blue solid line) and the dimer structure without the polymer layer is at the middle (blue dashed line). (**d–g**) Scattered field map of the modes D1, D2, D3, and D4 in the case of the nano-ridges with a thin polymer layer placed at the middle at wavelengths λ_{D1}~816 nm, λ_{D2}~975 nm, λ_{D3}~ 1483 nm, and λ_{D4}~1690 nm, respectively. (**h–j**) Same as in (**d–g**) but for the modes A1, A2, and A3 of the nano-ridges without the polymer layer at wavelengths λ_{A1}~735 nm, λ_{A2}~1005 nm, and λ_{A3}~1740 nm, respectively.

As a matter of comparison, we have simulated the scattering cross-section as a function of wavelength for $\theta = 55°$ for the dimer structure without the polymer layer inserted in the middle of the ridges (blue dashed line). The curve is overlaid with that of dimer (blue solid line) and monomer (red dashed line) structures with the polymer layer set in the middle as shown in Figure 3c. We can observe that the insertion of the thin polymer layers within the dimer ridges results in an improved quality factor for the modes. In addition, the spectrum is quite richer in the latter case as additional localized surface plasmons (with four modes) show up as compared to the case of no layers in the ridges (three modes). Furthermore, the localization of the field is significantly better with the polymer layer, as seen in Figure 3d–g, compared to the field maps in the case where no layer is inserted (Figure 3h–j).

To analyze the impact of geometrical parameters on the dimer modes and the physical origins of the latter, we examine how variations in the width of the dimer L_{dimer}, and the thickness of the polymer layer in the middle of the ridges, d_{pol}, affect the modes. The dimer width is defined as $L_{dimer} = 2\ W + d_m$, where W represents the single ridge width and d_m is the separation distance between the ridges. For the sake of simplicity, we alter L_{dimer} by changing only W and keeping d_m, which controls coupling intensity within the dimer, fixed. Also, all remaining parameters of the system are set equivalent to Figure 1a (as mentioned above). Figure 4a presents the scattering cross-section as a function of wavelength and L_{dimer}. We note a near-linear redshift in the resonance wavelengths as L_{dimer} is increased, which is similar to what we observed above in case of the monomer (Figure 2a) width increasing effect. This hints at the fact that modes D1–D4 of the dimer could be considered as plasmonic MIM-type standing waves confined in an effective plasmonic cavity. The latter is made of either the thin layers in the middle of the ridges plus the small space in between (like for mode D3) or of the gap layer underneath the dimer, where mode D4, for instance, is primarily localized. Let us call the latter cavity the lower cavity and the former the upper cavity. Note that some modes (D1, D2, and D4) are confined in both the lower and upper cavities (with a different rate) as well as in between the dimer. To confirm this, we consider that the MIM-type plasmonic standing waves corresponding to the modes are localized in either the lower or upper cavities (or both) of the nominal length, L_{dimer}, and hence, the resonance wavelength is given as follows:

$$\lambda_r = \frac{2n_{eff}\ L_{dimer}}{m} \tag{4}$$

Here, all the parameters are defined in an equivalent manner as in Equation (2) for the monomer. It should be stressed that the nominal length is replaced by $L_{eff} = L_{dimer} + \delta L_{dimer}$, where the latter parameter is a fitting parameter. Such a parameter is meant to account for the dimer mode extension off its outer edges as already explained.

As for the mode order, m, it is determined based on the localized field in the effective plasmonic cavity (lower or upper). From Figure 3d–g, we estimate m = 4 for the mode D2 and m = 2 for both modes D3 and D4. We should mention that because mode D1 has complex features like antenna-like components, with no neat/simple localization features in either the lower or upper cavity, we do not consider fitting its resonance wavelength. Figure 4b shows the fitting provided by Equation (4) (solid lines) to the positions of the modes D2, D3, and D4, resonance wavelengths (indicated by symbols) as functions of L_{dimer}. The fitting parameter δL_{dimer} has been determined for each mode as follows: 58.35 nm for mode D2, 11 nm for mode D3, and 286.76 nm for mode D4. We note that the fitting agrees perfectly with the FEM results for all the three modes, which clearly underscores the MIM character of such modes and that they correspond to MIM-type plasmon standing waves trapped in the effective lower/upper cavities.

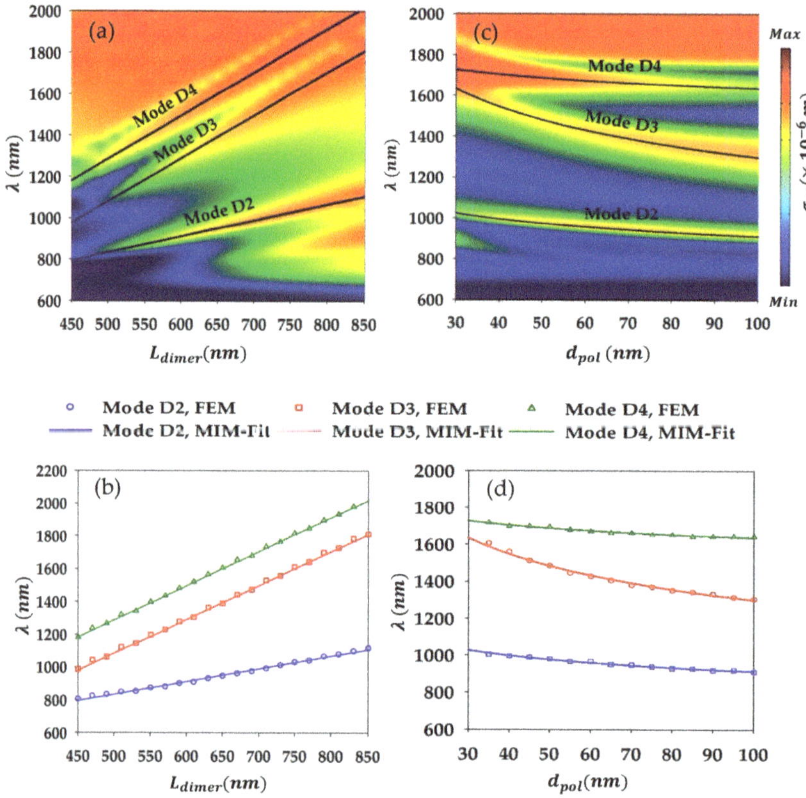

Figure 4. (**a**) Scattering cross-section of the dimer-based system (Figure 3a) as a function of the wavelength and L_{dimer}. (**b**) Plot of the resonance wavelengths of modes D2, D3, and D4 as a function of L_{dimer} (open symbols). The plot for each mode is fitted (solid lines) with Equation (4), giving the resonance wavelength of the effective (lower or upper) plasmonic cavity. Note that mode D3 is confined in the upper cavity while D4 is mostly trapped in the lower cavity and D2 is localized in the upper cavity and in the lower to a lesser extent. (**c**) Similar to (**a**), but as a function of the polymer layers thickness, d_{pol}. (**d**) Same as in (**b**), but as a function of d_{pol} and using Equation (5) as for the provided fitting to the modes.

We also explored the effect of varying the thickness of the polymer layers inserted in the middle of the ridges d_{pol}. Figure 4c shows the scattering cross-section as a function of wavelength and d_{pol}. This plot indicates that all modes are affected by changes in d_{pol}, but the impact is significantly more pronounced for mode D3, which is understandable since mode D3 is primarily localized within such layers. Following mode D3, we see that mode D2 is the most affected as compared to mode D4, which is consistent with their field localization as the latter is more confined beneath the ridges than in their middle while for the former it is the opposite. Note that for mode D1, due to its complex features, it is difficult to analyze its behavior versus changes in d_{pol}; however, it can be remarked that for $d_{pol} \geq 50$ nm, the mode is nearly cut off as its scattering section becomes quite weak. For $d_{pol} \leq 50$ nm, nevertheless, the mode shows a sensitivity with an increase in d_{pol} as its wavelength drops with a non-negligeable magnitude, reflecting its localization in the (middle) layers. Using a similar equation to Equation (4) and the same reasoning as in the case of the monomer, one can capture the modes' (D3, D2, and D4) behavior as d_{pol} is changed. Equation (3) can be written in case of the dimer as follows:

$$\lambda_r = \frac{2L_{eff}}{m}\frac{C}{d_{pol}} \tag{5}$$

Here, the ridge width W is simply replaced by the dimer effective width, L_{eff}, and C is a fitting parameter. Based on such equation, we provide the fitting of the three modes resonance wavelengths as a function of d_{pol} in Figure 4d, where we can observe the excellent agreement with the FEM results. Hence, we can be quite confident that modes D3, D2, and D4 arise out of the trapping of the MIM-type surface-plasmon-guided mode in the lower/upper effective cavity, giving rise to resonating MIM plasmonic standing waves.

4. Effective Volume of the Dimer Localized Plasmon Modes and Prospective Optomechanical Coupling

Here, we focus on the analysis of the effective mode volume of the localized surface plasmon modes of the dimer, aiming to classify these modes for an optimal enhanced optomechanical coupling with localized mechanical vibrations supported by the dimer. As a matter of fact, the present paper, which is about a thorough analysis of the plasmonic properties of the dimer-based device, comes after a previous study about the mechanical characteristics of the system [33]. The general objective is to achieve enhanced acousto-plasmonic coupling using surface acoustic waves (Figure 5a). In this framework, we would like to provide, in this last paragraph, a prospective qualitative analysis of enhanced optomechanical coupling harnessing the dimer plasmonic modes' effective volume. Note that the optical mode volume is one of the key parameters that govern light–matter interactions in general as it is related to field enhancement and spatial localization [38]. Having an estimation of the effective mode volume can help quantify the spatial extent of the electromagnetic energy confinement, directly impacting the strength of the interaction between light and matter. To get an estimate of the effective mode volumes V_m, we use the quasi-static approximation to calculate the mode volume integral for a leaky plasmonic cavity given [39–41] the following:

$$V_m = \frac{\int U\, dr^3}{\max(U)} \tag{6}$$

Here, the integral is performed over of the volume (surface times unit length) of both nano-ridges, U is the electromagnetic energy density, and $\max(U)$ is its maximum value in the simulation domain. For a dielectric medium, the energy density takes the simple form $U = \varepsilon\left|\vec{E}\right|^2$ while, in the case of a dispersive and lossy medium like a metal, an appropriate relation is as follows [37]:

$$U = (\varepsilon_1 + 2\omega\varepsilon_2/\Gamma)\left|\vec{E}\right|^2 \tag{7}$$

Here, $\varepsilon(\omega) = \varepsilon_1(\omega) + i\varepsilon_2(\omega)$ is the complex dielectric function of the metal. In our case, the latter is given by the Lorentz–Drude model (Equation (1)). Γ, on the other hand, is the mode damping taken as the linewidth of the respective LSP mode. As already mentioned, the simulation is performed in 2D due to symmetry along the y-axis, so to calculate the mode volume, one should perform surface integrals in the 2D cross-section system, and then multiply the result by a unit length along the y-direction. Figure 5b, shows the normalized effective mode volume V_m/V_0 calculated for each mode of the dimer (D1, D2, D3, and D4), where $V_0 = \left(\lambda/n_{pol}\right)^2 (\times 1\,[m])$ is the volume occupied by the associated resonance wavelength of the mode in the polymer layer. As can be seen, the effective volumes of modes are well below $\lambda/2n_{pol}$, reflecting the strong confinement of the dimer plasmonic modes far below the diffraction limit of light. From the values of the mode volumes, we observe that mode D1 has the largest value of 0.0117 $\left(\lambda/n_{pol}\right)^2$, indicating a relatively lower field localization rate compared to the other modes. This relates to such mode localization features like its tendency to radiate energy due to its antenna-like components associated with multipolar localizations at the edges. In contrast, mode D4, with the lowest value of 0.0016 $\left(\lambda/n_{pol}\right)^2$, is relatively the most strongly confined mode. Modes D2 and D3 fall in between with values 0.00306 $\left(\lambda/n_{pol}\right)^2$ and 0.00704 $\left(\lambda/n_{pol}\right)^2$, respectively. This result shows that for enhanced light–matter coupling purposes, mode D4 should offer the highest interaction rate as the coherent coupling rate between light and matter scales as $g \propto 1/\sqrt{V_m}$ [41].

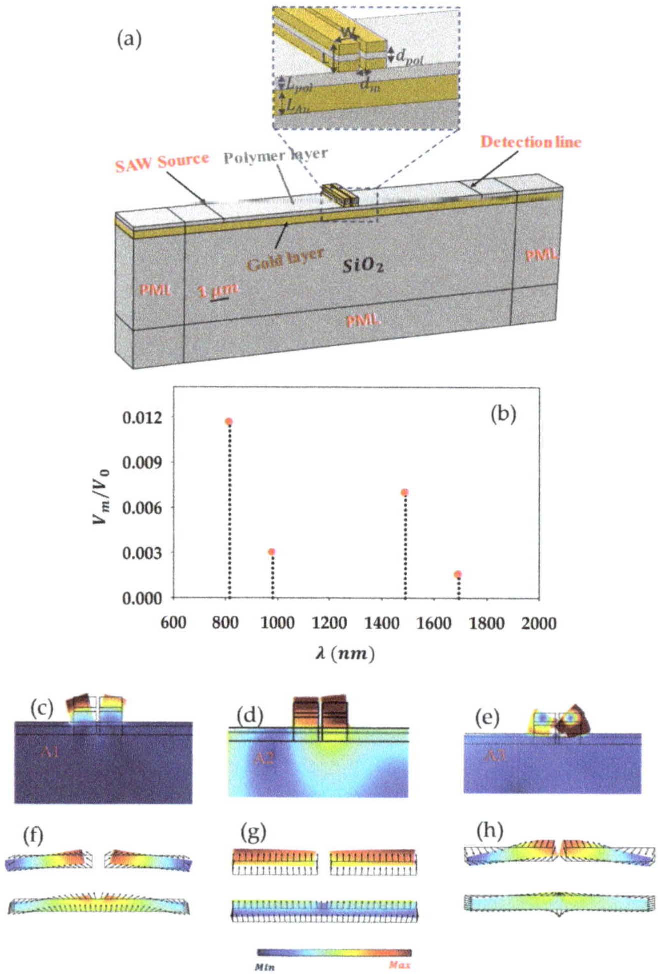

Figure 5. (**a**) Sketch of the system design used to excite confined localized mechanical vibrations in the dimer by means of surface acoustic waves (SAWs). The device consists of the dimer structure of Figure 3a deposited above an SiO_2 substrate. (**b**) Effective mode volume of the dimer LSP modes normalized to $V_0 = \left(\lambda/n_{pol}\right)^2 \times 1\ [m]$, representing the volume occupied by the associated resonance wavelengths in the polymer layer. (**c**–**e**) Zoom-in view on the mechanical displacement field in and around the dimer at its localized mechanical vibrations resonance frequencies, namely at $f_{A_1} = 0.238$ GHz for the symmetric flexural mode, $f_{A_2} = 0.456$ GHz for the in-phase compression mode, and $f_{A_3} = 0.619$ GHz for the high-order symmetric flexural mode. (**f**–**h**) Plots of the displacement field in the polymer layers, both underneath and in the middle of the ridges, associated with the mechanical vibration modes in (**c**–**e**), respectively.

Regarding the optomechanical coupling specifically, we can affirm that the acousto-plasmonic interaction is qualitatively related to the overlap degree between the plasmon and the mechanical vibration in a given cavity. This is estimated quantitively via the so-called optomechanical coefficient, whose expression is determined based on the photoelastic volume and the surface moving-boundary effects [16,42]. Given that the dimer plasmonic modes with the lowest volume magnitude V_m/V_0 are the most highly trapped plasmons, they should manifest the higher overlap rate with a given mechanical vibration localized in the dimer. As a result, such modes should lead to an optimal optomechanical coupling intensity. In Figure 5c–h, we recall the localized mechanical vibrations of the dimer that we deemed to be optomechanically active based on symmetry arguments in a previous work [33] and the design, enabling us to use surface acoustic waves to excite such mechanical vibrations. Considering, for instance, the in-phase compression mode A2 (Figure 5d), which deforms the polymer layers both under and in the middle of the ridges (see Figure 5g), where modes D2, D3, and D4 are localized, we can roughly classify such modes (based on their volume) in terms of the optomechanical coupling rate g as follows:

$$g(D4) \gtrsim g(D2) \gtrsim g(D3). \tag{8}$$

The same reasoning approximately applies if we consider another mechanical vibration of the dimer to classify the LSP modes in terms of acousto-plasmonic coupling magnitude. Obviously, a thorough study based on the rigorous simulations of the optomechanical coefficient g is needed to determine the best localized mechanical vibrations and plasmons that yield the highest coupling magnitude. However, this qualitative argument based on the modes volume already gives a good hint regarding the most promising plasmonic mode for the enhanced optomechanical purpose, namely mode D4, which shows the lowest volume or equivalently the highest confinement rate.

In summary, we investigated LSP modes in gold nano-ridges within a dimer-on-mirror structure, with nano-ridges sized at 320 nm in width and 300 nm in height. Previous studies on LSP resonances have predominantly focused on nanoparticles ranging from 1 to ~100 nm [9,17,21–25], with larger particle sizes remaining underexplored due to their experimental challenges and complex behavior [43]. While different shapes of nanoparticles, whether in monomer or dimer configurations, have been studied before, these studies have typically involved homogeneous nanoparticles with uniform material composition. These investigations have provided critical insights into how particle geometry and separation influence plasmonic behavior. In our work, we have explored the effect of introducing a thin dielectric layer at the center of the nano-ridges. This simple modification has been shown to significantly enhance the quality factor and field localization of the plasmonic modes compared to previous systems without such a layer [16]. Furthermore, our findings indicate that optical coupling in dimers leads to strongly confined plasmons with narrower linewidths and higher field localization than in monomers. Specifically, the dimer configuration yields LSP resonances with a quality factor of approximately threefold enhancement relative to a single nano-ridge.

Additionally, we have qualitatively classified the plasmonic modes of the dimer design in terms of optomechanical coupling intensity based on their effective volume. This classification provides valuable insights into the potential applications of these modes in optomechanical systems. Moreover, the further optimization and tuning of the design parameters, such as the thickness of the polymer layer and the dimensions of the nano-ridges, could lead to optimized enhancements in plasmonic performance.

5. Conclusions

In this paper, thorough finite-element-based numerical calculations have been carried-out to investigate localized surface plasmon resonances in a film-coupled nanoparticles system. The proposed design is made of two gold nano-ridges forming a dimer atop a metal plate in between which a thin polymer film acting as a spacer is inserted. We have first studied the optical scattering properties of a monomer ridge and found three

MIM-type localized surface plasmons in the frequency ranges of interest, namely the visible and near-infrared domains. Our results indicate that the presence of the polymer layer in the middle of the ridge gives rise to additional LSP modes confined within. In the dimer structure composed of two identical nano-ridges, we have observed that the plasmonic modes exhibit narrower linewidths and enhanced field localization compared to the monomer configuration. The enhanced localization is attributed to the strong coupling between the ridges leading to a higher rate of mode trapping. We have also shown that the inserted polymer layers within the ridges, in their middle, further enhance the modes' quality factor as compared to the ridges with no layers. The former case adds means to confine, even more, the dimer LSP modes (in the middle layers), which results in narrower linewidths. Also, in that case, a richer plasmonic spectrum based on the dimer is found as additional modes trapped in the ridges' middle layers show up. We have successfully demonstrated that the dimer/monomer modes arise out of the trapping of MIM-type guided modes in the polymer layers beneath or in the middle of the ridges, acting as plasmonic effective cavities.

Furthermore, based on the simulation of the LSP modes' effective volume of the dimer, we have been able to roughly class the modes in terms of their optomechanical coupling magnitude with a given mechanical vibration of the dimer. Specifically, mode D4, with its smallest normalized mode volume, stands out as the most promising LSP mode for enhanced acousto-plasmonic interaction. The showcased dimer device in this work could help design new acousto-optic components for high-frequency nanomechanical sensing or optical switching or to perform infrared/visible-light modulation operations.

Author Contributions: Conceptualization, A.N., E.H.E.B. and M.E.G.; methodology, M.E.G., M.A. and M.M.; validation, A.N., E.H.E.B. and M.E.G.; investigation, M.E.G., M.A. and M.M.; data curation, M.E.G. and A.N.; writing—original draft preparation, M.E.G. and A.N.; writing—review and editing, A.N., E.H.E.B. and M.E.G.; funding acquisition, E.H.E.B. and A.N. All authors have read and agreed to the published version of the manuscript.

Funding: This research received no external funding.

Institutional Review Board Statement: Not applicable.

Informed Consent Statement: Not applicable.

Data Availability Statement: No new data were created or analyzed in this study. Data sharing is not applicable to this article.

Acknowledgments: We would like to acknowledge the financial support under the framework of the Research Excellence Grants Program from the Moroccan National Center for Scientific and Technical Research (CNRST).

Conflicts of Interest: The authors declare no conflicts of interest.

References

1. Stockman, M.I.; Kneipp, K.; Bozhevolnyi, S.I.; Saha, S.; Dutta, A.; Ndukaife, J.; Kinsey, N.; Reddy, H.; Guler, U.; Shalaev, V.M. Roadmap on plasmonics. *J. Opt.* **2018**, *20*, 043001. [CrossRef]
2. Amrani, M.; Khattou, S.; Rezzouk, Y.; Mouadili, A.; Noual, A.; El Boudouti, E.H.; Djafari-Rouhani, B. Analytical and numerical study of T-shaped plasmonic demultiplexer based on Fano and induced transparency resonances. *J. Phys. D Appl. Phys.* **2021**, *55*, 075106. [CrossRef]
3. Duan, Q.; Liu, Y.; Chang, S.; Chen, H.; Chen, J.H. Surface plasmonic sensors: Sensing mechanism and recent applications. *Sensors* **2021**, *21*, 5262. [CrossRef] [PubMed]
4. Noual, A.; Akiki, R.; Pennec, Y.; El Boudouti, E.H.; Djafari-Rouhani, B. Surface acoustic waves-localized plasmon interaction in pillared phononic crystals. *Phys. Rev. Appl.* **2020**, *13*, 024077. [CrossRef]
5. Vasileiadis, T.; Noual, A.; Wang, Y.; Graczykowski, B.; Djafari-Rouhani, B.; Yang, S.; Fytas, G. Optomechanical hot-spots in metallic nanorod-polymer nanocomposites. *ACS Nano* **2022**, *16*, 20419–20429. [CrossRef] [PubMed]
6. Xia, S.; Zhai, X.; Wang, L.; Xiang, Y.; Wen, S. Plasmonically induced transparency in phase-coupled graphene nanoribbons. *Phys. Rev. B* **2022**, *106*, 075401. [CrossRef]
7. Xia, S.-X.; Zhai, X.; Wang, L.-L.; Wen, S.-C. Plasmonically induced transparency in double-layered graphene nanoribbons. *Photonics Res.* **2018**, *6*, 692–702. [CrossRef]

8. Zhang, D.; Xia, S.; Xu, W.; Zhai, X.; Wang, L. Topological plasmonically induced transparency in a graphene waveguide system. *Phys. Rev. B* **2024**, *109*, 245420. [CrossRef]
9. Koya, A.N.; Cunha, J.; Guerrero-Becerra, K.A.; Garoli, D.; Wang, T.; Juodkazis, S.; Proietti Zaccaria, R. Plasmomechanical systems: Principles and applications. *Adv. Funct. Mater.* **2021**, *31*, 2103706. [CrossRef]
10. Barzanjeh, S.; Xuereb, A.; Gröblacher, S.; Paternostro, M.; Regal, C.A.; Weig, E.M. Optomechanics for quantum technologies. *Nat. Phys.* **2022**, *18*, 15–24. [CrossRef]
11. Liu, X.; Liu, W.; Ren, Z.; Ma, Y.; Dong, B.; Zhou, G.; Lee, C. Progress of optomechanical micro/nano sensors: A review. *Int. J. Optomechatronics* **2021**, *15*, 120–159. [CrossRef]
12. Sansa, M.; Defoort, M.; Brenac, A.; Hermouet, M.; Banniard, L.; Fafin, A.; Gely, M.; Masselon, C.; Favero, I.; Jourdan, G. Optomechanical mass spectrometry. *Nat. Commun.* **2020**, *11*, 3781. [CrossRef]
13. Amendola, V.; Pilot, R.; Frasconi, M.; Maragò, O.M.; Iatì, M.A. Surface plasmon resonance in gold nanoparticles: A review. *J. Phys. Condens. Matter* **2017**, *29*, 203002. [CrossRef]
14. Ramos, D.; Malvar, O.; Davis, Z.J.; Tamayo, J.; Calleja, M. Nanomechanical plasmon spectroscopy of single gold nanoparticles. *Nano Lett.* **2018**, *18*, 7165–7170. [CrossRef] [PubMed]
15. Simone, G. Strong plasmon-mechanical coupling through standing acoustic waves and antenna enhancement, toward mass transduction. *J. Appl. Phys.* **2024**, *135*, 053152. [CrossRef]
16. Noual, A.; Akiki, R.; Lévêque, G.; Pennec, Y.; Djafari-Rouhani, B. Enhanced phonon-plasmon interaction in film-coupled dimer nanoridges mediated by surface acoustic waves. *Opt. Express* **2021**, *29*, 43104–43123. [CrossRef]
17. Ahmidayi, N.; d'Orsonnens, W.; Maurer, T.; Lévêque, G. Mechanical Enhancement of the Strain-Sensor Response in Dimers of Strongly Coupled Plasmonic Nanoparticles. *Ann. Der Phys.* **2023**, *535*, 2300319. [CrossRef]
18. Metcalfe, M. Applications of cavity optomechanics. *Appl. Phys. Rev.* **2014**, *1*, 031105. [CrossRef]
19. Firmansyah, T.; Wibisono, G.; Tjipto Rahardjo, E.; Kondoh, J. Reconfigurable localized surface plasmon resonance spectrum based on acousto-dynamic coupling in arrays gold nanoparticles induced by shear horizontal vibration. *Appl. Surf. Sci.* **2022**, *571*, 151331. [CrossRef]
20. Tsukanov, A.V. Optomechanical systems and quantum computing. *Russ. Microelectron.* **2011**, *40*, 333–342. [CrossRef]
21. Li, G.-C.; Zhang, Q.; Maier, S.A.; Lei, D. Plasmonic particle-on-film nanocavities: A versatile platform for plasmon-enhanced spectroscopy and photochemistry. *Nanophotonics* **2018**, *7*, 1865–1889. [CrossRef]
22. Maurer, T.; Adam, P.-M.; Lévêque, G. Coupling between plasmonic films and nanostructures: From basics to applications. *Nanophotonics* **2015**, *4*, 363–382. [CrossRef]
23. Goncalves, M.R. Plasmonic nanoparticles: Fabrication, simulation and experiments. *J. Phys. D Appl. Phys.* **2014**, *47*, 213001. [CrossRef]
24. Charchi, N.; Li, Y.; Huber, M.; Kwizera, E.A.; Huang, X.; Argyropoulos, C.; Hoang, T. Small mode volume plasmonic film-coupled nanostar resonators. *Nanoscale Adv.* **2020**, *2*, 2397–2403. [CrossRef] [PubMed]
25. Huang, S.; Ming, T.; Lin, Y.; Ling, X.; Ruan, Q.; Palacios, T.; Wang, J.; Dresselhaus, M.; Kong, J. Ultrasmall mode volumes in plasmonic cavities of nanoparticle-on-mirror structures. *Small* **2016**, *12*, 5190–5199. [CrossRef]
26. Deacon, W.M.; Lombardi, A.; Benz, F.; del Valle-Inclan Redondo, Y.; Chikkaraddy, R.; de Nijs, B.; Kleemann, M.-E.; Mertens, J.; Baumberg, J.J. Interrogating nanojunctions using ultraconfined acoustoplasmonic coupling. *Phys. Rev. Lett.* **2017**, *119*, 023901. [CrossRef]
27. Yeshchenko, O.A.; Kozachenko, V.V.; Naumenko, A.P.; Berezovska, N.I.; Kutsevol, N.V.; Chumachenko, V.A.; Haftel, M.; Pinchuk, A.O. Gold nanoparticle plasmon resonance in near-field coupled Au NPs layer/Al film nanostructure: Dependence on metal film thickness. *Photonics Nanostruct.-Fundam. Appl.* **2018**, *29*, 1–7. [CrossRef]
28. Armstrong, R.E.; Van Liempt, J.C.; Zijlstra, P. Effect of film thickness on the far-and near-field optical response of nanoparticle-on-film systems. *J. Phys. Chem. C* **2019**, *123*, 25801–25808. [CrossRef]
29. Mock, J.J.; Hill, R.T.; Degiron, A.; Zauscher, S.; Chilkoti, A.; Smith, D.R. Distance-dependent plasmon resonant coupling between a gold nanoparticle and gold film. *Nano Lett.* **2008**, *8*, 2245–2252. [CrossRef]
30. Lee, H.-Y.; Seong, H.K.; Yoo, J. Optimal shape design of the film-coupled nanoparticle using the phase field design method. *Int. J. Precis. Eng. Manuf.* **2016**, *17*, 453–460. [CrossRef]
31. Li, G.-C.; Zhang, Y.-L.; Lei, D.Y. Hybrid plasmonic gap modes in metal film-coupled dimers and their physical origins revealed by polarization resolved dark field spectroscopy. *Nanoscale* **2016**, *8*, 7119–7126. [CrossRef] [PubMed]
32. Kunert, B.; Guo, W.; Mols, Y.; Tian, B.; Wang, Z.; Shi, Y.; Van Thourhout, D.; Pantouvaki, M.; Van Campenhout, J.; Langer, R. III/V nano ridge structures for optical applications on patterned 300 mm silicon substrate. *Appl. Phys. Lett.* **2016**, *109*, 091101. [CrossRef]
33. El Ghafiani, M.; Noual, A.; Rezzouk, Y.; Moutaouekkil, M.; El Boudouti, E.H. Surface acoustic waves interaction with confined acoustic phonons in a coupled nanoridges dimer set atop of a multilayer design. *Arab J. Basic Appl. Sci.* **2023**, *30*, 593–600. [CrossRef]
34. Raman, A.; Fan, S. Photonic Band Structure of Dispersive Metamaterials Formulated as a Hermitian Eigenvalue Problem. *Phys. Rev. Lett.* **2010**, *104*, 087401. [CrossRef]
35. Palik, E.D. *Handbook of Optical Constants of Solids*; Academic Press: Cambridge, MA, USA, 1998; Volume 3.

36. Dorfmüller, J.; Vogelgesang, R.; Weitz, R.T.; Rockstuhl, C.; Etrich, C.; Pertsch, T.; Lederer, F.; Kern, K. Fabry-Pérot resonances in one-dimensional plasmonic nanostructures. *Nano Lett.* **2009**, *9*, 2372–2377. [CrossRef]
37. Maier, S.A. *Plasmonics: Fundamentals and Applications*; Springer: Berlin/Heidelberg, Germany, 2007; Volume 1.
38. Baumberg, J.J.; Aizpurua, J.; Mikkelsen, M.H.; Smith, D.R. Extreme nanophotonics from ultrathin metallic gaps. *Nat. Mater.* **2019**, *18*, 668–678. [CrossRef]
39. Koenderink, A.F. On the use of Purcell factors for plasmon antennas. *Opt. Lett.* **2010**, *35*, 4208–4210. [CrossRef]
40. Derom, S.; Vincent, R.; Bouhelier, A.; des Francs, G.C. Resonance quality, radiative/ohmic losses and modal volume of Mie plasmons. *Europhys. Lett.* **2012**, *98*, 47008. [CrossRef]
41. Srinivasan, K.; Borselli, M.; Painter, O.; Stintz, A.; Krishna, S. Cavity Q, mode volume, and lasing threshold in small diameter AlGaAs microdisks with embedded quantum dots. *Opt. Express* **2006**, *14*, 1094–1105. [CrossRef] [PubMed]
42. Ahmed, A.; Pelton, M.; Guest, J.R. Understanding how acoustic vibrations modulate the optical response of plasmonic metal nanoparticles. *ACS Nano* **2017**, *11*, 9360–9369. [CrossRef]
43. Kolwas, K. Optimization of Coherent Dynamics of Localized Surface Plasmons in Gold and Silver Nanospheres; Large Size Effects. *Materials* **2023**, *16*, 1801. [CrossRef] [PubMed]

photonics

MDPI

Review

The Use of Core-Shell Nanoparticles in Photovoltaics

Alexander Quandt [1,2,3,*], Daniel Wamwangi [1,4] and Sandile Kumalo [1]

1 School of Physics, University of the Witwatersrand, Private Bag 3, Wits 2050, South Africa; daniel.wamwangi@wits.ac.za (D.W.)
2 Istituto di Fotonica e Nanotecnologie-Consiglio Nazionale delle Ricerche (IFN-CNR), Characterization and Development of Materials for Photonics and Optoelectronics (CSMFO) Laboratory and Fondazione Bruno Kessler (FBK) Photonics Unit, Via Alla Cascata 56/C, 38123 Povo, TN, Italy
3 Photonics Laboratory of Angers LPhiA, SFR MATRIX, University of Angers, 2 Bd Lavoisier, 49045 Angers, France
4 Materials Physics Research Institute, University of the Witwatersrand, Private Bag 3, Wits 2050, South Africa
* Correspondence: alex.quandt@wits.ac.za

Abstract: The field of photovoltaics (PV) continually seeks innovative materials solutions to enhance the efficiency and the stability of their standard devices. Core-shell nanoparticles have emerged as a promising new technology with unique structural attributes and widely tunable properties. This paper reviews the use of plasmonic core-shell nanoparticles in PV applications through various experimental validations. We describe advancements in the design and in the control over the properties of core-shell nanoparticles and highlight their integration into various solar cells, based on their ability to finely tune optical, electronic, and chemical properties. We also discuss experimental results for organic, perovskite, and dye-sensitized solar cells, where core-shell nanoparticles have been successfully deployed. Additionally, we identify gaps in the current research, such as the need for scalable synthesis methods and long-term stability assessments, and we will point out promising new developments at the frontier of that field.

Keywords: photovoltaics; plasmonic; core-shell nanoparticles

1. Introduction

The global energy transition has highlighted the need for efficient, sustainable, and scalable renewable energy technologies [1]. Among these, photovoltaic (PV) systems, which convert sunlight directly into electricity, play a key role [2–4]. However, the cost of solar energy is not competitive with existing energy solutions. This is in part attributed to the costly fabrication of Si-based devices and the associated stringent material requirements. Therefore, the transition from bulk single crystal Si photoactive layers to thin films-based solar cells requires some practical trade-offs of the photo-absorption in the visible and near infra-red region. Even though there have been significant advancements in photoactive layer developments, challenges related to material properties, device physics, and operational limits continue to hinder their widespread adoption [5–8].

To address these challenges, researchers are exploring a range of strategies to improve light absorption, enhance the generation of excitons (bound electron-hole pairs) in materials where Coulombic attraction is significant, promote efficient exciton dissociation into free charge carriers, and facilitate the transport of these carriers in diverse photovoltaic devices. One promising approach is the incorporation of nanomaterials, particularly plasmonic nanoparticles (NPs), which exhibit tunable localized surface plasmon resonance (LSPR).

This unique optical phenomenon enables the manipulation of light at the nanoscale, enhancing the optoelectronic properties of the active layer and thereby the efficiency of PV devices [9–14].

An important advancement in this field is the development of core-shell NPs. These nanoparticles consist of a plasmonic core (typically a noble metal) surrounded by a dielectric shell. There is a lot of flexibility in the design of these nanoparticles, which allows for tunable optical properties, exciton generation, and improved charge transport, making them attractive for PV applications [15–17]. Another variant of the core-shell structure is the bimetallic core-shell nanoparticle, which consists of a metallic core and a different metallic shell. These core-shell structures have attracted extensive research interest due to their high photon trapping abilities and greater flexibility in modulating the plasmonic resonances through shell thickness variations [18,19].

This review paper is structured as follows: In the following section, we discuss the design principles and the most useful properties of plasmonic core-shell NPs, with a focus on their optical and electronic properties. Next, we examine their integration into various types of solar cells, including organic, perovskite, dye-sensitized, and inorganic PV devices. We also address the challenges associated with deploying these NPs in commercial applications and propose potential solutions. Finally, we conclude with a discussion on future research directions in this rapidly evolving field.

2. Design and Properties of Core-Shell Nanoparticles

2.1. Introduction to Core-Shell Nanoparticles

Nanotechnology has revolutionized material science by enabling the careful design and development of complex material systems through the control of atomic configurations, dimensionality, and patterning. As a result, research on nanostructures of varying geometry and aspect ratios, such as core-shell nanoparticles, has gained traction due to their unique, tunable properties and broad applicability in fields such as energy, electronics, and biomedicine. Core-shell nanoparticles consist of a core material surrounded by a shell that modifies their optical, electrical, and chemical characteristics [20]. By adjusting the core diameter and shell thickness, the LSPR characteristics, such as resonance wavelength, intensity, and field enhancement, can be precisely tuned, enabling more effective control of light–matter interactions compared to homogeneous nanoparticles. Additionally, core-shell architectures provide an inherent advantage in passivating surface defects, a challenge often encountered in homogeneous nanoparticles [21–23].

The unique properties of core-shell nanoparticles arise from the interaction between the core and the shell, which modifies light–matter interactions, resonance phenomena, and charge transport [24]. By designing the core-shell interface, researchers can achieve enhanced light absorption, reduced recombination losses, and improved charge transport. These features are particularly relevant for advanced optoelectronic applications, including solar cells.

2.2. Synthesis and Characterisation

2.2.1. Synthesis Methods

The formation of core-shell nanoparticles typically relies on multi-step synthesis routes, each offering unique advantages in controlling particle composition, morphology, and functionality. Common strategies include chemical reduction, seed-mediated growth, thermal decomposition, laser ablation, and templating techniques. These methods are often sequentially to achieve the desired core-shell architecture.

Chemical reduction is frequently used as an initial step to generate metallic nanoparticle seeds. It involves reducing metal precursors in the presence of stabilizing agents and allows good control over particle size and dispersion. While not sufficient for core-shell formation on its own, this step serves as the foundation for subsequent shell growth [25–30].

Building upon this, seed-mediated growth involves depositing a secondary material around preformed cores. This approach enables precise control over the shell thickness, composition, and overall particle geometry, making it particularly suitable for tuning optical and electronic properties [31–33].

Thermal decomposition, involving the breakdown of organometallic or inorganic precursors at elevated temperatures, produces uniform nanoparticles with high crystallinity. When applied in sequence, the shell precursor decomposes after core formation, and this method can be adapted for core-shell synthesis. It is especially valued for its scalability and reproducibility when integrated into multi-step protocols [34–37].

Pulsed laser ablation in liquids (PLAL) is a clean and versatile technique for synthesizing core-shell nanostructures. It involves focusing a high-energy femtosecond laser on a solid target submerged in liquid, generating a plasma plume that condenses into nanoparticles. Core-shell architectures can be achieved by ablating layered targets or using sequential ablation steps [38]. A recent study demonstrated the fabrication of Ti@W core-shell photocatalytic films using PLAL, where tungsten was deposited onto titanium substrates without the use of surfactants [39]. The resulting films exhibited superhydrophilic surfaces (contact angle $< 5°$) and enhanced visible-light photocatalytic activity, making them effective for self-cleaning applications.

Beyond discrete particle synthesis, creating ordered arrays of core-shell nanoparticles is critical for applications such as photonic crystals and sensors. Techniques like block copolymer lithography, often coupled with controlled shell growth, provide spatial patterning and uniformity over large areas. This method is particularly advantageous in device fabrication, where structural regularity is essential [40–42].

2.2.2. Characterization Techniques

The characterization of core-shell nanoparticles is crucial for understanding the relationship between their structural and optical properties, which are key to their performance in various applications. Several advanced techniques are used for this purpose, providing complementary insights into the morphology, composition, and optical behavior of these structures.

UV-Vis spectroscopy measurements are commonly employed to study optical absorption, particularly localized surface plasmon resonance (LSPR) and other optical properties when embedded in the photoactive or buffer layers. This method provides insights into how particle size, shape, and composition affect LSPR excitations and other modes [43–45]. The combination of UV-Vis spectroscopy with structural analysis helps establish correlations between the optical and structural properties of the nanoparticles.

Photoluminescence (PL) spectroscopy is a powerful tool for probing the electronic and optical properties of core-shell nanoparticles, particularly in the context of photovoltaic applications [46–48]. PL measurements, as shown in the spectra in Figure 1a, provide insights into exciton dynamics, recombination pathways, and energy transfer efficiency within the nanostructures. These properties are strongly influenced by the core-shell configuration, which can affect light absorption, emission peak positions and widths, and the spatial distribution of charge carriers [49–51]. For instance, spectral shifts observed in PL spectra can indicate changes in band alignment or enhanced exciton separation efficiency due to the presence of a shell layer [52–55]. Such analysis helps to distinguish between radiative recombination, non-radiative losses, and energy transfer mechanisms. Understanding and controlling these processes is essential for optimizing the design of core-shell nanoparticles to improve charge extraction and overall energy conversion efficiency in photovoltaic devices.

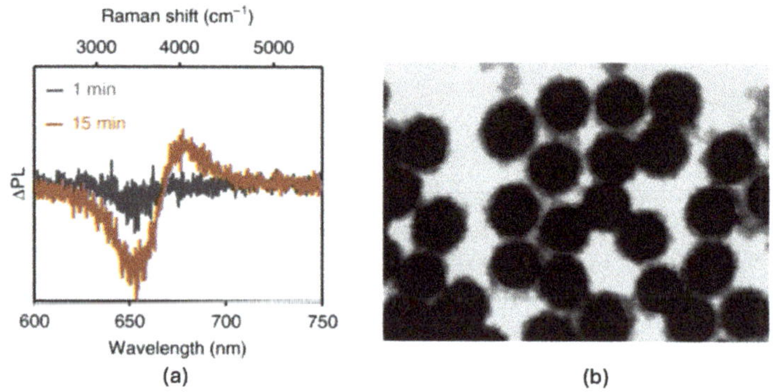

Figure 1. (**a**) Photoluminescence (ΔPL) spectra of photothermally grown Au@CeO_2 core@shell nanorod at t = 1 min (gray) and t = 15 min (brown). (**b**) Representative transmission electron microscopy (TEM) image showcasing the morphology of Au@CeO_2 core-shell NPs. Reproduced with permission from Ref. [55]. Copyright 2020, Springer Nature.

Transmission electron microscopy (TEM) is a standard method for structural analysis, offering high-resolution images of nanoparticle morphology. This technique, when applied to core-shell nanostructures, yields morphological information as shown in Figure 1b to enable precise correlation of optoelectronic properties with the geometrical measurements of core size, shell thickness, particle shape, and chemical composition. It also facilitates statistical analysis of particle size and shape distributions, which are critical for ensuring uniformity in nanoparticle synthesis and understanding their influence on performance [56–59].

2.3. Properties of Core-Shell Nanoparticles

2.3.1. Optical Properties

The optical absorption and scattering characteristics of core-shell nanoparticles are influenced by the aspect ratio, dielectric shell thickness, and core size. Adjusting the shell thickness (t) relative to the core radius (R_{core}) produces distinct optical effects. For thin shells ($t < R_{core}$), plasmonic behaviour is primarily dominated by the core, with the dielectric shell acting mainly as a spacer that isolates the core from environmental interactions, enhancing scattering cross-sections but offering limited influence over the plasmon resonance position. When the shell thickness approaches the core radius ($t \approx R_{core}$), strong coupling between core and shell materials occurs, altering both absorption and scattering behavior and providing a balanced enhancement of the optical response. For thick shells ($t > R_{core}$), the dielectric shell significantly modifies the surrounding electric field distribution, allowing fine-tuning of the plasmon resonance to align with specific wavelengths within the solar spectrum [60].

Core-shell nanoparticles offer significant advantages in PV applications. Their structural versatility allows for optimization of key processes, including light absorption, exciton generation, and charge transport. By surrounding the core with a dielectric shell, the nanoparticle's interaction with light can be finely controlled, enhancing the overall photovoltaic process [61–64]. Commonly used dielectric materials for shells include silica (SiO_2), titanium dioxide (TiO_2), zinc oxide (ZnO), and alumina (Al_2O_3). These materials not only passivate the core surface but also improve light scattering and charge transport, contributing to enhanced PV device performance [65–67].

In addition, rare-earth oxides, such as cerium oxide (CeO_2), yttrium oxide (Y_2O_3), and lanthanum oxide (La_2O_3), are emerging as promising shell materials. These oxides exhibit dynamic and active optical properties, making them ideal for advanced PV applications [68]. Their ability to facilitate down-conversion and up-conversion processes is particularly noteworthy. These processes involve converting high-energy photons into multiple low-energy photons or vice versa, effectively broadening the absorption spectrum and improving the overall efficiency of photovoltaic devices. Such advancements highlight the potential of core-shell nanoparticles to drive innovation and enhance performance in next-generation solar technologies [69–71].

2.3.2. Electrical and Chemical Properties

The diverse configurations of core-shell nanoparticles are shown in Figure 2. These architectures offer specific advantages depending on the target application. For instance, yolk–shell structures (Figure 2a) provide void space for energy storage or catalytic activity, while multilayered shells (Figure 2c) allow spectral tuning across multiple optical regimes. Surface-functionalized systems (Figure 2f) are crucial for biological or chemical sensing due to their customizable interfaces. The dielectric materials used in these NPs are typically insulators with low electrical conductivity but high polarizability suited for screening the electric field around the NP and to enhance plasmonic effects [72]. The quality of the interface between the metal and dielectric is critical, as smooth, well-defined interfaces minimize light scattering losses and allow for more efficient plasmonic resonance [73]. Some common dielectric materials include the following:

(i) **Silica (SiO_2):** Silica is chemically inert and transparent across the solar spectrum, making it ideal for encapsulating plasmonic cores like Au or Ag. The silica shell protects the core from oxidation and aggregation while fine-tuning the particle's SPR [74].

(ii) **Titanium Dioxide (TiO_2):** TiO_2 has a high refractive index, which enhances light scattering in core-shell nanoparticles. As a wide-bandgap semiconductor, it also facilitates electron transport in photovoltaic applications, especially in its anatase phase. Moreover, varying the shell thickness of the TiO_2 layer enables fine-tuning of the plasmon resonance wavelength and optical response, providing broader design flexibility compared to many other dielectric materials [75–77].

(iii) **Zinc Oxide (ZnO):** ZnO offers excellent electron mobility and is used as a shell material to enhance electron transport and light absorption in PV devices [78–80].

(iv) **Alumina (Al_2O_3):** Alumina is chemically resistant and acts as an insulating shell that prevents recombination, enhancing charge transport [81].

(v) **Polymeric Shells:** Flexible, organic-compatible polymers like polystyrene or polyvinyl alcohol are used in applications where dispersion within organic matrices is needed [82,83].

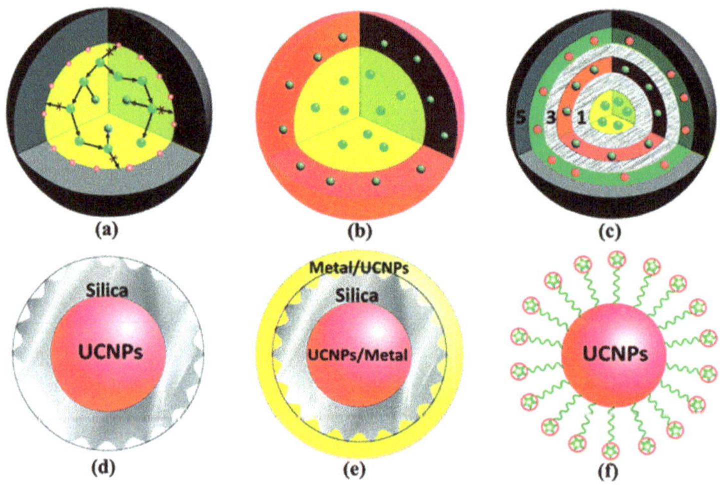

Figure 2. Schematic representations of different core–shell nanoparticle configurations: (**a**) Yolk–shell structures with a movable core inside a hollow shell, allowing for enhanced surface area and dynamic core–shell interactions; (**b**) Multicomponent core–shell with distinct compartmentalized regions for multifunctional applications; (**c**) Multilayered core–shell (layer-by-layer) with alternating materials offering tunable optical and electronic properties; (**d**) Classic core–shell with an upconversion nanoparticle (UCNP) core and silica shell for optical protection and surface passivation; (**e**) Composite core–shell with embedded metal/UCNP structures within silica for plasmon-enhanced upconversion; (**f**) Surface-functionalized UCNPs for targeted delivery or improved solubility via organic ligands or polymers. Reproduced with permission from Ref. [64]. Copyright 2015, The Royal Society of Chemistry.

3. Applications of Core-Shell Nanoparticles in Photovoltaics

3.1. *Organic Solar Cells*

Organic solar cells remain a cost effective and versatile PV technology that has gained rapid research attention [84–88]. Despite their enormous potential, they are limited by the short exciton diffusion length, which constrains carrier extraction. These diffusion lengths restrict the thickness of the photoactive layer to hundreds of nm, thereby limiting the photo-absorption and power conversion efficiency (PCE) in general [89,90]. A strategy to circumvent this limitation involves the use of plasmonic nanostructures to efficiently boost the light harvesting by the photoactive layer. Therefore, various core-shell NPs have been widely integrated into organic solar cells to enhance light absorption, exciton generation, and charge transport [91–93]. In the following sections, we present examples of enhanced performance in organic solar cells achieved by incorporating core-shell structures into the device architecture.

3.1.1. Augmentation of Active Layers

A promising strategy for enhancing photovoltaic performance involves embedding core-shell nanoparticles directly into the active layer of solar cells, as illustrated in Figure 3a. This approach has demonstrated improved light absorption in materials such as P3HT- or PTB7-based blends [94,95]. Embedding core-shell nanoparticles within the active layer significantly enhances light harvesting by amplifying localized electromagnetic fields [96]. This enhancement leads to improved absorption and exciton generation, enabling more efficient energy conversion. Studies by Zhang et al. have shown that incorporating core-shell nanoparticles into P3HT- or PTB7-based blends boosts the absorption of incident light, facilitating greater exciton generation and contributing to improved overall device performance [97–99].

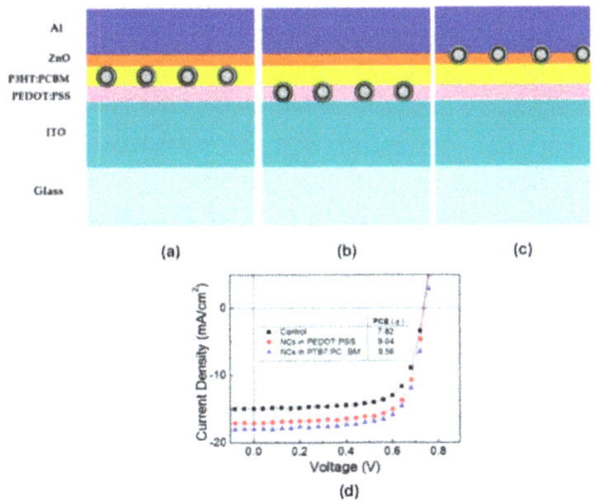

Figure 3. A schematic depicting the embedding of core-shell nanoparticles in (**a**) the active layer, (**b**) the hole transport layer, and (**c**) the electron transport layer of an organic solar cell. (**d**) Shows the J-V characteristics of devices with core-shell nanoparticles embedded in these layers. Reproduced with permission from Refs. [90,93]. Copyright 2016, The Royal Society of Chemistry and Copyright 2023, MDPI Micromachines.

3.1.2. Augmentation of Buffer Layers

Another strategy consists of incorporating core-shell NPs into the buffer layers through the near-field effect, such as the hole transport layer (HTL) or electron transport layer (ETL), shown in Figure 3b,c. The incorporation may involve either regular arrays of core-shell NPs or random embedding within these layers. For large core-shell structures, typically with core diameters of 100 nm or more, light trapping due to strong scattering effects dominate over absorption. These larger NPs can act as efficient scattering centers, extending the optical path length within the device and thus promoting greater light absorption in the active layer [100,101].

In contrast, smaller core-shell NPs with core diameters below 50 nm are more effective at generating localized near-field effects, which can increase exciton generation at the active layer interface. The proximity of these small NPs to the organic materials amplifies the near-field enhancement, benefiting charge separation and improving carrier extraction [102–105].

In the case of regular arrays, the periodicity and spacing of the nanostructures determine the nature of the LSPR excitation, and broadband multiple absorption peaks are typically observed. Arrays of core-shell NPs can also serve as diffractive elements, further enhancing light trapping within the active layer and ultimately increasing light-harvesting efficiency [106,107]. These configurations offer versatile pathways for selecting core-shell NPs tailored to enhance specific aspects of device performance as depicted in Figure 3d, depending on whether light trapping or near-field enhancement is prioritized.

3.2. Perovskite Solar Cells

While perovskite solar cells have demonstrated remarkable efficiency and stability in recent years, they still face certain limitations in their optical properties [108–110]. One major challenge is sub-optimal light absorption in the near-infrared (NIR) wavelengths, which reduces the overall photocurrent generation [111–116]. Additionally, the need to balance transparency with light harvesting in semitransparent perovskite solar cells restricts the thickness of the photoactive layer 150–300 nm, further limiting optical absorption and device efficiency [117–121]. Core-shell nanoparticles are being implemented to address these issues through the following:

3.2.1. Light Harvesting Enhancement

The incorporation of core-shell nanoparticles into photovoltaic devices as depicted in Figure 4a,b, particularly semitransparent perovskite solar cells, offers substantial advantages for light harvesting [122–125]. Metallic cores, such as Au or Ag, enhance local electromagnetic fields through localized surface plasmon resonance (LSPR) effects, leading to improved light absorption. These metallic cores enable light to be concentrated into sub-wavelength regions, amplifying the local electromagnetic field and optimizing light harvesting. This results in enhanced photocurrent generation, as presented in Figure 4c, and an overall increase in PCE, as illustrated in Figure 4d [126,127].

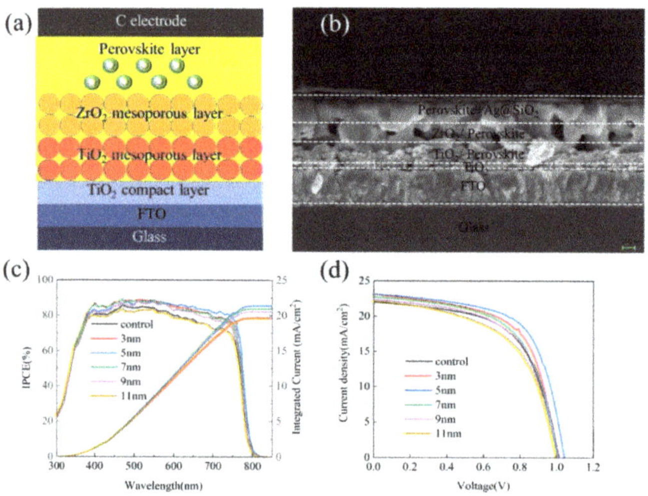

Figure 4. (**a**) Schematic representation of core-shell nanoparticles integrated into the perovskite layer. (**b**) SEM cross-sectional image of a perovskite solar cell (PSC) with core-shell NPs. (**c**) Incident photon to current conversion efficiency (IPCE) spectra of PSCs incorporating $Ag@SiO_2$ core-shell nanoparticles with varying shell thicknesses. (**d**) J-V characteristics of PSCs containing $Ag@SiO_2$ core-shell nanoparticles with different shell thicknesses. Reproduced with permission from Ref. [128]. Copyright 2020, MDPI Nanomaterials.

Despite these advantages, there are several practical challenges that must be addressed to fully realize the potential of core-shell NPs in perovskite solar cells. Issues such as material compatibility, achieving uniform dispersion of nanoparticles within the layers, and mitigating lead leaching from perovskite materials remain critical hurdles [129–131].

3.2.2. Stability Improvements

Dielectric shells provide critical stability enhancements for PSCs by shielding the sensitive perovskite layer from environmental stressors such as moisture, oxygen, and ultraviolet (UV) radiation. This protective barrier significantly enhances device longevity and operational stability, addressing one of the key challenges in perovskite solar cell technology [132–135].

Materials such as zinc oxide (ZnO) and titanium dioxide (TiO_2) are commonly employed as dielectric shells due to their dual functionality. These shells stabilize the perovskite layer while simultaneously improving electron mobility, facilitating efficient charge transport, and minimizing recombination losses. By integrating these dielectric shells, PSCs achieve enhanced electron transport and greater resilience against environmental degradation [136–139].

The ability of dielectric shells to mitigate degradation mechanisms is pivotal in extending the lifespan and reliability of PSCs. Their multifunctional role as both a stabilizing and performance-enhancing component underscores their importance in advancing the practical deployment of perovskite-based photovoltaic devices.

3.3. Dye-Sensitized Solar Cells

Core-shell NPs enhance the performance of dye-sensitized solar cells (DSSCs) by improving light absorption, electron injection, and transport and minimising recombination. Embedding these NPs in the photoanode as illustrated in Figure 5a,b or adding these core-shell nanoparticles (Figure 5c) to the electrolyte both increase the optical path length. They also enhance light scattering, which increases photon absorption [140]. Additionally, dielectric shells such as ZnO or SiO_2 improve electron transport, and they minimize recombination, leading to improved overall device performance, as shown in Figure 5d,e [141,142].

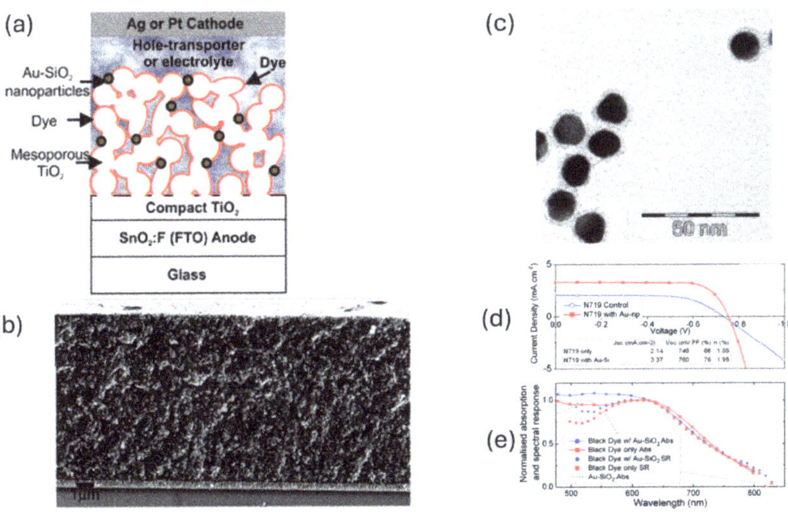

Figure 5. (**a**) Conceptual illustration of $Au@SiO_2$ core–shell nanoparticles embedded within the electrolyte of DSSCs, demonstrating their function in enhancing light absorption and facilitating charge transport. (**b**) Cross-sectional SEM image of a mesoporous TiO_2 photoanode incorporating $Au@SiO_2$ nanoparticles. (**c**) High-resolution depiction of core–shell nanoparticles featuring ~15 nm gold cores encapsulated by ~3 nm silica shells. (**d**) Current–voltage (J-V) characteristics of DSSCs showing the impact of $Au@SiO_2$ nanoparticle incorporation on device performance. (**e**) Comparison of normalized absorption and IPCE spectra for DSSCs with and without $Au@SiO_2$ integration. Reproduced with permission from Ref. [143]. Copyright 2010, American Chemical Society.

Energy level alignment is another important benefit, where the presence of core-shell NPs ensures more efficient electron transfer between the dye, the semiconductor, and the electrolyte [144]. DSSCs with integrated core-shell NPs have shown clear improvements in photocurrent and in PCE [145–149], where reductions in recombination losses, improved light harvesting, and better electron injections have led to PCE increases of 10–30%, thus contributing to much more efficient DSSCs [150–152].

3.4. Inorganic Solar Cells

The integration of core-shell NPs into inorganic solar cells has proven to be a promising approach for enhancing device efficiency. These nanoparticles play a critical role in improving light absorption, reducing reflection losses, and managing thermal effects in silicon-based and thin-film solar cells [153–155].

Enhanced Light Scattering and Trapping

Core-shell nanoparticles are instrumental in enhancing light absorption within silicon-based solar cells through effective light scattering and trapping mechanisms as demonstrated in Figure 6a,b. The dielectric shells of these nanoparticles scatter incoming light, increasing its optical path length within the active layer, which enhances photocurrent generation. This mechanism is particularly advantageous in reducing reflection losses, thereby allowing more light to be absorbed by the semiconductor material [156–158].

In thin-film photovoltaic devices, such as cadmium telluride (CdTe) and copper indium gallium selenide (CIGS) solar cells, core-shell nanoparticles facilitate improved light absorption in thinner active layers. By doing so, they enhance device efficiency without compromising charge transport properties [159,160]. Furthermore, core-shell nanoparticles can serve as anti-reflective coatings, decreasing surface reflectivity and maximizing light transmission into the active layer, thus boosting the overall efficiency of silicon solar cells [161].

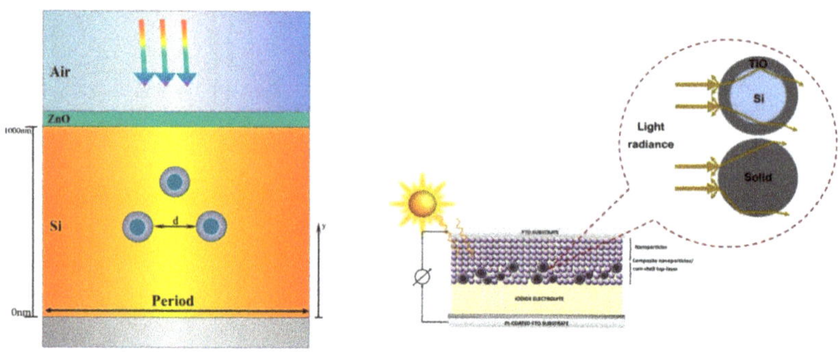

(a) (b)

Figure 6. (**a**) Schematic representation of a silicon solar cell incorporating core-shell nanoparticles. (**b**) Illustration of incident light interaction with SiO_2-TiO_2 core-shell structures compared to solid sphere structures, highlighting their role in light transmission and scattering. Reproduced with permission from Refs. [162,163]. Copyright 2019, Springer and Copyright 2019, Elsevier.

4. Challenges and Future Directions

4.1. Current Limitations

Despite their great potential outlined in this review, core-shell NPs face several challenges that must be addressed before their widespread adoption in photovoltaic (PV) technologies. These challenges include scalability issues, high manufacturing costs, limited plasmonic resonance lifetimes, and material stability [164,165]. Current synthesis methods, such as chemical reduction and seed-mediated growth, are costly and lack precise control over the general patterning and regularity of NP arrays. This lack of control compromises reproducibility and scalability for large-area production, which is critical for industrial applications [166].

A significant issue is the difficulty in patterning regular arrays of core-shell NPs for broadband optical absorption [167]. While conventional lithographic methods, such as electron-beam lithography and femtosecond laser processing, offer precise patterning, they remain costly, requiring specialized facilities, and hence are not scalable for large-area devices or modules [168,169]. Soft lithography techniques, including microcontact printing and nanoimprint lithography, provide a cost-effective and scalable alternative for large-area nanopatterning, making them promising for photovoltaic applications [170,171]. Block copolymer self-assembly has also emerged as a viable bottom-up approach for fabricating periodic nanostructures, offering tunability and large-area scalability, though challenges in defect control and domain alignment remain [172]. Additionally, advanced patterning methods used in OLED display manufacturing, such as inkjet printing and nanoimprint lithography, could offer new avenues for integrating core-shell NP arrays into scalable photovoltaic technologies [173].

Additionally, while dielectric shells are effective in protecting NPs, the long-term stability of these interfaces under real-world conditions remains a concern. Environmental factors, such as exposure to moisture, UV light, and temperature fluctuations, can degrade the metal-dielectric interface, leading to performance loss over time [174,175].

Finally, environmental and toxicity concerns, especially with materials like Ag, Au, and Pb, pose sustainability challenges that need to be addressed for broader adoption [176]. Additionally, these elements are increasingly being scrutinized as critical materials in the energy transition, with potential supply constraints due to their competing demand in sectors such as electronics, catalysis, and battery storage [177]. For instance, Ag is extensively used in photovoltaic modules and electronic circuits [178–183], while Au is crucial for high-performance electronic components [184–186]. Lead, despite its established toxicity, remains in use in perovskite solar cells, raising concerns about long-term environmental impact and regulatory restrictions [187,188]. While this challenge is primarily associated with the perovskite material itself rather than the core-shell structure, the integration of core-shell nanostructures into perovskite solar cells (fourth-generation solar cells) presents opportunities for enhanced stability and performance. In this context, colloidal core-shell structures of halide perovskites could also be considered, as they offer multiphoton excitations, which can be advantageous for improving light-harvesting efficiency and carrier dynamics in next-generation photovoltaic devices [189]. Exploring alternative materials or recycling strategies will be essential for ensuring the sustainable deployment of core-shell NPs in PV applications.

4.2. Some Solutions

To address these challenges, targeted approaches are needed that provide direct solutions to the high cost and low scalability problems. Among them are the following:

1. Scalable Synthesis Techniques
    - **Advanced Synthesis Methods**: Continuous flow synthesis and template-assisted approaches offer better scalability and control over particle uniformity. For example, continuous flow systems provide consistent reaction conditions, improving monodispersity and reproducibility in large-scale production [190,191]. From the point of view of large-scale device fabrication, these methods are not yet industry-standard due to high initial setup costs and technical challenges in scaling to multi-ton production levels [192,193].

- **Self-assembly with Directed Control**: Self-assembly techniques combined with external fields (electric or magnetic) can pattern nanoparticles into ordered arrays, improving optical performance while reducing fabrication costs. Large-scale fabrication would require consistent control of nanoparticle positioning across large areas, which remains challenging and limits industrial adoption [194–197].

2. Integration into PV Devices
    - **Efficient Deposition Techniques**: Advanced deposition methods such as spin-coating, spray-coating, and inkjet printing of core-shell nanoparticles enable the incorporation of nanoparticles into large-area solar modules with little extra costs. For example, inkjet printing allows precise deposition of core-shell NPs into pre-designed patterns, ensuring optimal placement for enhancing light trapping and charge transport [198,199]. Despite all the progress made in recent years, there remain some serious obstacles for large-scale implementations, such as issues with nozzle clogging and ink formulation, which need to be addressed to improve reliability and throughput [200,201].
    - **Optimized Placement**: The placement of core-shell NPs within the active or transport layers of PV devices should be optimized based on the target performance metric, such as light absorption or charge extraction. While simulations provide valuable insights, translating these into practical fabrication methods requires standardized protocols and advanced computational design tools to address inconsistencies [202–204].
3. Improving Long-term Stability
    - **Robust Shell Materials**: Using advanced shell materials, such as alumina or zirconia, can enhance chemical and thermal stability. These materials resist environmental degradation and maintain performance under real-world conditions [205,206]. Large-scale fabrication would require the development of cost-effective and scalable methods for depositing such shells, as current approaches like atomic layer deposition are expensive and need further refinement [207].
    - **Encapsulation Techniques**: Encapsulation of the entire solar cell, using barrier films or coatings, can protect core-shell NPs from moisture, oxygen, and UV light, extending the operational lifetime of the device [208–210]. This additional step adds to manufacturing costs, which industries are reluctant to absorb unless significant performance gains are demonstrated [211].
4. Environmental Sustainability
    - **Non-toxic Materials**: Developing non-toxic alternatives to toxic materials (e.g., aluminium-based plasmonic nanoparticles instead of lead-based materials) is essential for sustainable adoption [212]. Despite all the progress made in developing these alternatives, achieving comparable optical properties with non-toxic materials remains a significant technical challenge [213–215].
    - **Lifecycle Assessment**: Comprehensive lifecycle assessments can help quantify the environmental impact of core-shell NPs, identifying areas where improvements can be made in material selection and processing [216,217]. These assessments are still in their infancy for many emerging nanoparticle technologies, hindering their industrial acceptance [218].
5. Advanced Material Design
    - **Bimetallic Core-Shell Structures:** Recent theoretical studies highlight the potential of bimetallic plasmonic core-shell structures with limited surface segregation.

Combining metals with complementary properties (e.g., Au-Ag or Cu-Al) can result in core-shell NPs with tunable optical and electronic properties, tailored for specific PV applications. These designs leverage fundamental material properties, such as differences in cohesive energy and enthalpy of mixing, to achieve optimal performance [219–222]. From the point of view of large-scale device fabrication, scalable and cost-effective synthesis of these structures remains an obstacle [223].

5. Conclusions

Core-shell nanoparticles represent a disruptive advancement in photovoltaic technologies, offering innovative solutions to major challenges like light absorption and charge transport. These NPs can be integrated into a wide range of PV technologies, including organic, perovskite, dye-sensitized, and inorganic solar cells, where they lead to significant improvements in efficiency and performance of these devices. A summary of typical experimental results is shown in Table 1. The data refer to standard photovoltaic devices and reflect the current benchmarks in the field.

Table 1. Summary of performance enhancements with plasmonic core-shell nanoparticles in photovoltaic devices.

Device	Type of Core-Shell Nanoparticles	Improvements in Efficiency	Reference
Organic Solar Cells	Ag@TiO_2	From 7.82% to 9.56%	[93]
Perovskite Solar Cells	Au@SiO_2	From 11.44% to 14.57%	[122]
Dye-Sensitized Solar Cells	Ag@PVP	From 7.04% to 7.9%	[224]

State-of-the-art research in this area continues to address the challenges of scalability, material compatibility, and environmental sustainability, where core-shell NPs hold the potential to drive the next generation of high-efficient, durable, and cost-effective solar cells. Mastering these new technologies will make solar energy more affordable and accessible worldwide, where advanced solar technologies have already become an essential part of a global transition towards sustainable and renewable energy production and consumption.

Author Contributions: Conceptualization, A.Q., D.W. and S.K.; Literature Review and Investigation A.Q., D.W. and S.K.; resources, S.K. and D.W; Data Curation, S.K.; Writing—Original Draft Preparation, S.K.; Writing—Review and Editing, A.Q., D.W. and S.K.; Supervision, A.Q. and D.W.; Project Administration, A.Q.; Funding Acquisition, A.Q. and D.W. All authors have read and agreed to the published version of the manuscript.

Funding: DW would like to thank the financial support through the National Equipment program of the NRF (grant UID 116181). This research received financial support by the DSI-NRF Centre of Excellence in Strong Materials (CoE-SM) in terms of materials, major equipment and an MSc and PhD bursary for SK.

Acknowledgments: AQ would like to thank the Mandelstam Institute for Theoretical Physics at the University of the Witwatersrand for support, Maurizio Ferrari and Alessandro Chiasera for their kind invitation to the IFN-CNR in Trento, and Bouchta Sahraoui and his group at the Photonics Laboratory of the University of Angers LPhiA as his main hosts during his Sabbatical leave in 2024/2025.

Conflicts of Interest: The authors declare no conflict of interest.

References

1. Gielen, D.; Boshell, F.; Saygin, D.; Bazilian, M.D.; Wagner, N.; Gorini, R. The role of renewable energy in the global energy transformation. *Energy Strategy Rev.* **2019**, *24*, 50. [CrossRef]
2. Hassan, Q.; Algburi, S.; Sameen, A.Z.; Salman, H.M.; Jaszczur, M. A review of hybrid renewable energy systems: Solar and wind-powered solutions: Challenges, opportunities, and policy implications. *Results Eng.* **2023**, *20*, 101621. [CrossRef]
3. Hasan, M.M.; Hossain, S.; Mofijur, M.; Kabir, Z.; Badruddin, I.A.; Yunus Khan, T.M.; Jassim, E. Harnessing solar power: A review of photovoltaic innovations, solar thermal systems, and the dawn of energy storage solutions. *Energies* **2023**, *16*, 6456. [CrossRef]
4. Maka, A.O.; Ghalut, T.; Elsaye, E. The pathway toward decarbonisation and net-zero emissions by 2050: The role of solar energy technology. *Green Technol. Sustain.* **2024**, *2*, 100107. [CrossRef]
5. Machín, A.; Márquez, F. Advancements in photovoltaic cell materials: Silicon, Organic, and Perovskite Solar cells. *Materials* **2024**, *17*, 1165. [CrossRef]
6. Verlinden, P.; Young, D.L.; Xiong, G.; Reese, M.O.; Mansfield, L.M.; Powalla, M.; Haegel, N.M. Photovoltaic device innovation for a solar future. *Device* **2023**, *1*, 100013. [CrossRef]
7. Gerold, E.; Antrekowitsch, H. Advancements and Challenges in Photovoltaic Cell Recycling: A Comprehensive Review. *Sustainability* **2024**, *16*, 2542. [CrossRef]
8. Noman, M.; Khan, Z.; Jan, S.T. A comprehensive review on the advancements and challenges in perovskite solar cell technology. *RSC Adv.* **2024**, *14*, 5131. [CrossRef]
9. Lu, L.; Luo, Z.; Xu, T.; Yu, L. Cooperative plasmonic effect of Ag and Au nanoparticles on enhancing performance of polymer solar cells. *Nano Lett.* **2013**, *13*, 64. [CrossRef]
10. Omrani, M.; Keshavarzi, R.; Abdi-Jalebi, M.; Gao, P. Impacts of plasmonic nanoparticles incorporation and interface energy alignment for highly efficient carbon-based perovskite solar cells. *Sci. Rep.* **2022**, *12*, 5367. [CrossRef]
11. Smith, J.G.; Faucheaux, J.A.; Jain, P.K. Plasmon resonances for solar energy harvesting: A mechanistic outlook. *Nano Today* **2015**, *10*, 80. [CrossRef]
12. Babicheva, V.E. Optical processes behind plasmonic applications. *Nanomaterials* **2023**, *13*, 1270. [CrossRef]
13. Arinze, E.S.; Qiu, B.; Nyirjesy, G.; Thon, S.M. Plasmonic nanoparticle enhancement of solution-processed solar cells: Practical limits and opportunities. *ACS Photonics* **2016**, *3*, 173. [CrossRef]
14. Erwin, W.R.; Zarick, H.F.; Talbert, E.M.; Bardhan, R. Light trapping in mesoporous solar cells with plasmonic nanostructures. *Energy Environ. Sci.* **2016**, *9*, 1601. [CrossRef]
15. Zhang, Y.J.; Radjenovic, P.M.; Zhou, X.S.; Zhang, H.; Yao, J.L.; Li, J.F. Plasmonic core–shell nanomaterials and their applications in spectroscopies. *Adv. Mater.* **2021**, *33*, 2005900. [CrossRef] [PubMed]
16. Li, J.F.; Zhang, Y.J.; Ding, S.Y.; Panneerselvam, R.; Tian, Z.Q. Core–shell nanoparticle-enhanced Raman spectroscopy. *Chem. Rev.* **2017**, *117*, 5069. [CrossRef]
17. Wei, S.; Wang, Q.; Zhu, J.; Sun, L.; Lin, H.; Guo, Z. Multifunctional composite core–shell nanoparticles. *Nanoscale* **2011**, *3*, 4502. [CrossRef]
18. Ghosh Chaudhuri, R.; Paria, S. Core/shell nanoparticles: Classes, properties, synthesis mechanisms, characterization, and applications. *Chem. Rev.* **2012**, *112*, 2433. [CrossRef] [PubMed]
19. Gawande, M.B.; Goswami, A.; Asefa, T.; Guo, H.; Biradar, A.V.; Peng, D.L.; Varma, R.S. Core–shell nanoparticles: Synthesis and applications in catalysis and electrocatalysis. *Chem. Soc. Rev.* **2015**, *44*, 7590. [CrossRef]
20. Li, S.; Lin, H.; Chu, C.; Martin, C.; MacSwain, W.; Meulenberg, R.W.; Zheng, W. Interfacial B-site ion diffusion in all-inorganic core/shell perovskite nanocrystals. *ACS Nano* **2023**, *17*, 22477. [CrossRef]
21. Rycenga, M.; Cobley, C.M.; Zeng, J.; Li, W.; Moran, C.H.; Zhang, Q.; Xia, Y. Controlling the synthesis and assembly of silver nanostructures for plasmonic applications. *Chem. Rev.* **2011**, *111*, 3669. [CrossRef]
22. Cortés, E.; Wendisch, F.J.; Sortino, L.; Mancini, A.; Ezendam, S.; Saris, S.; Maier, S.A. Optical metasurfaces for energy conversion. *Chem. Rev.* **2022**, *122*, 15082. [CrossRef] [PubMed]
23. Alessandri, I.; Lombardi, J.R. Enhanced Raman scattering with dielectrics. *Chem. Rev.* **2016**, *116*, 14921. [CrossRef] [PubMed]
24. Kluczyk-Korch, K.; Antosiewicz, T.J. Hot carrier generation in a strongly coupled molecule–plasmonic nanoparticle system. *Nanophotonics* **2023**, *12*, 1722. [CrossRef] [PubMed]
25. Cao, K.; Zhu, Q.; Shan, B.; Chen, R. Controlled synthesis of Pd/Pt core shell nanoparticles using area-selective atomic layer deposition. *Sci. Rep.* **2015**, *5*, 8470. [CrossRef]
26. Chatterjee, K.; Sarkar, S.; Rao, K.J.; Paria, S. Core/shell nanoparticles in biomedical applications. *Adv. Colloid Interface Sci.* **2014**, *209*, 8. [CrossRef]
27. El-Toni, A.M.; Habila, M.A.; Labis, J.P.; ALOthman, Z.A.; Alhoshan, M.; Elzatahry, A.A.; Zhang, F. Design, synthesis and applications of core–shell, hollow core, and nanorattle multifunctional nanostructures. *Nanoscale* **2016**, *8*, 2510. [CrossRef]

28. Purbia, R.; Paria, S. Yolk/shell nanoparticles: Classifications, synthesis, properties, and applications. *Nanoscale* **2015**, *7*, 19789. [CrossRef]
29. Altammar, K.A. A review on nanoparticles: Characteristics, synthesis, applications, and challenges. *Front. Microbiol.* **2023**, *14*, 1155622. [CrossRef]
30. Nyabadza, A.; McCarthy, É.; Makhesana, M.; Heidarinassab, S.; Plouze, A.; Vazquez, M.; Brabazon, D. A review of physical, chemical and biological synthesis methods of bimetallic nanoparticles and applications in sensing, water treatment, biomedicine, catalysis and hydrogen storage. *Adv. Colloid Interface Sci.* **2023**, *321*, 103010. [CrossRef]
31. Yao, Q.; Yuan, X.; Fung, V.; Yu, Y.; Leong, D.T.; Jiang, D.E.; Xie, J. Understanding seed-mediated growth of gold nanoclusters at molecular level. *Nat. Commun.* **2017**, *8*, 927. [CrossRef] [PubMed]
32. Kwizera, E.A.; Chaffin, E.; Wang, Y.; Huang, X. Synthesis and properties of magnetic-optical core–shell nanoparticles. *RSC Adv.* **2017**, *7*, 17153. [CrossRef]
33. Han, G.H.; Kim, K.Y.; Nam, H.; Kim, H.; Yoon, J.; Lee, J.H.; Yu, T. Facile direct seed-mediated growth of AuPt bimetallic shell on the surface of Pd nanocubes and application for direct H_2O_2 synthesis. *Catalysts* **2020**, *10*, 650. [CrossRef]
34. Nguyen, M.D.; Tran, H.V.; Xu, S.; Lee, T.R. Fe_3O_4 nanoparticles: Structures, synthesis, magnetic properties, surface functionalization, and emerging applications. *Appl. Sci.* **2021**, *11*, 11301. [CrossRef]
35. Brollo, M.E.F.; López-Ruiz, R.; Muraca, D.; Figueroa, S.J.; Pirota, K.R.; Knobel, M. Compact Ag@ Fe3O4 core-shell nanoparticles by means of single-step thermal decomposition reaction. *Sci. Rep.* **2014**, *4*, 6839. [CrossRef]
36. Baaziz, W.; Pichon, B.P.; Liu, Y.; Grenèche, J.M.; Ulhaq-Bouillet, C.; Terrier, E.; Begin-Colin, S. Tuning of synthesis conditions by thermal decomposition toward core–Shell $Co_xFe_{1-x}O@Co_yFe_{3-y}O_4$ and $CoFe_2O_4$ nanoparticles with spherical and cubic shapes. *Chem. Mater.* **2014**, *26*, 5063. [CrossRef]
37. Reimhult, E.; Schroffenegger, M.; Lassenberger, A. Design principles for thermoresponsive core–shell nanoparticles: Controlling thermal transitions by brush morphology. *Langmuir* **2019**, *35*, 7092. [CrossRef]
38. Riedel, R.; Mahr, N.; Yao, C.; Wu, A.; Yang, F.; Hampp, N. Synthesis of gold–silica core–shell nanoparticles by pulsed laser ablation in liquid and their physico-chemical properties towards photothermal cancer therapy. *Nanoscale* **2020**, *12*, 3007. [CrossRef]
39. Cui, H.; Fang, X.; Qi, X.; Wang, Y.; Wang, H.; Zhai, Z.; Liu, J. Laser-fabricated core-shell Ti/W based photocatalytic films with superhydrophilic self-cleaning properties. *Surf. Interfaces* **2025**, *61*, 106147. [CrossRef]
40. Lermusiaux, L.; Plissonneau, M.; Bertry, L.; Drisko, G.L.; Buissette, V.; Le Mercier, T.; Tréguer-Delapierre, M. Seeded growth of ultrathin gold nanoshells using polymer additives and microwave radiation. *Sci. Rep.* **2021**, *11*, 17831. [CrossRef]
41. Barad, H.N.; Kwon, H.; Alarcón-Correa, M.; Fischer, P. Large area patterning of nanoparticles and nanostructures: Current status and future prospects. *ACS Nano* **2021**, *15*, 5861. [CrossRef] [PubMed]
42. Jones, M.R.; Osberg, K.D.; Macfarlane, R.J.; Langille, M.R.; Mirkin, C.A. Templated techniques for the synthesis and assembly of plasmonic nanostructures. *Chem. Rev.* **2011**, *111*, 3736. [CrossRef] [PubMed]
43. Kamarudheen, R.; Kumari, G.; Baldi, A. Plasmon-driven synthesis of individual metal@ semiconductor core@ shell nanoparticles. *Nat. Commun.* **2020**, *11*, 3957. [CrossRef]
44. Mayer, K.M.; Hafner, J.H. Localized surface plasmon resonance sensors. *Chem. Rev.* **2011**, *111*, 3828. [CrossRef]
45. Petryayeva, E.; Krull, U.J. Localized surface plasmon resonance: Nanostructures, bioassays and biosensing—A review. *Anal. Chim. Acta* **2011**, *706*, 8. [CrossRef] [PubMed]
46. Chen, G.; Shen, J.; Ohulchanskyy, T.Y.; Patel, N.J.; Kutikov, A.; Li, Z.; Han, G. (α-NaYbF4: Tm3+)/CaF2 core/shell nanoparticles with efficient near-infrared to near-infrared upconversion for high-contrast deep tissue bioimaging. *ACS Nano* **2012**, *6*, 8280. [CrossRef]
47. Canham, L. Introductory lecture: Origins and applications of efficient visible photoluminescence from silicon-based nanostructures. *Faraday Discuss.* **2020**, *222*, 10. [CrossRef]
48. Guidelli, E.J.; Baffa, O.; Clarke, D.R. Enhanced UV emission from silver/ZnO and gold/ZnO core-shell nanoparticles: Photoluminescence, radioluminescence, and optically stimulated luminescence. *Sci. Rep.* **2015**, *5*, 14004. [CrossRef]
49. Lee, S.H.; Kim, Y.; Jang, H.; Min, J.H.; Oh, J.; Jang, E.; Kim, D. The effects of discrete and gradient mid-shell structures on the photoluminescence of single InP quantum dots. *Nanoscale* **2019**, *11*, 23251. [CrossRef]
50. Shin, T.; Cho, K.S.; Yun, D.J.; Kim, J.; Li, X.S.; Moon, E.S.; Jung, T.S. Exciton recombination, energy-, and charge transfer in single-and multilayer quantum-dot films on silver plasmonic resonators. *Sci. Rep.* **2016**, *6*, 26204. [CrossRef]
51. Feng, L.; Li, W.; Bao, J.; Zheng, Y.; Li, Y.; Ma, Y.; Wu, A. Synthesis and luminescence properties of core-shell-shell composites: SiO2@ PMDA-Si-Tb@ SiO2 and SiO2@ PMDA-Si-Tb-phen@ SiO2. *Nanomaterials* **2019**, *9*, 189. [CrossRef] [PubMed]
52. Wang, S.; Jarrett, B.R.; Kauzlarich, S.M.; Louie, A.Y. Core/shell quantum dots with high relaxivity and photoluminescence for multimodality imaging. *J. Am. Chem. Soc.* **2007**, *129*, 3848. [CrossRef] [PubMed]

53. Fan, W.; Leung, M.K. Recent development of plasmonic resonance-based photocatalysis and photovoltaics for solar utilization. *Molecules* **2016**, *21*, 180. [CrossRef]
54. Liu, X.; Qiu, J. Recent advances in energy transfer in bulk and nanoscale luminescent materials: From spectroscopy to applications. *Chem. Soc. Rev.* **2015**, *44*, 8714. [CrossRef]
55. Chen, C.; Zheng, S.; Song, H. Photon management to reduce energy loss in perovskite solar cells. *Chem. Soc. Rev.* **2021**, *50*, 7250. [CrossRef] [PubMed]
56. Bijelić, L.; Ruiz-Zepeda, F.; Hodnik, N. The role of high-resolution transmission electron microscopy and aberration corrected scanning transmission electron microscopy in unraveling the structure-property relationships of Pt-based fuel cells electrocatalysts. *Inorg. Chem. Front.* **2024**, *11*, 323. [CrossRef]
57. Malatesta, M. Transmission electron microscopy as a powerful tool to investigate the interaction of nanoparticles with subcellular structures. *Int. J. Mol. Sci.* **2021**, *22*, 12789. [CrossRef]
58. Lee, B.; Yoon, S.; Lee, J.W.; Kim, Y.; Chang, J.; Yun, J.; Lee, J.H. Statistical characterization of the morphologies of nanoparticles through machine learning based electron microscopy image analysis. *ACS Nano* **2020**, *14*, 17125. [CrossRef]
59. Rice, S.B.; Chan, C.; Brown, S.C.; Eschbach, P.; Han, L.; Ensor, D.S.; Grulke, E.A. Particle size distributions by transmission electron microscopy: An interlaboratory comparison case study. *Metrologia* **2013**, *50*, 663. [CrossRef]
60. Logsdail, A.J.; Johnston, R.L. Predicting the optical properties of core–shell and Janus segregated Au–M nanoparticles (M= Ag, Pd). *J. Phys. Chem. C* **2012**, *116*, 23628. [CrossRef]
61. Zhang, W.E.I.; Saliba, M.; Stranks, S.D.; Sun, Y.; Shi, X.; Wiesner, U.; Snaith, H.J. Enhancement of perovskite-based solar cells employing core–shell metal nanoparticles. *Nano Lett.* **2013**, *13*, 4510. [CrossRef] [PubMed]
62. Selopal, G.S.; Zhao, H.; Wang, Z.M.; Rosei, F. Core/shell quantum dots solar cells. *Adv. Funct. Mater.* **2020**, *30*, 1908762. [CrossRef]
63. Li, W.; Elzatahry, A.; Aldhayan, D.; Zhao, D. Core–shell structured titanium dioxide nanomaterials for solar energy utilization. *Chem. Soc. Rev.* **2018**, *47*, 8237. [CrossRef]
64. Chen, G.; Ågren, H.; Ohulchanskyy, T.Y.; Prasad, P.N. Light upconverting core–shell nanostructures: Nanophotonic control for emerging applications. *Chem. Soc. Rev.* **2015**, *44*, 1713. [CrossRef] [PubMed]
65. Ju, S.; Kim, H.; Kwak, H.; Kang, C.; Jung, I.; Oh, S.; Lee, K.T. Dielectric light-trapping nanostructure for enhanced light absorption in organic solar cells. *Sci. Rep.* **2023**, *13*, 20649. [CrossRef] [PubMed]
66. Dolatyari, M.; Tahmasebi, M.; Dolatyari, S.; Rostami, A.; Zarghami, A.; Yadav, A.; Klein, A. Core/Shell ZnO/TiO_2, SiO_2/TiO_2, Al_2O_3/TiO_2, and Al1. 9Co0. $1O_3/TiO_2$ Nanoparticles for the Photodecomposition of Brilliant Blue E-4BA. *Inorganics* **2024**, *12*, 281. [CrossRef]
67. Yu, P.; Yao, Y.; Wu, J.; Niu, X.; Rogach, A.L.; Wang, Z. Effects of plasmonic metal core-dielectric shell nanoparticles on the broadband light absorption enhancement in thin film solar cells. *Sci. Rep.* **2017**, *7*, 7696. [CrossRef]
68. Hamed, M.S.; Adedeji, M.A.; Mola, G.T. Rare-earth metal-induced plasmon resonances for enhanced photons harvesting in inverted thin film organic solar cell. *Energy Fuels* **2021**, *35*, 15017. [CrossRef]
69. Li, B.; Tian, F.; Cui, X.; Xiang, B.; Zhao, H.; Zhang, H.; Wang, D. Review for rare-earth-modified perovskite materials and optoelectronic applications. *Nanomaterials* **2022**, *12*, 1773. [CrossRef]
70. Zhong, Y.; Dai, H. A mini-review on rare-earth down-conversion nanoparticles for NIR-II imaging of biological systems. *Nano Res.* **2020**, *13*, 1281. [CrossRef]
71. Wang, B.; Zhu, X.; Li, S.; Chen, M.; Liu, N.; Yang, H.; Yang, Y. Enhancing the photovoltaic performance of perovskite solar cells using plasmonic Au@ Pt@ Au core-shell nanoparticles. *Nanomaterials* **2019**, *9*, 1263. [CrossRef] [PubMed]
72. Wang, L.; Hasanzadeh Kafshgari, M.; Meunier, M. Optical properties and applications of plasmonic-metal nanoparticles. *Adv. Funct. Mater.* **2020**, *30*, 2005400. [CrossRef]
73. Zhang, X.; Chen, Y.L.; Liu, R.S.; Tsai, D.P. Plasmonic photocatalysis. *Rep. Prog. Phys.* **2013**, *76*, 046401. [CrossRef]
74. Bogireddy, N.K.R.; Pal, U.; Gomez, L.M.; Agarwal, V. Size controlled green synthesis of gold nanoparticles using Coffea arabica seed extract and their catalytic performance in 4-nitrophenol reduction. *RSC Adv.* **2018**, *8*, 24826. [CrossRef] [PubMed]
75. Ghosh, S.; Das, A.P. Modified titanium oxide (TiO_2) nanocomposites and its array of applications: A review. *Toxicol. Environ. Chem.* **2015**, *97*, 514. [CrossRef]
76. Taleb, A.; Mesguich, F.; Hérissan, A.; Colbeau-Justin, C.; Yanpeng, X.; Dubot, P. Optimized TiO_2 nanoparticle packing for DSSC photovoltaic applications. *Sol. Energy Mater. Sol. Cells* **2016**, *148*, 59. [CrossRef]
77. Bist, A.; Pant, B.; Ojha, G.P.; Acharya, J.; Park, M.; Saud, P.S. Novel materials in perovskite solar cells: Efficiency, stability, and future perspectives. *Nanomaterials* **2023**, *13*, 1724. [CrossRef] [PubMed]
78. Vittal, R.; Ho, K.C. Zinc oxide based dye-sensitized solar cells: A review. *Renew. Sustain. Energy Rev.* **2017**, *70*, 935. [CrossRef]
79. Zhang, P.; Wu, J.; Zhang, T.; Wang, Y.; Liu, D.; Chen, H.; Li, S. Perovskite solar cells with ZnO electron-transporting materials. *Adv. Mater.* **2018**, *30*, 1703737. [CrossRef]

80. Liu, C.; Xiao, C.; Li, W. Zinc oxide nanoparticles as electron transporting interlayer in organic solar cells. *J. Mater. Chem. C* **2021**, *9*, 14114. [CrossRef]
81. Lee, W.; Park, S.J. Porous anodic aluminum oxide: Anodization and templated synthesis of functional nanostructures. *Chem. Rev.* **2014**, *114*, 7556. [CrossRef] [PubMed]
82. Ramli, R.A.; Laftah, W.A.; Hashim, S. Core–shell polymers: A review. *RSC Adv.* **2013**, *3*, 15565. [CrossRef]
83. Xia, X.; Chao, D.; Qi, X.; Xiong, Q.; Zhang, Y.; Tu, J.; Fan, H.J. Controllable growth of conducting polymers shell for constructing high-quality organic/inorganic core/shell nanostructures and their optical-electrochemical properties. *Nano Lett.* **2013**, *13*, 4568. [CrossRef]
84. Lazaroiu, A.C.; Gmal Osman, M.; Strejoiu, C.V.; Lazaroiu, G. A comprehensive overview of photovoltaic technologies and their efficiency for climate neutrality. *Sustainability* **2023**, *15*, 16297. [CrossRef]
85. Yu, J.; Zheng, Y.; Huang, J. Towards high performance organic photovoltaic cells: A review of recent development in organic photovoltaics. *Polymers* **2014**, *6*, 2473. [CrossRef]
86. Wang, K.; Liu, C.; Meng, T.; Yi, C.; Gong, X. Inverted organic photovoltaic cells. *Chem. Soc. Rev.* **2016**, *45*, 2937. [CrossRef]
87. Roncali, J. Molecular bulk heterojunctions: An emerging approach to organic solar cells. *Acc. Chem. Res.* **2009**, *42*, 1719. [CrossRef]
88. Wang, G.; Adil, M.A.; Zhang, J.; Wei, Z. Large-area organic solar cells: Material requirements, modular designs, and printing methods. *Adv. Mater.* **2019**, *31*, 1805089. [CrossRef]
89. Firdaus, Y.; Le Corre, V.M.; Karuthedath, S.; Liu, W.; Markina, A.; Huang, W.; Anthopoulos, T.D. Long-range exciton diffusion in molecular non-fullerene acceptors. *Nat. Commun.* **2020**, *11*, 5220. [CrossRef]
90. van Bavel, S.; Sourty, E.; de With, G.; Frolic, K.; Loos, J. Relation between photoactive layer thickness, 3D morphology, and device performance in P3HT/PCBM bulk-heterojunction solar cells. *Macromolecules* **2009**, *42*, 7396. [CrossRef]
91. Gattu Subramanyam, P.; Krishnaswamy, N.; Guha, K.; Iannacci, J.; Ude, E.N.; Muniswamy, V. Enhanced optical management in organic solar cells by virtue of square-lattice triple core-shell nanostructures. *Micromachines* **2023**, *14*, 1574. [CrossRef] [PubMed]
92. Lee, J.M.; Kim, S.O. Enhancing organic solar cells with plasmonic nanomaterials. *ChemNanoMat* **2016**, *2*, 19. [CrossRef]
93. Liu, S.; Jiang, R.; You, P.; Zhu, X.; Wang, J.; Yan, F. Au/Ag core–shell nanocuboids for high-efficiency organic solar cells with broadband plasmonic enhancement. *Energy Environ. Sci.* **2016**, *9*, 898. [CrossRef]
94. Anrango-Camacho, C.; Pavón-Ipiales, K.; Frontana-Uribe, B.A.; Palma-Cando, A. Recent advances in hole-transporting layers for organic solar cells. *Nanomaterials* **2022**, *12*, 443. [CrossRef]
95. Chou, C.H.; Chen, F.C. Plasmonic nanostructures for light trapping in organic photovoltaic devices. *Nanoscale* **2014**, *6*, 8458. [CrossRef] [PubMed]
96. Yao, K.; Zhong, H.; Liu, Z.; Xiong, M.; Leng, S.; Zhang, J.; Jen, A.K.Y. Plasmonic metal nanoparticles with core–bishell structure for high-performance organic and perovskite solar cells. *ACS Nano* **2019**, *13*, 5397. [CrossRef]
97. Kosco, J.; Bidwell, M.; Cha, H.; Martin, T.; Howells, C.T.; Sachs, M.; McCulloch, I. Enhanced photocatalytic hydrogen evolution from organic semiconductor heterojunction nanoparticles. *Nat. Mater.* **2020**, *19*, 559. [CrossRef]
98. Wu, B.; Wu, X.; Guan, C.; Fai Tai, K.; Yeow, E.K.L.; Jin Fan, H.; Sum, T.C. Uncovering loss mechanisms in silver nanoparticle-blended plasmonic organic solar cells. *Nat. Commun.* **2013**, *4*, 2004. [CrossRef]
99. Zhang, R.; Zhou, Y.; Peng, L.; Li, X.; Chen, S.; Feng, X.; Huang, W. Influence of SiO_2 shell thickness on power conversion efficiency in plasmonic polymer solar cells with Au nanorod@ SiO_2 core-shell structures. *Sci. Rep.* **2016**, *6*, 25036. [CrossRef]
100. Chen, J.D.; Zhou, L.; Ou, Q.D.; Li, Y.Q.; Shen, S.; Lee, S.T.; Tang, J.X. Enhanced light harvesting in organic solar cells featuring a biomimetic active layer and a self-cleaning antireflective coating. *Adv. Energy Mater.* **2014**, *4*, 1301777. [CrossRef]
101. Thrithamarassery Gangadharan, D.; Xu, Z.; Liu, Y.; Izquierdo, R.; Ma, D. Recent advancements in plasmon-enhanced promising third-generation solar cells. *Nanophotonics* **2017**, *6*, 175. [CrossRef]
102. Ai, B.; Fan, Z.; Wong, Z.J. Plasmonic–perovskite solar cells, light emitters, and sensors. *Microsyst. Nanoeng.* **2022**, *8*, 5. [CrossRef]
103. Mohammadi, M.H.; Fathi, D.; Eskandari, M. Light trapping in perovskite solar cells with plasmonic core/shell nanorod array: A numerical study. *Energy Rep.* **2021**, *7*, 1415. [CrossRef]
104. Alkhalayfeh, M.A.; Aziz, A.A.; Pakhuruddin, M.Z. An overview of enhanced polymer solar cells with embedded plasmonic nanoparticles. *Renew. Sustain. Energy Rev.* **2021**, *141*, 110726. [CrossRef]
105. Ha, M.; Kim, J.H.; You, M.; Li, Q.; Fan, C.; Nam, J.M. Multicomponent plasmonic nanoparticles: From heterostructured nanoparticles to colloidal composite nanostructures. *Chem. Rev.* **2019**, *119*, 12278. [CrossRef]
106. Wang, W.; Qi, L. Light management with patterned micro-and nanostructure arrays for photocatalysis, photovoltaics, and optoelectronic and optical devices. *Adv. Funct. Mater.* **2019**, *29*, 1807275. [CrossRef]
107. Han, G.S.; Chung, H.S.; Kim, B.J.; Kim, D.H.; Lee, J.W.; Swain, B.S.; Jung, H.S. Retarding charge recombination in perovskite solar cells using ultrathin MgO-coated TiO_2 nanoparticulate films. *J. Mater. Chem. A* **2015**, *3*, 9164. [CrossRef]

108. Liu, H.; Xiang, L.; Gao, P.; Wang, D.; Yang, J.; Chen, X.; Zhang, Y. Improvement strategies for stability and efficiency of perovskite solar cells. *Nanomaterials* **2022**, *12*, 3295. [CrossRef]
109. Cao, Y.; Liu, H.; Gao, F.; Li, D.; Xiang, L.; Gao, J.; Li, S. Interface optimization and growth control for high efficiency wide bandgap perovskite solar cells. *Surf. Interfaces* **2023**, *37*, 102680. [CrossRef]
110. Cao, Y.; Wang, X.; Sun, J.; Xiang, L.; Li, D.; He, L.; Li, S. Synergistic ion-anchoring passivation for perovskite solar cells with efficiency exceeding 24% and ultra-ambient stability. *ACS Appl. Mater. Interfaces* **2023**, *15*, 40032–40041. [CrossRef]
111. Forgács, D.; Gil-Escrig, L.; Pérez-Del-Rey, D.; Momblona, C.; Werner, J.; Niesen, B.; Bolink, H.J. Efficient monolithic perovskite/perovskite tandem solar cells. *Adv. Energy Mater.* **2017**, *7*, 1602121. [CrossRef]
112. Wali, Q.; Elumalai, N.K.; Iqbal, Y.; Uddin, A.; Jose, R. Tandem perovskite solar cells. *Renew. Sustain. Energy Rev.* **2018**, *84*, 110. [CrossRef]
113. Song, Z.; Chen, C.; Li, C.; Awni, R.A.; Zhao, D.; Yan, Y. Wide-bandgap, low-bandgap, and tandem perovskite solar cells. *Semicond. Sci. Technol.* **2019**, *34*, 093001. [CrossRef]
114. Lal, N.N.; Dkhissi, Y.; Li, W.; Hou, Q.; Cheng, Y.B.; Bach, U. Perovskite tandem solar cells. *Adv. Energy Mater.* **2017**, *7*, 1602761. [CrossRef]
115. Li, H.; Zhang, W. Perovskite tandem solar cells: From fundamentals to commercial deployment. *Chem. Rev.* **2020**, *120*, 9950. [CrossRef]
116. Ho-Baillie, A.W.; Zheng, J.; Mahmud, M.A.; Ma, F.J.; McKenzie, D.R.; Green, M.A. Recent progress and future prospects of perovskite tandem solar cells. *Appl. Phys. Rev.* **2021**, *8*, 041307. [CrossRef]
117. Xue, Q.; Xia, R.; Brabec, C.J.; Yip, H.L. Recent advances in semi-transparent polymer and perovskite solar cells for power generating window applications. *Energy Environ. Sci.* **2018**, *11*, 1709. [CrossRef]
118. Tai, Q.; Yan, F. Emerging semitransparent solar cells: Materials and device design. *Adv. Mater.* **2017**, *29*, 1700192. [CrossRef]
119. Shi, B.; Duan, L.; Zhao, Y.; Luo, J.; Zhang, X. Semitransparent perovskite solar cells: From materials and devices to applications. *Adv. Mater.* **2020**, *32*, 1806474. [CrossRef]
120. Liao, J.F.; Wu, W.Q.; Jiang, Y.; Zhong, J.X.; Wang, L.; Kuang, D.B. Understanding of carrier dynamics, heterojunction merits and device physics: Towards designing efficient carrier transport layer-free perovskite solar cells. *Chem. Soc. Rev.* **2020**, *49*, 381. [CrossRef]
121. Rahmany, S.; Etgar, L. Semitransparent perovskite solar cells. *ACS Energy Lett.* **2020**, *5*, 1531. [CrossRef]
122. Chandrasekhar, P.S.; Dubey, A.; Reza, K.M.; Hasan, M.N.; Bahrami, B.; Komarala, V.K.; Qiao, Q. Higher efficiency perovskite solar cells using Au@ SiO_2 core–shell nanoparticles. *Sustain. Energy Fuels* **2018**, *2*, 2267.
123. Fard, A.H.M.; Matloub, S. Design and simulation of bifacial perovskite solar cell with high efficiency using cubic plasmonic nanoparticles. *Sol. Energy* **2024**, *280*, 112871.
124. Cheng, P.; An, Y.; Jen, A.K.Y.; Lei, D. New Nanophotonics Approaches for Enhancing the Efficiency and Stability of Perovskite Solar Cells. *Adv. Mater.* **2024**, *36*, 2309459. [CrossRef]
125. Xu, L.; Yuan, S.; Ma, L.; Zhang, B.; Fang, T.; Li, X.; Song, J. All-inorganic perovskite quantum dots as light-harvesting, interfacial, and light-converting layers toward solar cells. *J. Mater. Chem. A* **2021**, *9*, 18947. [CrossRef]
126. Li, X.; Zhu, J.; Wei, B. Hybrid nanostructures of metal/two-dimensional nanomaterials for plasmon-enhanced applications. *Chem. Soc. Rev.* **2016**, *45*, 3187.
127. Tan, D.Q. The search for enhanced dielectric strength of polymer-based dielectrics: A focused review on polymer nanocomposites. *J. Appl. Polym. Sci.* **2020**, *137*, 49379. [CrossRef]
128. He, Z.; Zhang, C.; Meng, R.; Luo, X.; Chen, M.; Lu, H.; Yang, Y. Influence of Ag@ SiO_2 with different shell thickness on photoelectric properties of hole-conductor-free perovskite solar cells. *Nanomaterials* **2020**, *10*, 2364. [CrossRef]
129. Bai, Y.; Liu, C.; Shan, Y.; Chen, T.; Zhao, Y.; Yu, C.; Pang, H. Metal-organic frameworks nanocomposites with different dimensionalities for energy conversion and storage. *Adv. Energy Mater.* **2022**, *12*, 2100346. [CrossRef]
130. Cayre, O.J.; Chagneux, N.; Biggs, S. Stimulus responsive core-shell nanoparticles: Synthesis and applications of polymer based aqueous systems. *Soft Matter* **2011**, *7*, 2234. [CrossRef]
131. Datye, A.K.; Votsmeier, M. Opportunities and challenges in the development of advanced materials for emission control catalysts. *Nat. Mater.* **2021**, *20*, 1059. [CrossRef]
132. Wang, Q.; Phung, N.; Di Girolamo, D.; Vivo, P.; Abate, A. Enhancement in lifespan of halide perovskite solar cells. *Energy Environ. Sci.* **2019**, *12*, 865. [CrossRef]
133. Ma, S.; Yuan, G.; Zhang, Y.; Yang, N.; Li, Y.; Chen, Q. Development of encapsulation strategies towards the commercialization of perovskite solar cells. *Energy Environ. Sci.* **2022**, *15*, 13. [CrossRef]
134. Bogachuk, D.; Zouhair, S.; Wojciechowski, K.; Yang, B.; Babu, V.; Wagner, L.; Hinsch, A. Low-temperature carbon-based electrodes in perovskite solar cells. *Energy Environ. Sci.* **2020**, *13*, 3880. [CrossRef]

135. Li, J.; Xia, R.; Qi, W.; Zhou, X.; Cheng, J.; Chen, Y.; Zhang, X. Encapsulation of perovskite solar cells for enhanced stability: Structures, materials and characterization. *J. Power Sources* **2021**, *485*, 229313. [CrossRef]
136. Ekeocha, J.; Ellingford, C.; Pan, M.; Wemyss, A.M.; Bowen, C.; Wan, C. Challenges and opportunities of self-healing polymers and devices for extreme and hostile environments. *Adv. Mater.* **2021**, *33*, 2008052. [CrossRef]
137. Rao, H.; Sun, W.; Ye, S.; Yan, W.; Li, Y.; Peng, H.; Huang, C. Solution-processed CuS NPs as an inorganic hole-selective contact material for inverted planar perovskite solar cells. *ACS Appl. Mater. Interfaces* **2016**, *8*, 7805. [CrossRef]
138. Najafi, M.; Di Giacomo, F.; Zhang, D.; Shanmugam, S.; Senes, A.; Verhees, W.; Andriessen, R. Highly efficient and stable flexible perovskite solar cells with metal oxides nanoparticle charge extraction layers. *Small* **2018**, *14*, 1702775. [CrossRef]
139. Wang, Q.; Chueh, C.C.; Zhao, T.; Cheng, J.; Eslamian, M.; Choy, W.C.; Jen, A.K.Y. Effects of self-assembled monolayer modification of nickel oxide nanoparticles layer on the performance and application of inverted perovskite solar cells. *ChemSusChem* **2017**, *10*, 3803. [CrossRef]
140. Tsai, J.K.; Hsu, W.D.; Wu, T.C.; Meen, T.H.; Chong, W.J. Effect of compressed TiO_2 nanoparticle thin film thickness on the performance of dye-sensitized solar cells. *Nanoscale Res. Lett.* **2013**, *8*, 6. [CrossRef]
141. Tiwana, P.; Docampo, P.; Johnston, M.B.; Snaith, H.J.; Herz, L.M. Electron mobility and injection dynamics in mesoporous ZnO, SnO_2, and TiO_2 films used in dye-sensitized solar cells. *ACS Nano* **2011**, *5*, 5166. [CrossRef]
142. Ding, B.; Lee, B.J.; Yang, M.; Jung, H.S.; Lee, J.K. Surface-Plasmon Assisted Energy Conversion in Dye-Sensitized Solar Cells. *Adv. Energy Mater.* **2011**, *1*, 421. [CrossRef]
143. Brown, M.D.; Suteewong, T.; Kumar, R.S.S.; D'Innocenzo, V.; Petrozza, A.; Lee, M.M.; Snaith, H.J. Plasmonic dye-sensitized solar cells using core− shell metal− insulator nanoparticles. *Nano Lett.* **2011**, *11*, 438. [CrossRef]
144. Mélinon, P.; Begin-Colin, S.; Duvail, J.L.; Gauffre, F.; Boime, N.H.; Ledoux, G.; Warot-Fonrose, B. Engineered inorganic core/shell nanoparticles. *Phys. Rep.* **2014**, *543*, 197. [CrossRef]
145. Feng, H.P.; Tang, L.; Zeng, G.M.; Zhou, Y.; Deng, Y.C.; Ren, X.; Yu, J.F. Core-shell nanomaterials: Applications in energy storage and conversion. *Adv. Colloid Interface Sci.* **2019**, *267*, 46. [CrossRef]
146. Wooh, S.; Lee, Y.G.; Tahir, M.N.; Song, D.; Meister, M.; Laquai, F.; Char, K. Plasmon-enhanced photocurrent in quasi-solid-state dye-sensitized solar cells by the inclusion of gold/silica core–shell nanoparticles in a TiO_2 photoanode. *J. Mater. Chem. A* **2013**, *1*, 12634. [CrossRef]
147. Xu, Q.; Liu, F.; Meng, W.; Huang, Y. Plasmonic core-shell metal-organic nanoparticles enhanced dye-sensitized solar cells. *Opt. Express* **2012**, *20*, A907. [CrossRef]
148. Zheng, Y.Z.; Tao, X.; Zhang, J.W.; Lai, X.S.; Li, N. Plasmonic enhancement of light-harvesting efficiency in tandem dye-sensitized solar cells using multiplexed gold core/silica shell nanorods. *J. Power Sources* **2018**, *376*, 32. [CrossRef]
149. Chandrasekhar, P.S.; Chander, N.; Anjaneyulu, O.; Komarala, V.K. Plasmonic effect of Ag@ TiO_2 core–shell nanocubes on dye-sensitized solar cell performance based on reduced graphene oxide–TiO_2 nanotube composite. *Thin Solid Film.* **2015**, *594*, 55. [CrossRef]
150. Devadiga, D.; Selvakumar, M.; Shetty, P.; Santosh, M.S. Recent progress in dye sensitized solar cell materials and photo-supercapacitors: A review. *J. Power Sources* **2021**, *493*, 229698. [CrossRef]
151. Gong, J.; Sumathy, K.; Qiao, Q.; Zhou, Z. Review on dye-sensitized solar cells (DSSCs): Advanced techniques and research trends. *Renew. Sustain. Energy Rev.* **2017**, *68*, 246. [CrossRef]
152. Zhang, W.; Wu, Y.; Bahng, H.W.; Cao, Y.; Yi, C.; Saygili, Y.; Grätzel, M. Comprehensive control of voltage loss enables 11.7% efficient solid-state dye-sensitized solar cells. *Energy Environ. Sci.* **2018**, *11*, 1787. [CrossRef]
153. Nguyen, H.T.; Nguyen, T.T.; Tran, T.T.; Bang, J.; Kumar, M.; Kim, J.; Yun, J.H. Colour-Passive radiative cooling in optoelectronics with Silver/Quantum dot decorated silica multifunctional hybrid structures. *Chem. Eng. J.* **2024**, *488*, 150840. [CrossRef]
154. Mirnaziry, S.R.; Shameli, M.A.; Yousefi, L. Design and analysis of multi-layer silicon nanoparticle solar cells. *Sci. Rep.* **2022**, *12*, 13259. [CrossRef]
155. Ali, A.; El-Mellouhi, F.; Mitra, A.; Aïssa, B. Research progress of plasmonic nanostructure-enhanced photovoltaic solar cells. *Nanomaterials* **2022**, *12*, 788. [CrossRef]
156. Tabrizi, A.A.; Pahlavan, A. Efficiency improvement of a silicon-based thin-film solar cell using plasmonic silver nanoparticles and an antireflective layer. *Opt. Commun.* **2020**, *454*, 124437. [CrossRef]
157. Marangi, F.; Lombardo, M.; Villa, A.; Scotognella, F. New strategies for solar cells beyond the visible spectral range. *Opt. Mater. X* **2021**, *11*, 100083. [CrossRef]
158. Sheverdin, A.; Valagiannopoulos, C. Core-shell nanospheres under visible light: Optimal absorption, scattering, and cloaking. *Phys. Rev. B* **2019**, *99*, 075305. [CrossRef]
159. Devasahayam, S.; Hussain, C.M. Thin-film nanocomposite devices for renewable energy current status and challenges. *Sustain. Mater. Technol.* **2020**, *26*, e00233. [CrossRef]

160. Manthina, V.; Correa Baena, J.P.; Liu, G.; Agrios, A.G. ZnO–TiO_2 nanocomposite films for high light harvesting efficiency and fast electron transport in dye-sensitized solar cells. *J. Phys. Chem. C* **2012**, *116*, 23870. [CrossRef]
161. Huang, J.Y.; Wang, Y.; Fei, G.T.; Xu, S.H.; Wang, B.; Zeng, Z. Dual-functional antireflection and down-shifting coating for Si solar cells. *Colloids Surf. A Physicochem. Eng. Asp.* **2022**, *652*, 129907. [CrossRef]
162. Mokari, G.; Heidarzadeh, H. Efficiency enhancement of an ultra-thin silicon solar cell using plasmonic coupled core-shell nanoparticles. *Plasmonics* **2019**, *14*, 1041. [CrossRef]
163. Zaine, S.N.A.; Mohamed, N.M.; Khatani, M.; Samsudin, A.E.; Shahid, M.U. Improving the light scattering efficiency of photoelectrode dye-sensitized solar cell through optimization of core-shell structure. *Mater. Today Commun.* **2019**, *19*, 220. [CrossRef]
164. Mondal, K.; Sharma, A. Recent advances in the synthesis and application of photocatalytic metal–metal oxide core–shell nanoparticles for environmental remediation and their recycling process. *RSC Adv.* **2016**, *6*, 83612. [CrossRef]
165. Langer, J.; Jimenez de Aberasturi, D.; Aizpurua, J.; Alvarez-Puebla, R.A.; Auguié, B.; Baumberg, J.J.; Liz-Marzán, L.M. Present and future of surface-enhanced Raman scattering. *ACS Nano* **2019**, *14*, 117. [CrossRef]
166. Zhang, Y.; Zhang, L.; Cui, K.; Ge, S.; Cheng, X.; Yan, M.; Liu, H. Flexible electronics based on micro/nanostructured paper. *Adv. Mater.* **2018**, *30*, 1801588. [CrossRef]
167. Yadav, A.S.; Tran, D.T.; Teo, A.J.; Dai, Y.; Galogahi, F.M.; Ooi, C.H.; Nguyen, N.T. Core–shell particles: From fabrication methods to diverse manipulation techniques. *Micromachines* **2023**, *14*, 497. [CrossRef]
168. Srivastava, K.; Jacobs, T.S.; Ostendorp, S.; Jonker, D.; Brzesowsky, F.A.; Susarrey-Arce, A.; Odijk, M. Alternative nano-lithographic tools for shell-isolated nanoparticle enhanced Raman spectroscopy substrates. *Nanoscale* **2024**, *16*, 7582. [CrossRef]
169. Stokes, K.; Clark, K.; Odetade, D.; Hardy, M.; Goldberg Oppenheimer, P. Advances in lithographic techniques for precision nanostructure fabrication in biomedical applications. *Discov. Nano* **2023**, *18*, 153. [CrossRef]
170. Biswas, A.; Bayer, I.S.; Biris, A.S.; Wang, T.; Dervishi, E.; Faupel, F. Advances in top–down and bottom–up surface nanofabrication: Techniques, applications future prospects. *Adv. Colloid Interface Sci.* **2012**, *170*, 2. [CrossRef]
171. Ye, X.; Qi, L. Two-dimensionally patterned nanostructures based on monolayer colloidal crystals: Controllable fabrication, assembly, and applications. *Nano Today* **2011**, *6*, 608. [CrossRef]
172. Bandaru, S.; Arora, D.; Ganesh, K.M.; Umrao, S.; Thomas, S.; Bhaskar, S.; Chakrabortty, S. Recent Advances in Research from Nanoparticle to Nano-Assembly: A Review. *Nanomaterials* **2024**, *14*, 1387. [CrossRef]
173. Cai, Z.; Li, Z.; Ravaine, S.; He, M.; Song, Y.; Yin, Y.; Zhang, A.O. From colloidal particles to photonic crystals: Advances in self-assembly and their emerging applications. *Chem. Soc. Rev.* **2021**, *50*, 5898. [CrossRef]
174. Bellani, S.; Bartolotta, A.; Agresti, A.; Calogero, G.; Grancini, G.; Di Carlo, A.; Bonaccorso, F. Solution-processed two-dimensional materials for next-generation photovoltaics. *Chem. Soc. Rev.* **2021**, *50*, 11965.
175. Jankiewicz, B.J.; Jamiola, D.; Choma, J.; Jaroniec, M. Silica–metal core–shell nanostructures. *Adv. Colloid Interface Sci.* **2012**, *170*, 47. [CrossRef]
176. Shilpa, G.; Kumar, P.M.; Kumar, D.K.; Deepthi, P.R.; Sadhu, V.; Sukhdev, A.; Kakarla, R.R. Recent advances in the development of high efficiency quantum dot sensitized solar cells (QDSSCs): A review. *Mater. Sci. Energy Technol.* **2023**, *6*, 546. [CrossRef]
177. Raabe, D. The materials science behind sustainable metals and alloys. *Chem. Rev.* **2023**, *123*, 2436. [CrossRef]
178. Alaaeddin, M.H.; Sapuan, S.M.; Zuhri, M.Y.M.; Zainudin, E.S.; Al-Oqla, F.M. Photovoltaic applications: Status and manufacturing prospects. *Renew. Sustain. Energy Rev.* **2019**, *102*, 318. [CrossRef]
179. Zarmai, M.T.; Ekere, N.N.; Oduoza, C.F.; Amalu, E.H. A review of interconnection technologies for improved crystalline silicon solar cell photovoltaic module assembly. *Appl. Energy* **2015**, *154*, 173. [CrossRef]
180. Luo, W.; Khoo, Y.S.; Hacke, P.; Naumann, V.; Lausch, D.; Harvey, S.P.; Ramakrishna, S. Potential-induced degradation in photovoltaic modules: A critical review. *Energy Environ. Sci.* **2017**, *10*, 43. [CrossRef]
181. Ogbomo, O.O.; Amalu, E.H.; Ekere, N.N.; Olagbegi, P.O. A review of photovoltaic module technologies for increased performance in tropical climate. *Renew. Sustain. Energy Rev.* **2017**, *75*, 1225. [CrossRef]
182. Farrell, C.C.; Osman, A.I.; Doherty, R.; Saad, M.; Zhang, X.; Murphy, A.; Rooney, D.W. Technical challenges and opportunities in realising a circular economy for waste photovoltaic modules. *Renew. Sustain. Energy Rev.* **2020**, *128*, 109911. [CrossRef]
183. Fuentes, M.; Nofuentes, G.; Aguilera, J.; Talavera, D.L.; Castro, M. Application and validation of algebraic methods to predict the behaviour of crystalline silicon PV modules in Mediterranean climates. *Sol. Energy* **2007**, *81*, 1396. [CrossRef]
184. Miozzo, L.; Yassar, A.; Horowitz, G. Surface engineering for high performance organic electronic devices: The chemical approach. *J. Mater. Chem.* **2010**, *20*, 2513. [CrossRef]
185. Si, W.; Yan, C.; Chen, Y.; Oswald, S.; Han, L.; Schmidt, O.G. On chip, all solid-state and flexible micro-supercapacitors with high performance based on MnO x/Au multilayers. *Energy Environ. Sci.* **2013**, *6*, 3218. [CrossRef]
186. Pearton, S.J.; Ren, F.; Zhang, A.P.; Lee, K.P. Fabrication and performance of GaN electronic devices. *Mater. Sci. Eng. R Rep.* **2000**, *30*, 55. [CrossRef]

187. Schileo, G.; Grancini, G. Lead or no lead? Availability, toxicity, sustainability and environmental impact of lead-free perovskite solar cells. *J. Mater. Chem. C* **2021**, *9*, 67. [CrossRef]
188. Urbina, A. The balance between efficiency, stability and environmental impacts in perovskite solar cells: A review. *J. Phys. Energy* **2020**, *2*, 022001. [CrossRef]
189. Chen, W.; Bhaumik, S.; Veldhuis, S.A.; Xing, G.; Xu, Q.; Grätzel, M.; Sum, T.C. Giant five-photon absorption from multidimensional core-shell halide perovskite colloidal nanocrystals. *Nat. Commun.* **2017**, *8*, 15198. [CrossRef]
190. Seifert, J.S.; Nees, N.; Khan, H.; Traoré, N.E.; Drobek, D.; Peukert, W.; Taylor, R.N.K. Continuous flow synthesis and simulation-supported investigation of tunable plasmonic gold patchy nanoparticles. *Nanoscale* **2024**, *16*, 19284. [CrossRef]
191. Fischer, K.; Marlow, P.; Manger, F.; Sprau, C.; Colsmann, A. Microfluidics: Continuous-Flow Synthesis of Nanoparticle Dispersions for the Fabrication of Organic Solar Cells. *Adv. Mater. Technol.* **2022**, *7*, 2200297. [CrossRef]
192. Baig, N.; Kammakakam, I.; Falath, W. Nanomaterials: A review of synthesis methods, properties, recent progress, and challenges. *Mater. Adv.* **2021**, *2*, 1821. [CrossRef]
193. Mabrouk, M.; Das, D.B.; Salem, Z.A.; Beherei, H.H. Nanomaterials for biomedical applications: Production, characterisations, recent trends and difficulties. *Molecules* **2021**, *26*, 1077. [CrossRef]
194. Chai, Z.; Childress, A.; Busnaina, A.A. Directed assembly of nanomaterials for making nanoscale devices and structures: Mechanisms and applications. *ACS Nano* **2022**, *16*, 17641. [CrossRef] [PubMed]
195. Yang, Z.; Wei, J.; Giżynski, K.; Song, M.G.; Grzybowski, B.A. Interference-like patterns of static magnetic fields imprinted into polymer/nanoparticle composites. *Nat. Commun.* **2017**, *8*, 1. [CrossRef]
196. Grzelczak, M.; Vermant, J.; Furst, E.M.; Liz-Marzán, L.M. Directed self-assembly of nanoparticles. *ACS Nano* **2010**, *4*, 3591. [CrossRef]
197. Ambhulkar, S.; Ravichandran, D.; Zhu, Y.; Thippanna, V.; Ramanathan, A.; Patil, D.; Song, K. Nanoparticle Assembly: From Self-Organization to Controlled Micropatterning for Enhanced Functionalities. *Small* **2024**, *20*, 2306394. [CrossRef]
198. Karunakaran, S.K.; Arumugam, G.M.; Yang, W.; Ge, S.; Khan, S.N.; Lin, X.; Yang, G. Recent progress in inkjet-printed solar cells. *J. Mater. Chem. A* **2019**, *7*, 13873. [CrossRef]
199. Razza, S.; Castro-Hermosa, S.; Di Carlo, A.; Brown, T.M. Research Update: Large-area deposition, coating, printing, and processing techniques for the upscaling of perovskite solar cell technology. *Apl Mater.* **2016**, *4*, 91508. [CrossRef]
200. Zavanelli, N.; Kim, J.; Yeo, W.H. Recent advances in high-throughput nanomaterial manufacturing for hybrid flexible bioelectronics. *Materials* **2021**, *14*, 2973. [CrossRef]
201. Pirela, S.V.; Martin, J.; Bello, D.; Demokritou, P. Nanoparticle exposures from nano-enabled toner-based printing equipment and human health: State of science and future research needs. *Crit. Rev. Toxicol.* **2017**, *47*, 683. [CrossRef] [PubMed]
202. Kalfagiannis, N.; Karagiannidis, P.G.; Pitsalidis, C.; Panagiotopoulos, N.T.; Gravalidis, C.; Kassavetis, S.; Logothetidis, S. Plasmonic silver nanoparticles for improved organic solar cells. *Sol. Energy Mater. Sol. Cells* **2012**, *104*, 165. [CrossRef]
203. Cabrini, S.; Kawata, S. (Eds.) *Nanofabrication Handbook*; CRC Press: Boca Raton, FL, USA, 2012; Volume 180.
204. Ortiz-Perez, A.; van Tilborg, D.; van der Meel, R.; Grisoni, F.; Albertazzi, L. Machine learning-guided high throughput nanoparticle design. *Digit. Discov.* **2024**, *3*, 1280. [CrossRef]
205. Li, M.; Zhou, S.; Guan, Q.; Li, W.; Li, C.; Bouville, F.; Saiz, E. Robust Underwater Oil-Repellent Biomimetic Ceramic Surfaces: Combining the Stability and Reproducibility of Functional Structures. *ACS Appl. Mater. Interfaces* **2022**, *14*, 46077. [CrossRef] [PubMed]
206. Ramakrishna, S.; Santhosh Kumar, K.S.; Mathew, D.; Reghunadhan Nair, C.P. A robust, melting class bulk superhydrophobic material with heat-healing and self-cleaning properties. *Sci. Rep.* **2015**, *5*, 18510. [CrossRef]
207. Ribes, À.; Sánchez-Cabezas, S.; Hernández-Montoto, A.; Villaescusa, L.A.; Aznar, E.; Martínez-Máñez, R.; Cuenca-Bustos, A. Lab and pilot-scale synthesis of MxOm@ SiC core–shell nanoparticles. *Materials* **2020**, *13*, 649. [CrossRef]
208. Rolston, N.; Printz, A.D.; Hilt, F.; Hovish, M.Q.; Brüning, K.; Tassone, C.J.; Dauskardt, R.H. Improved stability and efficiency of perovskite solar cells with submicron flexible barrier films deposited in air. *J. Mater. Chem. A* **2017**, *5*, 22975. [CrossRef]
209. Uddin, A.; Upama, M.B.; Yi, H.; Duan, L. Encapsulation of organic and perovskite solar cells: A review. *Coatings* **2019**, *9*, 65. [CrossRef]
210. Cao, M.; Ji, W.; Chao, C.; Li, J.; Dai, F.; Fan, X. Recent Advances in UV-Cured Encapsulation for Stable and Durable Perovskite Solar Cell Devices. *Polymers* **2023**, *15*, 3911. [CrossRef]
211. Irede, E.L.; Awoyemi, R.F.; Owolabi, B.; Aworinde, O.R.; Kajola, R.O.; Hazeez, A.; Ifijen, I.H. Cutting-edge developments in zinc oxide nanoparticles: Synthesis and applications for enhanced antimicrobial and UV protection in healthcare solutions. *RSC Adv.* **2024**, *14*, 20992–21034. [CrossRef]
212. Choudhary, A.; Halder, A.; Aggarwal, P.; Govind Rao, V. Plasmonic chemistry for sustainable ammonia production. *Commun. Mater.* **2024**, *5*, 69. [CrossRef]

213. Zhang, X.; Ye, S.; Zhang, X.; Wu, L. Optical properties of SiO 2@ M (M= Au, Pd, Pt) core–shell nanoparticles: Material dependence and damping mechanisms. *J. Mater. Chem. C* **2015**, *3*, 2282. [CrossRef]
214. Chiozzi, V.; Rossi, F. Inorganic–organic core/shell nanoparticles: Progress and applications. *Nanoscale Adv.* **2020**, *2*, 5090. [CrossRef]
215. Chambon, S.; Schatz, C.; Sébire, V.; Pavageau, B.; Wantz, G.; Hirsch, L. Organic semiconductor core–shell nanoparticles designed through successive solvent displacements. *Mater. Horiz.* **2014**, *1*, 431. [CrossRef]
216. Nizam, N.U.M.; Hanafiah, M.M.; Woon, K.S. A content review of life cycle assessment of nanomaterials: Current practices, challenges, and future prospects. *Nanomaterials* **2021**, *11*, 3324. [CrossRef]
217. Saavedra, E.L.; Osma, J.F. Impact of Nanoparticle Additions on Life Cycle Assessment (LCA) of Ceramic Tiles Production. *Nanomaterials* **2024**, *14*, 910. [CrossRef] [PubMed]
218. Usmani, S.M.; Bremer-Hoffmann, S.; Cheyns, K.; Cubadda, F.; Dumit, V.I.; Escher, S.E.; Haase, A. Review of new approach methodologies for application in risk assessment of nanoparticles in the food and feed sector: Status and challenges. *EFSA Support. Publ.* **2024**, *21*, 8826E.
219. Eom, N.; Messing, M.E.; Johansson, J.; Deppert, K. General trends in core–shell preferences for bimetallic nanoparticles. *ACS Nano* **2021**, *15*, 8883. [CrossRef]
220. Mahmud, S.; Satter, S.S.; Singh, A.K.; Rahman, M.M.; Mollah, M.Y.A.; Susan, M.A.B.H. Tailored engineering of bimetallic plasmonic Au@ Ag Core@ Shell nanoparticles. *ACS Omega* **2019**, *4*, 18061. [CrossRef]
221. Gargiulo, J.; Herran, M.; Violi, I.L.; Sousa-Castillo, A.; Martinez, L.P.; Ezendam, S.; Cortés, E. Impact of bimetallic interface design on heat generation in plasmonic Au/Pd nanostructures studied by single-particle thermometry. *Nat. Commun.* **2023**, *14*, 3813. [CrossRef]
222. Saha, A.; Khalkho, B.R.; Deb, M.K. Au–Ag core–shell composite nanoparticles as a selective and sensitive plasmonic chemical probe for L-cysteine detection in Lens culinaris (lentils). *RSC Adv.* **2021**, *11*, 20380. [CrossRef] [PubMed]
223. Wang, C.; Shi, Y.; Qin, D.; Xia, Y. Bimetallic core–shell nanocrystals: Opportunities and challenges. *Nanoscale Horiz.* **2023**, *8*, 1194. [CrossRef] [PubMed]
224. Amiri, O.; Salavati-Niasari, M.; Bagheri, S.; Yousefi, A.T. Enhanced DSSCs efficiency via Cooperate co-absorbance (CdS QDs) and plasmonic core-shell nanoparticle (Ag@ PVP). *Sci. Rep.* **2016**, *6*, 25227. [CrossRef] [PubMed]

photonics

MDPI

Review

Stimulated Raman Scattering Microscopy: A Review

Rajeev Ranjan [1,2] and Luigi Sirleto [1,*]

1 National Research Council (CNR), Institute of Applied Sciences and Intelligent Systems (ISASI), Via Pietro Castellino 111, 80131 Napoli, Italy
2 CNRS, Centrale Marseille, Institut Fresnel, Aix Marseille Univ, 13005 Marseille, France
* Correspondence: luigi.sirleto@cnr.it

Abstract: Stimulated Raman scattering (SRS) microscopy is a high-speed imaging modality based on intrinsic molecular vibrations, producing chemical maps in living systems. Such capability, allowing for direct visualization without the perturbation of biological processes, has enabled a plethora of biological and medical applications. In this review, after introducing the basic theory and competitive effects of SRS, some crucial features for SRS microscopy implementations, such as noise, spectral bandwidth, speed, chemical sensitivity, spatial resolution, and quantum enhancement, are discussed. Finally, some SRS applications in biological and medical imaging are described. Even if certainly not exhaustive, we aimed to offer a broad overview, providing guidance for newcomers and hinting at a more detailed investigation to interested researchers in this rapidly growing field.

Keywords: Raman imaging; nonlinear optics; optical microscopy; stimulated Raman scattering; ultrafast optics

1. Introduction

Today, fluorescence microscopy is used worldwide, even if it suffers from drawbacks such as photobleaching, phototoxicity, and challenging quantification. Most small biomolecules, such as nucleosides, amino acids, fatty acids, choline, glucose, and cholesterol, are intrinsically nonfluorescent, so fluorescent tags are mandatory. Unluckily, the latter, such as organic dyes, fluorescent proteins, or quantum dots, are all relatively larger than small biomolecules; consequently, they can severely falsify their native biochemical or biophysical properties. Therefore, label-free imaging, with high sensitivity and high chemical selectivity of unlabeled living cells, is preferable [1–3].

In this framework, vibrational microscopy techniques have emerged as a powerful approach based on a chemically label-free selective contrast due to intrinsic molecule vibrations. Among them, stimulated Raman scattering (SRS) microscopy, which was adapted to microscopy only in the last two decades, is one of the most promising techniques. SRS microscopy is sensitive to the same molecular vibrations probed in spontaneous Raman spectroscopy but exhibits a nonlinear dependence on the incoming light fields, producing coherent radiation. Because of nonlinear excitation, taking advantage of SRS microscopy, it is possible to achieve images faster than conventional Raman microscopes, with typical acquisition times of a few seconds. Other advantages of SRS microscopy are high sensitivity, spatial and spectral resolution, and 3D sectioning capability. Since the invention of SRS microscopy, it has been extensively employed to study composition, structure, chemical distribution, molecular transport, metabolic conversion in living cells, and disease in biological systems in a label-free manner. Applications of SRS have generated new insights in many fields, including lipid droplet biology, cell metabolism, drug delivery, neurobiology, tumor biology, and developmental biology [4–10].

In Raman amplification, two laser pulses, i.e., a pump and a probe (Stokes) beams, at frequencies and ω_s (with $\omega_p > \omega_s$) are incident on a material sample. If their difference frequency matches a particular molecular vibrational frequency, the energy from the intense pump beam is transferred to the weaker signal beam. The SRS effect occurs in the form of a gain of the Stokes beam (stimulated Raman gain, SRG) and a loss of the pump beam (stimulated Raman loss, SRL). Because of its coherent nature, the molecular bonds oscillate in phases and interfere constructively inside the focus area of the laser beam so that the SRS signal can be orders of magnitude greater than spontaneous Raman scattering. Although it is impossible to detect a single vibrational mode at room temperature, detecting a macromolecule with thousands of identical vibrational modes that interfere coherently is feasible [11–26].

This review article is organized as follows. In the next paragraph, some crucial features of SRS implementations, such as noise, spectral bandwidth, speed, chemical sensitivity, spatial resolution, and quantum enhancement, are discussed. These features are important for at least two reasons. First, there are significant efforts to improve them in many directions. Second, these features allow for the matching of the potentials of SRS microscopes with the requirements imposed by applications. Finally, in the last section, some life science applications of SRS are reported. Of course, covering all the application fields would be impossible, so only some recent results have been reported.

2. Features of SRS Microscopes

Almost all SRS microscopes reported in the literature are home-built with highly different laser sources, optics, and detection mechanisms. They present a large variety in terms of main features, such as noise, spectral bandwidth, speed, chemical sensitivity, and spatial resolution. As a consequence, there is a lack of standard procedures or reference materials for SRS, which has important implications for reliability, reproducibility, and consistency [27]. This section analyzes and discusses the main features of SRS microscopes.

2.1. Noise and Competitive Effects

We note that the SRS signal is generated at the frequency of incident beams acting as local oscillators. Such heterodyne detection boosts the signal level and removes the non-resonant background, which limits CARS microscopy. Therefore, initially, it was suggested that SRS microscopy was free of spurious background signals. However, the SRS signal is generally overlaid with parasitic signals stemming from various linear (scattering and absorption) and nonlinear optical effects whose sources are ubiquitous [4–26].

Due to other nonlinear optical processes, unwanted signals, which can overlay with SRS, can be classified into two groups: nonlinear transient absorption [28,29] and nonlinear transient scattering processes [30,31] (see Figure 1). Both heterodyne modalities produce a spectrally overlapped background with the SRS signal. Nonlinear transient absorption (TA) is a pump–probe process involving electronic-energy-level excitations, which may be classified into excited-state absorption, induced fluorescence, and ground-state photobleaching processes [28]. The resulting attenuation of the probe beam can be misinterpreted as an SRS signal [29] but has no chemical specificity. Nonlinear transient scattering includes two main effects: cross-phase modulation (XPM) and thermal lensing (TL). In cross-phase modulation (XPM), the pump beam (ω_p) changes the nonlinear refractive index at the focus through the optical Kerr effect, thus affecting the probe beam (ω_{pr}) propagation and frequency. [30]. TL is induced by local heating due to the absorption of pump photons, which changes the refractive index at the focus, thus affecting the propagation of the probe beam [31]. Both effects can be reduced by using collection optics with a large numerical aperture, i.e., employing an objective collecting lens with a numerical aperture that exceeds that of the excitations.

We note that all these nonlinear optical (NLO) techniques, SRS, nonlinear transient absorption, and nonlinear transient scattering processes, are based on the detection of small intensity changes (of the order of 10^{-5}–10^{-4}) of large signals, in which the signal wavelengths do not depend on the nonlinear processes. The signal exists at the same wavelengths as the excitation pulses, and the spectral separation of the relatively weak signal from the excitation beams is impossible. In these techniques, the signal, i.e., a small alternating voltage (AC) at the sub-microvolt level, must be discriminated from the laser *1/f* noise. For this reason, most of the investigations relied on high-frequency optical modulation schemes and phase-sensitive detection [32–34].

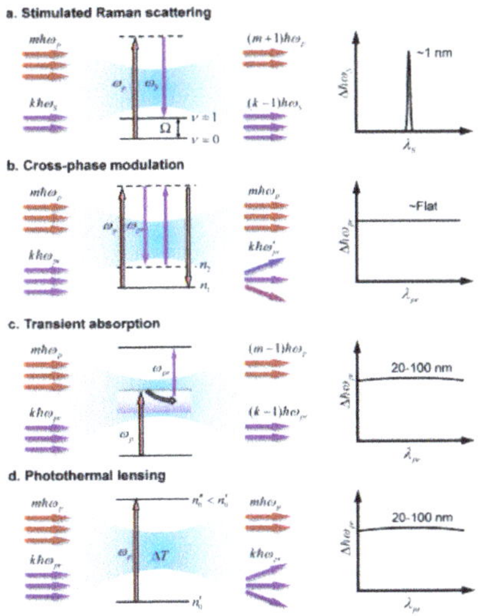

Figure 1. Basic principles of pump–probe modalities and their spectral characteristics: (**a**), stimulated Raman scattering (SRS); (**b**), cross-phase modulation (XPM); (**c**), transient absorption (TA); and (**d**), thermal lensing. Adapted from [35] Copyright 2013 Optica Publishing Group.

We note that XPM involves only energy states, so the XPM signal can be considered wavelength-independent for a small spectral range. Transient absorption and photothermal effects are based on electronic transitions and, therefore, have weak wavelength dependence. In contrast, the SRS signal originates from vibrational transitions, which have very sharp spectral features. As a consequence, one may selectively detect the SRS process by modulating a laser beam's wavelength within a few nm. Based on this concept, frequency-modulated SRS is one of the most successful methods proposed to reduce the background signal [35]. In this method, either the pump or Stokes input of the laser pulse is frequency-modulated so that their difference is modulated on and off the vibrational resonance. This results in a signal at the Stokes frequency that is modulated between two values, one corresponding to purely non-resonant interaction and the other to both resonant interactions. The difference between these two signals effectively removes the average background contribution and gives much-improved image contrast.

In [35], a simple but robust FM-SRS (Frequency Modulation-SRS) microscopy strategy was based on a commercial laser system and standard optics. Effective suppression of backgrounds, including non-resonant cross-phase modulation and an electronic background from two-photon absorption or the pump–probe process, was demonstrated on a variety of Raman modes.

It is worth noting that due to the weak Raman cross-section of biomolecules, SRS microscopy often suffers from a low signal-to-noise ratio (SNR), and SRS applications in biological/biomedical imaging could be compromised. In the literature on SRS, noise investigations are often overlooked, even if the signal-to-noise ratio (SNR) is crucial for spectroscopy [36,37] and biological imaging [38–45]. A safer way would be to evaluate both the stimulated Raman loss and the stimulated Raman gain, as these two contributions should be equal in the absence of artifacts. In ref [45], an SRS scheme, SRGAL (stimulated Raman gain and loss), was presented. The SRGAL concept modulates the pump at frequency f1 and the Stokes at frequency f2 while simultaneously demodulating the pump at f2 to access SRL and the Stokes at f1 to access SRG. In the SRGAL scheme, the SRL and SRG images are collected simultaneously but within separate signal channels. As a result, the SRGAL scheme allows for (1) faster imaging for a given signal-to-noise ratio (SNR), as the SRG and SRL channels can be added to double the SRS signal, and (2) the semi-quantitative evaluation of the SRS artifacts by directly comparing the magnitude of the SRG and SRL channels at each pixel level.

One key SRGAL advantage is to improve the SNR ratio by summating the SRL and SRG images. In Figure 2, the SNR improvement is demonstrated for an onion surface thick slice (about 150 microns). The SNR was measured within the delineated region of interest (ROI) (loss: dashed green ROI, gain: dashed red ROI). By summing the SRL and SRG images of the onion cell, a 2-fold improvement of the SNRSRGAL (black curve) compared to SNRSRL (green curve) and SNRSRG (red curve) was obtained. The 2-fold improvement is further shown in Figure 2 through the ratio 2 SNRSRGAL/(SNRSRL+SNRSRG° that is constant along the scan direction and equal to 2 [45].

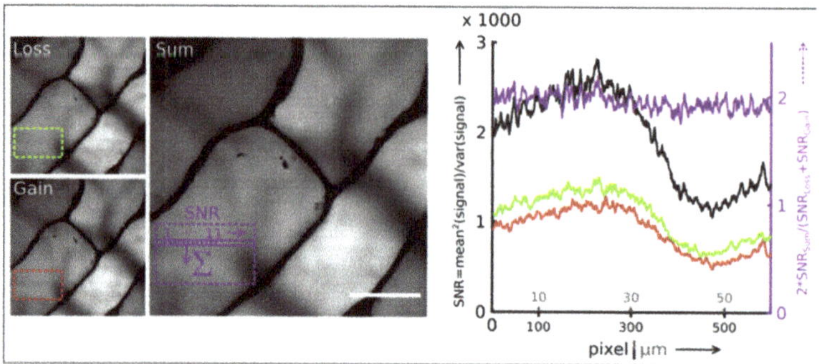

Figure 2. SRGAL signal-to-noise (*SNRSRGAL*) improvement. Scale bar: 50 µm. Pixel integration time: 20 ms. Adapted from [45] Copyright 2020 Optica Publishing Group.

Visualizing heterogeneous morphology, segmentation, and quantifying image features is crucial for nonlinear optics microscopy applications. The identification and quantification of the structure and the spatial distributions of chemicals and organelles are necessary in order to accurately compare images and make objective conclusions about an experiment. On this line of argument, recently, there has been a sudden increase in applications of quantitative biological imaging, a field that has been advanced by developments in optical imaging technology, as well as in computational image analysis techniques [46,47]. However, many pitfalls are associated with quantification due to complications such as sample scattering and a non-Raman background. A useful and detailed discussion of SRS-based quantitative analysis is reported in reference [47].

2.2. Spectral Bandwidth and Speed

Label-free stimulated Raman scattering microscopy allows for the imaging of a variety of molecular species, targeting their intrinsic chemical bonds in the C-H region (2800–3100 cm^{-1}), the silent region (1800–2800 cm^{-1}), and the fingerprint region (<1800 cm^{-1}). Almost all biomolecules contain carbon and hydrogen, so the CH region is the most used in SRS microscopy. The two Raman bands, CH_2, near 2845 cm^{-1}, and CH_3, near 2930 cm^{-1}, corresponding to lipids and proteins, respectively, are typically investigated. Due to the difference between peaks (about 95 cm^{-1}) and large spectral shapes (about 100 cm^{-1}), the CH_2 and CH_3 Raman bands partially overlap. In the C-H region, because the density of CH bonds is high, the SRS signal level is high, while the SRS molecular specificity is low [48–51].

SRS imaging can be demonstrated in three modalities: single-frequency, hyperspectral, and multiplex.

Single-frequency SRS microscopy is implemented by using a pair of Fourier transform-limited (FTL) picosecond laser sources. This approach ensures a high spectral resolution (*10 cm^{-1}) that is very useful in the fingerprint region (~800 and 1800 cm^{-1}), where Raman peaks are narrow, closely spaced, and may be in abundance for a particular chemical [48,49]. Utilizing a picosecond (ps) pump and Stokes lasers, single-color SRS has reached video-rate speeds, opening doors for a plethora of biological and medical applications [52,53]. The drawback is that ps pulses show a low peak intensity, thus needing high laser power for imaging.

Two-color stimulated Raman scattering (SRS) microscopy has succeeded in label-free digital histology, with diagnostic results similar to the hematoxylin and eosin stain. In ref. [54], the pulse profiles of Stokes beams were precisely engineered to detect two Raman frequencies simultaneously through a dual-phase lock-in amplifier in-phase and quadrature output channels. Such a method could reach the maximum speed as in single-color SRS.

High-throughput-stimulated Raman scattering (SRS) microscopy is highly desired for large tissue imaging. However, in most cases, SRS microscopy was limited to a very small field of view (FOV) of about 300 μm or to mapping a relatively large tissue in a very long time [55,56]. In [57], an inertia-free acousto-optic deflector (AOD)-based high-speed large-field stimulated Raman scattering microscopy was presented. The immune system of the mechanical response time ensures both speed and integration time. An SRS imaging of a 12 × 8 mm^2 mouse brain slice in only 8 min at an image resolution of approximately 1 μm and 32 slices from a whole brain in 12 h were achieved.

We note that while single-frequency with ps coherent Raman scattering guarantees the best spectral resolution, the optimal image contrast and signal intensity ratio are obtained by matching the spectral resolution with the width of the Raman lines under consideration (5–100 cm^{-1}) [58]. Because picosecond pulse excitation can only match the linewidths in the fingerprint region (5–20 cm^{-1}), the question is raised as to whether broader bandwidth femtosecond (fs) pulses can be suited to the optimal excitation of CH stretching vibrations [59,60] In addition, we note that when femtosecond (fs) pulses are used, the SRS signal is improved by one order of magnitude [61]. On the other hand, low spectral selectivity is obtained using fs sources, and multi-band excitation can occur, which does not allow for, in principle, the separation of lipids and proteins.

Hyperspectral SRS (hsSRS) imaging can be obtained by changing the wavelength of one or both lasers. Different Raman bands, corresponding to different chemical components of the sample, can be imaged by tuning the frequency either of the pump or the Stokes beams in sequential scans, allowing for the detection of one chemical contrast for each scanning. Thus, CH_2 (2845 cm^{-1}) and CH_3 (2940 cm^{-1}) stretching signals can be collected at one Raman shift at a time, leading, in principle, to mapping the distributions of the lipid and protein contents. Nevertheless, such improvement comes at the price of reduced imaging speed. When fs pulses are used, SRS images acquired at 2845 cm^{-1} can be attributed to the lipids, while in order to obtain protein content, SRS images acquired at 2940 cm^{-1} contain both lipids and protein signals. Therefore, the calibration of individual components is mandatory [59,60,62].

Hyperspectral SRS (hsSRS) imaging, i.e., the spectral scanning of a narrowband laser pulse and collection of images at a series of Raman shifts, has reached the speed of a few seconds per stack [63–67]. Hyperspectral SRS (hsSRS) imaging can also be achieved by chirping pump and Stokes lasers and tuning the time delay between them. In this technique, called spectral focusing [64], each temporal delay between the chirped pulses corresponds to a Raman shift. SRS spectroscopic imaging can be obtained by scanning the temporal delay of one of the pulses and recording a series of images. A spectral-focusing scheme has been demonstrated with a total acquisition time of several tens of seconds [64,68–70], where the spectral acquisition speed was limited by the waiting time for the stabilization and communication of a motorized translational stage used for delay tuning. In [71], microsecond-scale SRS spectroscopic imaging was presented by temporally tuning two spectrally focused pulses through a resonant delay line. The platform was able to form a spectral image of 40,000 pixels within 3.3 s. The SRS spectrum at each pixel was acquired within 83 μs, covering the 200 cm^{-1} spectral window with 25 cm^{-1} spectral resolution.

Another possibility is given by femtosecond pulse-shaping technology [72], which is able to obtain a spectral resolution of about 10 cm^{-1}. In [73], the development of a unique spatial light-modulated stimulated Raman scattering (SLM-SRS) microscopy, tailoring the broadband excitation beam for multiplexed vibrational imaging with higher sensitivity, is reported. Pulse-shaping is achieved using a sparse-sampling mask designed to allow for the collective excitation of the predominant spectral windows against the overlapping spectra. SLM-SRS microscopy permits rapid multiplexed SRS imaging of polystyrene and polymethyl methacrylate beads in Brownian motion in dimethyl sulfoxide (DMSO) at 70 ms intervals without motion artifacts.

In multiplex SRS imaging, the SRS spectrum at each pixel is captured simultaneously over a large spectral window by using broadband excitation at the cost of more pixel integration time [74,75]. In multiplex SRS imaging, by replacing the lock-in amplifier with a gate array, lock-in-free single-color-line-scan SRS imaging that acquired 20 frames per second was obtained [76]. Using a 32-channel tuned amplifier array, Zhang et al. [77] obtained an acquisition speed of 5 μs per spectrum (~200 cm^{-1}).

Several computational methods have been proposed to break the speed limit of Raman imaging bounded by signal integration time. In [78], the sparsity of natural images was used to record a Raman image. This work reported three-dimensional sparsely sampled spectroscopic SRS imaging that measures ~20% of pixels throughout the stack. An acquisition speed of 0.8 s per image stack, with 50 frames in the spectral domain and 40,000 pixels in the spatial domain, was achieved.

2.3. Improvement in Chemical Sensitivity

Strategies aimed at enhancing the visualization and quantitation of biomolecules by SRS have focused on the introduction of small Raman-active tags, which present their Raman band in the cellular silent region of the Raman spectrum, where the background interference from intrinsic cellular components is minimal, and the signal contrast is maximized. Two main strategies are employed; the former concerns using Raman-active groups, which are either inherent to the molecule under investigation or selectively introduced for imaging purposes. The tiny size and good biocompatibility of vibrational tags enable the nonperturbative tracing of small molecules [79], which is difficult to achieve with fluorescence because of the bulky fluorophores.

Furthermore, by virtue of much narrower vibrational peaks (≈10 cm^{-1}) compared to fluorescence peaks (≈500 cm^{-1}), Raman imaging offers scalable multiplexity and breaks the "palette number barrier" of fluorescence, which can typically only image 4–5 colors simultaneously. By combining electronic pre-resonance spectroscopy with stimulated Raman scattering (SRS) microscopy (i.e., epr- SRS), the Raman cross-sections of electronically coupled vibrational modes in light-absorbing dyes can be enhanced by 10^{13}-fold [80–83]. This drastic enhancement achieved the nanomolar sensitivity of Raman-active dyes.

2.4. Super-Resolution

Because SRS is an optical technique using near-infrared or visible light, its spatial resolution is governed by the diffraction limit (≈300 nm).

In order to break the diffraction limit, the approach is focused on reducing the point spread function (PSD) or increasing the cutoff frequency of the optical transfer function (OFT). It includes the following three different directions: (a) the effective PSF reduction based on nonlinear optical processes such as stimulated emission depletion (STED) microscopy [84] and saturation excitation (SAX) microscopy [85]; (b) OTF expansion by Moire effects, such as structured illumination microscopy (SIM) [86]; and (c) PSF, which is suppressed by focal volume engineering, such as superoscillation [87].

For direction (b), we note that the Moiré fringe is a wide-field technique, while CRS is a nonlinear optical process requiring a tightly focused laser beam. Although SIM-based super-resolution CARS microscopy was theoretically proposed [88,89], the related experiment has not yet been demonstrated. Therefore, the feasible approaches for super-resolution CRS microscopy should be approach (a) [90–98] and approach (c) [99–101]. However, approach (c) can only improve the resolution up to 2-fold, while approach (a) provides, in principle, unlimited resolution, so, in recent years, it has attracted much more attention.

Note that the CRS process is the 3rd-order nonlinear process; for super-resolution CRS imaging, the nonlinear optical process of even higher orders (n > 3) should be involved [102]. The key to achieving super-resolution CRS imaging is to properly detect the higher-order nonlinear process with a relatively high excitation power without damaging the sample. In approach (a), several theoretical schemes are proposed based on either the saturation [98] or suppression [93,96,98] of the CRS signal. To apply super-resolution CRS techniques in biological samples, saturated SRS (SSRS) microscopy and higher-order CARS (HO-CARS) are proposed and demonstrated experimentally [90,91].

In reference [91], saturated stimulated-Raman-scattering (SSRS) microscopy was presented to break the diffraction limit for far-field super-resolution imaging in biological systems. An experimental implementation of SSRS microscopy was developed to enhance the spatial resolution of SRS imaging in samples (e.g., plant cells, live HeLa cells) with a spectral-focusing hyperspectral SRS system. With saturation occurring in coherent Raman processes under a certain level of excitation powers, it was demonstrated that the SSRS technique has the ability to become an appealing tool for label-free far-field super-resolution SRS imaging in biological and biomedical systems (see Figure 3).

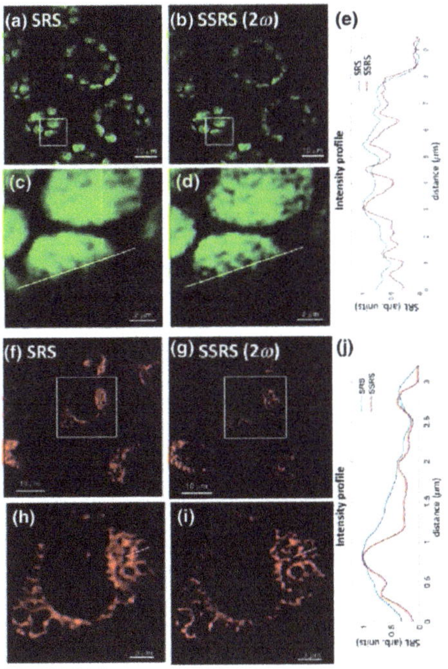

Figure 3. Comparison of the conventional SRS and super-resolution SSRS images of chloroplast and Raman-tagged mitochondria in live HeLa cells. (**a**,**b**) SRS image and super-resolution SSRS image of chloroplasts in leaf sheath cells of rice. The Raman shift is 1530 cm^{-1}. (**c**,**d**) Enlarged SRS and SSRS images of the selected areas. (**e**) Intensity profiles across the lines are indicated in (**c**,**d**). (**f**,**g**) SRS image and second-order super-resolution SSRS image of mitochondria in Hela cells. The Raman shift is 2216 cm^{-1}. (**h**,**i**) Enlarged SRS and SSRS images of the selected areas. (**j**), Intensity profiles across the lines indicated in (**h**,**i**). Adapted from [91] with the permission number RNP/24/MAY/078995 of the American Physical Society.

Over the last half-decade, a sample-oriented nanoscale imaging technique known as expansion microscopy (ExM) has emerged as a powerful technique for nanoscale optical imaging through the physical magnification of the tissue samples embedded in a water-swellable polymer gel [103–105].

By combining SRS imaging with this framework, a new technique called molecule anchorable gel-enabled nanoscale imaging of fluorescence and stimulated Raman scattering microscopy (MAGNIFIERS), integrating SRS microscopy with expansion microscopy (ExM), has been described in ref [105]. MAGNIFIERS offers chemical-specific nanoscale imaging with sub-50 nm resolution and has scalable multiplexity when combined with multiplex Raman probes and fluorescent labels. MAGNIFIERS has been used to visualize nanoscale features in a label-free manner with the C–H vibration of proteins, lipids, and DNA in a broad range of biological specimens [105].

2.5. Quantum Enhancement

It is known that quantum correlations can be used to extract more information per photon used in an optical measurement [106–109]. This breaks the trade-off between the signal-to-noise ratio and intensity [110]. For this reason, quantum correlations are now used routinely to improve many optical measurements in proof-of-principle experiments [111]. While several different quantum states of light can, in principle, be used to provide such a quantum advantage, so far, it is only the ubiquitous squeezed states of light that have demonstrably been shown to provide a real practical advantage [112] due to their generation simplicity and robustness to loss. Squeezed light, first observed in the 1980s [113], is a unique type of quantum state of light in which the amplitude or the phase has a lower quantum uncertainty than that of a coherent state, but the uncertainty principle is not violated. Squeezed vacuum states can be generated by means of parametric down-conversion (PDC), which converts a pump photon of frequency o0 into a signal photon of frequency o1 and an idler photon of frequency o2. The generated photons are quantum-mechanically correlated in their photon number and quadratures. As a result of the quadrature correlations, the joint measurement of the photons yields quadrature noise reduction, that is, squeezing. Because squeezed vacuum states have an electric field with zero mean value, they are not directly suitable for SRS, but its mean value must be coherently amplified to empower the squeezed vacuum state.

In SRS, the sensitivity and imaging speed depend on the background noise of the probe beam. This background noise is fundamentally limited by shot noise when the probe beam is in a coherent state, produced by a conventional laser [114]. Still, in principle, it can be arbitrarily improved by increasing the input beams' power. However, in biological systems, especially in living systems, the power must be kept low to avoid changing the biological dynamics of the specimens and, in particular, to avoid damage due to excessive heating. For SRS, photodamage places acute constraints on sensitivity and imaging speed [115], presenting a roadblock for powerful prospective applications such as label-free spectrally multiplexed imaging. State-of-the-art coherent Raman microscopes are already limited by shot noise. Consequently, the roadblock cannot be overcome through improvements in instrumentation.

Although quantum-enhanced stimulated Raman scattering (QE-SRS) helps to improve sensitivity by reducing the shot noise, it is still challenging to exceed the sensitivity of classical SRS microscopy. This is mainly because of the difficulty in applying high-power squeezed light pulses (SRS pump or Stokes) in QE-SRS. The typical average power of laser pulses for SRS imaging is several tens of milliwatts. Therefore, to exceed the sensitivity of classical SRS, developing a QE-SRS system that can operate at a high-power regime is crucial.

To improve the signal-to-noise ratio (SNR), the squeezed light-enhanced SRS (SQ-SRS) has been demonstrated in both SRS spectroscopy [116] and SRS microscopy [117]. In [116], SRS spectroscopy was demonstrated with quantum-enhanced balanced detection (QE-BD). In Figure 4, the QE-SRS spectra are measured by observing the output signal of the LIA on an oscilloscope (OSC), which is shown [116]. The noise became stronger or weaker depending on anti-squeezing (ASQ) or squeezing (SQ). The histogram of the voltage signals in Figure 4b is plotted to estimate the noise level quantitatively. Applying Gaussian fitting, the standard deviation σ was derived as 0.378, 0.139, and 0.109 V for ASQ, shot noise limited (SNL), and SQ-SRS traces, respectively. This corresponds to the quantum enhancement of 2.11 dB.

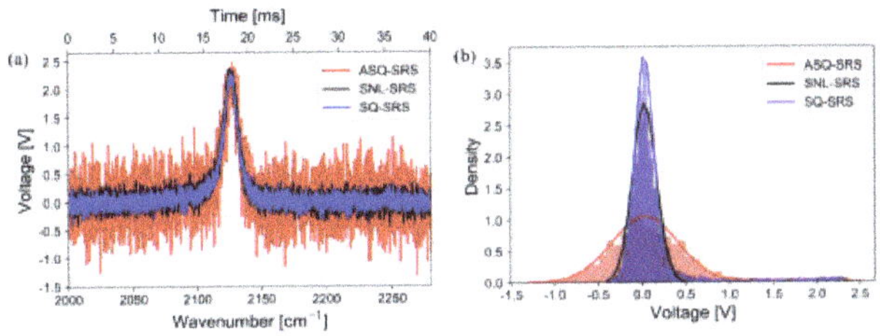

Figure 4. The QE-SRS spectra were acquired with a lock-in amplifier and an oscilloscope. (**a**) The measured SRS spectra with and without SQV. (**b**) Noise distribution with Gaussian fitting (σASQ = 0.40 V, σSNL = 0.152 V, σSQ = 0.116 V). Adapted from [116] Copyright 2022 Optica Publishing Group.

In [118], the quantum enhancement of continuous-wave (CW) SRS spectroscopy was demonstrated by spectroscopically measuring the carbon–hydrogen (C-H) vibrations of polymethylmethacrylate (PMMA) and polydimethylsiloxane (PDMS) with a sensitivity improvement of approximately 56% relative to shot-noise-limited Raman spectroscopy.

In [119], it was experimentally shown that quantum correlations allow for a signal-to-noise ratio beyond the photodamage limit of conventional microscopy. The correlations allow for the imaging of molecular bonds within a cell with a 35-percent improved signal-to-noise ratio compared with conventional microscopy, corresponding to a 14-percent improvement in concentration sensitivity.

In [120], the deconvoluted SQ-SRS (Dcv-SQ-SRS) super-resolution SRS imaging technique, in which the resolution of the deconvoluted image is further enhanced by using quantum light, was presented. Firstly, SQ-SRS was adopted to improve the SNR of conventional SRS beyond the shot noise limit. Then, the deconvolution method was applied to retrieve Dcv-SQ-SRS images beyond the diffraction limit. Thanks to a higher SNR, with the Dcv-SQ-SRS technique, $\sim$a 2.2-fold improvement in spatial resolution compared to conventional SRS microscopy was proved.

Several super-resolution methods have been proposed to break the diffraction limit in coherent Raman scattering microscopy. Typically, these methods require a higher excitation power than conventional SRS imaging, leading to a high risk of photodamage to the tissue samples. On the other hand, the deconvolution methods could extract information beyond the diffraction limit without increasing the laser power. Still, its performance is limited by SNR [121].

3. Life Science Applications

Tremendous work has been conducted on cellular imaging, and the developments and applications of SRS in this area have been thoroughly discussed and reviewed [4–10,122]. Mammalian cells store excess lipid molecules in specialized intracellular organelles called lipid droplets (LDs). The sizes of LDs vary from tens of nm to tens of μm in diameter. LDs are ubiquitously conserved from yeast to mammals and are involved in maintaining lipid homeostasis through lipid synthesis, metabolism, and transportation. Based on controlling these important cellular functions, LDs are closely associated with many human diseases but also with membrane trafficking, cell signaling, proliferation, and apoptosis [123]. This is why great attention has been reserved for LD biology in the last decade.

Although intracellular lipid accumulation has been observed in human cancer tissues and cells, it has not been widely used as a prognostic factor or therapeutic target due to a limited understanding of lipid metabolism in cancer. In particular, the role of lipid accumulation in cancer progression remains elusive, partly because of the lack of tools for mapping lipid species at a single-cell level. Quantitative analysis of lipid content at a single-cell level in human patient cancerous tissue by coupling confocal Raman microscopy with SRS microscopy enabled the identification of the metabolic signature of aggressive human prostate cancer [124].

One advantage of label-free imaging is the capability of quantitative long-duration imaging. This is especially important when studying biomolecule dynamics and tracking changes during ongoing processes, such as embryonic development and injury. Dou et al. performed time-lapse SRS imaging of a developing Drosophila embryo to track single-lipid droplet motion within large populations of droplets (Figure 5). By analyzing the velocity and turning frequency of each droplet, the mathematical model for the LD movement was developed to show the key regulatory point of LD dynamics in the developmental process. This work offers the potential of using the SRS imaging technique to study lipid trafficking in living cells and organisms [125].

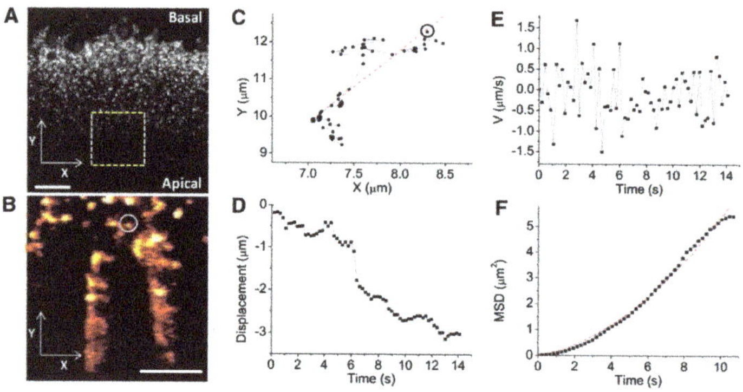

Figure 5. Droplet bidirectional motion is characterized by single-droplet tracking. (**A**) A snapshot of the embryo cortex. (Yellow box) Imaged region for droplet tracking. Acquisition time, 2 µs per pixel. Bar, 10 µm. (**B**) A typical trafficking droplet tracked over 10 s. Acquisition time, 2 µs per pixel (4∼5 frames/s). Bar, 5 µm. (**C**) The trajectory of the droplet is marked in phase B (dashed line) least-squares linear fitting that characterizes the major axis of displacement. (**D**) Droplet displacement (along the dashed line in phase C, where segments of continuous runs were extracted). Decreasing ordinate values indicates apically directed motion. (**E**,**F**) The instantaneous velocity and MSD curve of the tracked droplet. An overall quadratic shape of the MSD curve corresponds to directed transport. The fitting of this MSD curve led to a diffusion coefficient (**D**) of 0.032 ± 0.003 µm^2/s and an average velocity (V) of 0.201 ± 0.041 µm/s. It is reproduced by [125] under license number 5711300529639 with permission under the Elsevier agreement.

SRS is severely limited in imaging depth due to the turbidity and heterogeneity of tissue in both transmissive and epi-mode imaging. The studies, which explicitly report the achieved imaging depths of coherent Raman microscopy in tissue, suggest the limit to be around 20–100 µm, with varying tissue types being the cause of the large variance. Hill et al. [126] investigated maximum signal sizes, scattering lengths, and achievable imaging depths as a function of tissue type and sample thickness using a simultaneous SRS epi/transmissive imaging approach in four different ex vivo murine tissues: brain, lung, kidney, and liver. As one would expect, their findings demonstrated that epi (transmissive) SRS signal size and imaging depth increase (decrease) as tissue thickness grows, reflecting an increase (reduction) in backscattered (transmitted) photons. When tissue thicknesses reach 2 mm, epi and transmissive signal sizes are roughly equivalent, and epi images provide slightly higher penetration depths than transmissive images.

Hematoxylin and eosin (H&E) staining has been the "gold standard" for pathologies like cancer, but it requires extensive time and labor, making it not ideal in the operation room where quick decisions must be made. Fast and accurate pathological diagnoses require real-time assessment of excised tissue. Two-color SRS microscopy has been used for virtual histology by imaging at 2845 cm^{-1} and 2930 cm^{-1}, respectively, for lipid and protein and has achieved excellent results comparable to H&E staining [127].

Lu et al. [128] compared the large-scale SRS imaging of brain tumor tissue resection from patients with H&E staining and found that SRS microscopy could capture not only fundamental diagnostic hallmarks (such as cell nuclei and cell density) for tumor classification but also features not detectable by H&E approaches, including abundant lipid droplets within glioma cells, collagen deposition in gliosarcoma, and the disruption of myelinated fibers.

In addition to tumors, SRS imaging has been applied to detect a variety of brain diseases, such as Alzheimer's disease (AD) and amyotrophic lateral sclerosis (ALS). Ji et al. [129] applied multicolor narrowband SRS microscopy to detect Aβ plaques in the APP/PS1 AD mouse model ex vivo brain tissue. They differentiated misfolded Aβ plaques from normal proteins and lipids in frozen and fresh brain tissues based on the spectral shift (~10 cm^{-1}) of the amide-I band of SRS spectra.

Studying amyotrophic lateral sclerosis (ALS) and exploring potential interventions would be easier if we could more effectively visualize motor axon degeneration. Tian et al. [130] demonstrated in their research that SRS microscopy could be employed to monitor peripheral nerve degeneration sensitively in ALS mouse models (Figure 6). Their findings showed that significant degeneration of peripheral nerves could be detected concurrently with the earliest observable signs of muscle denervation, occurring before any measurable decline in motor function. Moreover, through serial imaging, they could observe the disease's long-term progression and the effects of drugs in living animals.

Acetylcholine (ACh) is a neurotransmitter that plays a crucial role in the central nervous system. Fu et al. [131] reported the first label-free imaging of ACh in frog muscle ex vivo by employing frequency-modulated spectral-focusing SRS microscopy and using the vibrational signature of ACh at 720 cm^{-1} (intrinsic C-N bonds stretching).

The miniaturization of the SRS microscopy system is desired for in vivo and in situ imaging of live animals and in the clinic. A fiber-delivered handheld SRS imaging system was reported in [132]. Background-free in situ and in vivo imaging was achieved on normal and tumorigenic dog brain tissues and live human skin. However, achieving real-time SRS imaging for intraoperative assessment is still a challenge.

The high attrition rates of candidate drugs during clinical development remain a common feature within the pharmaceutical industry. Incorporating imaging into complex drug-screening models has the potential to improve the robustness of preclinical studies of drug uptake, retention, and metabolism.

However, imaging the uptake of drugs and drug carriers presents a significant challenge. Thus, spectroscopically bio-orthogonal Raman-active tags have been adopted, which improve detection sensitivity. The use of isotopes has recently been demonstrated as an alternative Raman-tagging strategy. Deuterium and carbon-13 are useful isotopic substitutions for Raman spectroscopy. The principal benefit of this strategy is that isotopes have almost the identical size and electronic characteristics as the parent compounds. Still, they may display favorable Raman spectral characteristics, particularly as C–D bands fall into the cellular silent region (~2100 cm^{-1}). Replacing the hydrogen atoms in a molecule with deuterium may even improve metabolic stability, and the resultant deuterated molecules are typically safe to handle and dose to patients. When coupled with biorthogonal-labeling strategies, live-cell Raman imaging is a promising imaging modality applicable to drug discovery and medicine. However, the low sensitivity of the Raman scattering process remains a significant limitation of the approach, where, in many cases, typical detection limits are above physiologically relevant drug concentrations. Several studies using Raman imaging to probe intracellular drug distribution have been reviewed by Tipping et al. [133].

Antimicrobial resistance threatens the effective treatment of infectious diseases caused by bacteria and fungi. Rapid antimicrobial susceptibility testing (AST) plays a significant role in suppressing the emergence of antimicrobial resistance to combat this crisis. Current gold standard AST methods are quite time-consuming and lead to the emergence and spread of antimicrobial resistance. The rapid AST of infectious pathogens is urgently needed to guide appropriate antibiotic prescriptions in a timely manner before clinical treatment.

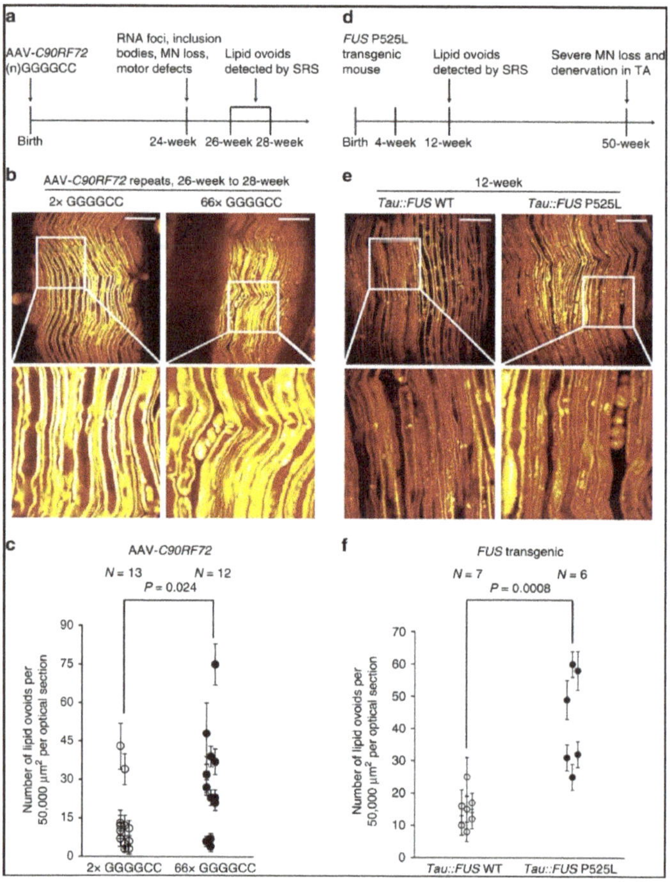

Figure 6. SRS imaging of non-SOD1 mouse models has demonstrated its broad applicability in the study of ALS. The information below is adapted from [130] and is used under a Creative Commons CC-BY license. (**a**) The progression of the disease in the AAV-C9ORF72 repeat expansion overexpression mouse model of ALS is illustrated. (**b**) Representative SRS images of the AAV-C9ORF72 mouse model showcase the healthy nerve fiber morphology for 2 repeats (2Rs) and the deposition of lipid ovoids for 66 repeats (66Rs). The scale bar is set at 50 μm. (**c**) The quantification of lipid ovoids in the AAV-C9ORF72 mouse model of ALS is presented. The data are expressed as mean±s.e.m., with error bars indicating the standard error of the mean. (**d**) The progression of disease-related phenotypes in the FUSP525L mouse ALS model is depicted. (**e**) Ex vivo SRS images of FUSP525L versus FUS WT at 12 weeks of age are provided, with a scale bar of 50 μm. (**f**) The quantification of lipid ovoids and the corresponding statistical significance analysis for the FUSP525L mouse model of ALS are included.

Hong et al. [134] reviewed the utility of SRS for rapidly determining the antimicrobial susceptibility of bacteria and fungi by single-cell metabolic imaging. Rapid AST was achieved in functional analysis by single-cell SRS imaging of the deuterated glucose metabolic activity, D_2O metabolic incorporation, and de novo lipogenesis. Unlike fluorescent labels, the deuterium substitution does not affect glucose metabolism inside the cell. By mapping the signal arising from the unnatural C-D bond, one is able to determine the concentration and the speed at which glucose or D_2O is consumed by a bacterial or fungal cell under normal or antimicrobial-laden conditions. SRS microscopy was further substantiated as a feasible platform for the rapid detection of antifungal susceptibility by the direct imaging of lipogenesis activity. The transformative aspect of SRS metabolic imaging lies in its real-time analysis capability, enabling the determination of antimicrobial susceptibility within a single cell cycle. This functionality facilitates the timely identification of suitable antimicrobial agents for precise treatment.

4. Conclusions

In the last two decades, significant biological and medical insights have been made possible by the chemical specificity and nonperturbative nature of SRS microscopy. However, improvements in SRS detection sensitivity and spatial resolution are always desired.

Super-resolution and quantum enhancement have recently opened the door to new possibilities for SRS spectroscopy and microscopy. Super-resolution could allow for the resolution of many crucial nanoscale organizations in cells and tissues. On the other hand, using squeezed light to enhance the sensitivity of the stimulated Raman signal enables studies of biological samples with a lower risk of damage and biophysical effects that may not be visible using the standard classical approaches. With further development of advanced quantum light generation and detection schemes in the near future, the QE-SRS technique with much-reduced shot noise can benefit many aspects of bioimaging in terms of its sensitivity, biomolecular specificity, imaging speed, and spatial resolution beyond the limit of classical light [10,118,119].

Author Contributions: The authors participated equally in this work. All authors have read and agreed to the published version of the manuscript.

Funding: This research received no external funding.

Institutional Review Board Statement: Not applicable.

Informed Consent Statement: Not applicable.

Data Availability Statement: Data are contained within the article.

Conflicts of Interest: The authors declare no conflicts of interest.

References

1. Streets, A.M.; Li, A.; Chen, T.; Huang, Y. Imaging without fluorescence: Nonlinear optical microscopy for quantitative cellular imaging. *Anal. Chem.* **2014**, *86*, 8506–8513. [CrossRef] [PubMed]
2. Min, W.; Freudiger, C.W.; Lu, S.; Xie, X.S. Coherent nonlinear optical imaging: Beyond fluorescence microscopy. *Annu. Rev. Phys. Chem.* **2011**, *62*, 501–530. [CrossRef]
3. Popp, J.; Kiefer, W. Raman scattering, fundamentals. In *Encyclopedia of Analytical Chemistry*; Meyers, R.A., Ed.; Wiley: Hoboken, NJ, USA, 2000; pp. 13104–13142.
4. Ferrara, M.A.; Ranjan, R.; Righini, G.C.; Sirleto, L. Stimulated Raman scattering: Towards applications in nano and biophotonics. In *Woodhead Publishing Series in Electronic and Optical Materials, Advances in Nonlinear Photonics, Woodhead Publishing*; Righini, G.C., Sirleto, L., Eds.; Woodhead Publishing: Thorston, UK, 2023; pp. 489–515, ISBN 9780323983846. [CrossRef]
5. Zumbusch, A.; Langbein, A.; Borri, P. Nonlinear vibrational microscopy applied to lipid biology. *Prog. Lipid Res.* **2013**, *52*, 615–632. [CrossRef]
6. Alfonso-García, A.; Mittal, R.; Lee, E.S.; Potma, E.O. Biological imaging with coherent Raman scattering microscopy: A tutorial. *J. Biomed. Opt.* **2014**, *19*, 071407. [CrossRef] [PubMed]
7. Cheng, J.X.; Xie, X.S. Vibrational spectroscopic imaging of living systems: An emerging platform for biology and medicine. *Science* **2015**, *350*, aaa10670.
8. Lee, H.J.; Cheng, J.-X. Imaging chemistry inside living cells by stimulated Raman scattering microscopy. *Methods* **2017**, *128*, 119–128. [CrossRef]

9. Hill, A.H.; Fu, D. Cellular Imaging Using Stimulated Raman Scattering Microscopy. *Anal. Chem.* **2019**, *91*, 9333–9342. [CrossRef] [PubMed]
10. Shi, L.; Fung, A.A.; Zhou, A. Advances in stimulated Raman scattering imaging for tissues and animals. *Quant. Imaging Med. Surg.* **2021**, *11*, 1078–1101. [CrossRef]
11. McClung, F.J.; Hellwarth, R.W. Giant Optical Pulsations from ruby. *J. Appl. Phys.* **1962**, *33*, 828–829. [CrossRef]
12. Eckhardt, G.; Hellwarth, R.W.; McClung, F.J.; Schwarz, S.E.; Weiner, D.; Woodbury, E.J. Stimulated Raman scattering from organic liquids. *Phys. Rev. Lett.* **1962**, *9*, 455–457. [CrossRef]
13. Eckhardt, G.; Bortfeld, D.P.; Geller, M. Stimulated emission of stokes and anti-stokes raman lines from diamond, calcite, and α-sulfur single crystals. *Appl. Phys. Lett.* **1963**, *3*, 137–138. [CrossRef]
14. Lallemand, P.; Simova, P.; Bret, G. Pressure-induced line shift and collisional narrowing in hydrogen gas determined by stimulated Raman emission. *Phys. Rev. Lett.* **1966**, *17*, 1239–1241. [CrossRef]
15. Boyd, R.W. *Nonlinear Optics*, 2nd ed.; Academic Press: Boston, MA, USA, 2003.
16. Shen, Y.R. *The Principles of Nonlinear Optics*; John Wiley & Sons, Inc.: Hoboken, NJ, USA, 2003.
17. Yariv, A. *Quantum Electronics*; John Wiley & Sons, Inc.: Hoboken, NJ, USA, 1967.
18. Armstrong, J.A.; Bloembergen, N.; Ducuing, J.; Pershan, P.S. Interactions between light waves in a nonlinear dielectric. *Phys. Rev.* **1962**, *127*, 1918–1939. [CrossRef]
19. Bloembergen, N.; Shen, Y.R. Coupling between vibrations and light waves in Raman Laser Media. *Phys. Rev. Lett.* **1964**, *12*, 504–507. [CrossRef]
20. Bloembergen, N.; Shen, Y.R. Multimode effects in stimulated Raman emission. *Phys. Rev. Lett.* **1964**, *13*, 720–724. [CrossRef]
21. Shen, Y.R.; Bloembergen, N. Theory of stimulated Brillouin and Raman scattering. *Phys. Rev.* **1965**, *137*, 6A. [CrossRef]
22. Bloembergen, N. The Stimulated Raman Effect. *Am. J. Phys.* **1967**, *35*, 989–1022. [CrossRef]
23. Wang, C.S. Theory of Stimulated Raman Scattering. *Phys. Rev.* **1969**, *182*, 484–494. [CrossRef]
24. Hellwarth, R.W. Theory of Stimulated Raman Scattering. *Phys. Rev.* **1963**, *130*, 1850–1852. [CrossRef]
25. Raymer, M.G.; Mostowski, J. Stimulated raman scattering: Unified treatment of spontaneous initiation and spatial propagation. *Phys. Rev. A* **1981**, *24*, 1980–1993. [CrossRef]
26. Raymer, M.G.; Walmsley, I.A.; Mostowski, J.; Sobolewska, B. Quantum theory of spatial and temporal coherence properties of stimulated Raman scattering. *Phys. Rev. A* **1985**, *32*, 332–344. [CrossRef] [PubMed]
27. Tsikritsis, D.; Legge, E.J.; Belsey, N.A. Practical considerations for quantitative and reproducible measurements with stimulated Raman scattering microscopy. *Analyst* **2022**, *147*, 4642–4656. [CrossRef] [PubMed]
28. Fu, D.; Ye, T.; Matthews, T.E.; Yurtsever, G.; Warren, W.S. Two-color, two-photon, and excited-state absorption microscopy. *J. Biomed. Opt.* **2007**, *12*, 054004. [CrossRef] [PubMed]
29. Isobe, K.; Kawano, H.; Suda, A.; Kumagai, A.; Miyawaki, A.; Midorikawa, K. Simultaneous imaging of two-photon absorption and stimulated Raman scattering by spatial overlap modulation nonlinear optical microscopy. *Biomed. Opt. Express* **2013**, *4*, 1548–1558. [CrossRef] [PubMed]
30. Wilson, J.W.; Samineni, P.; Warren, W.S.; Fischer, M.C. Cross-phase modulation spectral shifting: Nonlinear phase contrast in a pump-probe microscope. *Biomed. Opt. Express* **2012**, *3*, 854–862. [CrossRef] [PubMed]
31. Uchiyama, K.; Hibara, A.; Kimura, H.; Sawada, T.; Kitamori, T. Thermal lens microscope. *Jpn. J. Appl. Phys.* **2000**, *39*, 5316–5322. [CrossRef]
32. D'Arco, A.; Ferrara, M.A.; Indolfi, M.; Tufano, V.; Sirleto, L. Label-free imaging of small lipid droplets by femtosecond-stimulated Raman scattering microscopy. *J. Nonlinear Opt. Phys. Mater.* **2017**, *26*, 1750052. [CrossRef]
33. Ranjan, R.; Indolfi, M.; Ferrara, M.A.; Sirleto, L. Implementation of a nonlinear microscope based on stimulated Raman scattering. *J. Vis. Exp.* **2019**, *149*, e59614.
34. Ranjan, R.; Ferrara, M.A.; Filograna, A.; Valente, C.; Sirleto, L. Femtosecond stimulated Raman microscopy: Home-built realization and a case study of biological imaging. *J. Instrum.* **2019**, *14*, P09008. [CrossRef]
35. Zhang, D.; Slipchenko, M.N.; Leaird, D.E.; Weiner, A.M.; Cheng, J.-X. Spectrally modulated stimulated Raman scattering imaging with an angle-to-wavelength pulse shaper. *Opt. Express* **2013**, *21*, 13864–13874. [CrossRef]
36. Ozeki, Y.; Dake, F.; Kajiyama, S.; Fukui, K.; Itoh, K. Analysis and experimental assessment of the sensitivity of stimulated Raman scattering microscopy. *Opt. Express* **2009**, *17*, 3651. [CrossRef] [PubMed]
37. Dietze, D.R.; Mathies, R.A. Femtosecond stimulated Raman spectroscopy. *ChemPhysChem* **2016**, *17*, 1224–1251. [CrossRef] [PubMed]
38. Audier, X.; Heuke, S.; Volz, P.; Rimk, I.; Rigneault, H. Noise in stimulated Raman scattering measurement: From basics to practice. *APL Photonics* **2020**, *5*, 011101. [CrossRef]
39. Nose, K.; Ozeki, Y.; Kishi, T.; Sumimura, K.; Nishizawa, N.; Fukui, K.; Kanematsu, Y.; Itoh, K. Sensitivity enhancement of fiber-laser-based stimulated Raman scattering microscopy by collinear balanced detection technique. *Opt. Express* **2012**, *20*, 13958. [CrossRef] [PubMed]
40. Zada, L.; Fokker, B.; Leslie, H.A.; Vethaak, A.D.; de Boer, J.F.; Ariese, F. Stimulated Raman scattering simulation for imaging optimization. *J. Eur. Opt. Soc. Rapid Publ.* **2017**, *17*, 1.
41. Ranjan, R.; Costa, G.; Ferrara, M.A.; Sansone, M.; Sirleto, L. Noises investigations and image denoising in femtosecond stimulated Raman scattering microscopy. *J. Biophotonics* **2022**, *15*, e202100379. [CrossRef] [PubMed]

42. Ranjan, R.; Costa, G.; Ferrara, M.A.; Sansone, M.; Sirleto, L. Noise Measurements and Noise Statistical Properties Investigations in a Stimulated Raman Scattering Microscope Based on Three Femtoseconds Laser Sources. *Photonics* **2022**, *9*, 910. [CrossRef]
43. Berto, P.; Andresen, E.R.; Rigneault, H. Background-free stimulated Raman spectroscopy and microscopy. *Phys. Rev. Lett.* **2014**, *112*, 053905. [CrossRef]
44. Ranjan, R.; D'arco, A.; Ferrara, M.A.; Indolfi, M.; Larobina, M.; Sirleto, L. Integration of stimulated Raman gain and stimulated Raman losses detection modes in a single nonlinear microscope. *Opt. Express* **2018**, *26*, 26317. [CrossRef]
45. Heuke, S.; Lombardini, A.; Büttner, E.; Rigneault, H. Simultaneous stimulated Raman gain and loss detection (SRGAL). *Opt. Express* **2020**, *28*, 29619–29630. [CrossRef]
46. D'Arco, A.; Brancati, N.; Ferrara, M.A.; Indolfi, M.; Frucci, M.; Sirleto, L. Subcellular chemical and morphological analysis by stimulated Raman scattering microscopy and image analysis techniques. *Biomed. Opt. Express* **2016**, *7*, 1853. [CrossRef] [PubMed]
47. Manifold, B.; Fu, D. Quantitative stimulated Raman scattering microscopy: Promises and pitfalls. *Annu. Rev. Anal. Chem.* **2022**, *15*, 269–289. [CrossRef] [PubMed]
48. Saar, B.G.; Freudiger, C.W.; Reichman, J.; Stanley, C.M.; Holtom, G.R.; Xie, X.S. Video-rate molecular imaging in vivo with stimulated Raman scattering. *Science* **2010**, *330*, 1369–1370. [CrossRef]
49. Freudiger, C.W.; Min, W.; Saar, B.G.; Lu, S.; Holtom, G.R.; He, C.; Tsai, J.C.; Kang, J.X.; Xie, X.S. Label-Free Biomedical Imaging with High Sensitivity by Stimulated Raman Scattering Microscopy. *Science* **2008**, *322*, 1857–1861. [CrossRef]
50. Ranjan, R.; Antonietta Ferrara, M.; Sirleto, L. *Femtosecond Stimulated Raman Microscopy in C-H Region of Raman Spectra of Biomolecules and Its Extension to Silent and Fingerprint Regions [Internet]. Novel Imaging and Spectroscopy*; IntechOpen: London, UK, 2020. [CrossRef]
51. Ferrara, M.A.; Filograna, A.; Ranjan, R.; Corda, D.; Valente, C.; Sirleto, L. Threedimensional label-free imaging throughout adipocyte differentiation by stimulated Raman microscopy. *PLoS ONE* **2019**, *14*, e0216811. [CrossRef]
52. Fu, D.; Zhou, J.; Zhu, W.S.; Manley, P.W.; Wang, Y.K.; Hood, T.; Wylie, A.; Xie, X.S. Imaging the intracellular distribution of tyrosine kinase inhibitors in living cells with quantitative hyperspectral stimulated Raman scattering. *Nat. Chem.* **2014**, *6*, 614–622. [CrossRef]
53. Orringer, D.A.; Pandian, B.; Niknafs, Y.S.; Hollon, T.C.; Boyle, J.; Lewis, S.; Garrard, M.; Hervey-Jumper, S.L.; Garton, H.J.; Maher, C.O.; et al. Rapid intraoperative histology of unprocessed surgical specimens via fibre-laser-based stimulated Raman scattering microscopy. *Nat. Biomed. Eng.* **2017**, *1*, 0027. [CrossRef]
54. He, R.; Xu, Y.; Zhang, L.; Ma, S.; Wang, X.; Ye, D.; Ji, M. Dual-phase stimulated Raman scattering microscopy for real-time two-color imaging. *Optica* **2017**, *4*, 44–47. [CrossRef]
55. Galli, R.; Uckermann, O.; Temme, A.; Leipnitz, E.; Meinhardt, M.; Koch, E.; Schackert, G.; Steiner, G.; Kirsch, M. Assessing the Efficacy of Coherent Anti-Stokes Raman Scattering Microscopy for the Detection of Infiltrating Glioblastoma in Fresh Brain Samples. *J. Biophotonics* **2017**, *10*, 404–414. [CrossRef]
56. Pekmezci, M.; Morshed, R.A.; Chunduru, P.; Pandian, B.; Young, J.; Villanueva-Meyer, J.E.; Tihan, T.; Sloan, E.A.; Aghi, M.K.; Molinaro, A.M.; et al. Detection of Glioma Infiltration at the Tumor Margin Using Quantitative Stimulated Raman Scattering Histology. *Sci. Rep.* **2021**, *11*, 12162. [CrossRef]
57. Yan, S.; Li, Y.; Huang, Z.; Yuan, X.; Wang, P. High-Speed Stimulated Raman Scattering Microscopy Using Inertia-Free AOD Scanning. *J. Phys. Chem. B* **2023**, *127*, 4229–4234. [CrossRef] [PubMed]
58. Pegoraro, A.F.; Ridsdale, A.; Moffatt, D.J.; Jia, Y.W.; Pezacki, J.P.; Stolow, A. Optimally chirped multimodal CARS microscopy based on a single Ti:sapphire oscillator. *Opt. Express* **2009**, *17*, 2984–2996. [CrossRef] [PubMed]
59. Francis, A.; Berry, K.; Chen, Y.; Figueroa, B.; Fu, D. Label-free pathology by spectrally sliced femtosecond stimulated Raman scattering (SRS) microscopy. *PLoS ONE* **2017**, *12*, e0178750. [CrossRef] [PubMed]
60. Sirleto, L.; Ranjan, R.; Ferrara, M.A. Analysis of pulses bandwidth and spectral resolution in femtosecond stimulated Raman scattering microscopy. *Appl. Sci.* **2019**, *11*, 3903. [CrossRef]
61. Zhang, D.; Slipchenko, M.N.; Cheng, J.X. Highly Sensitive Vibrational Imaging by Femtosecond Pulse Stimulated Raman Loss. *J. Phys. Chem. Microsc. Lett.* **2011**, *2*, 1248–1253. [CrossRef] [PubMed]
62. Fu, D.; Lu, F.K.; Zhang, X.; Freudiger, C.; Pernik, D.R.; Holtom, G.; Xie, X.S. Quantitative Chemical Imaging with Multiplex Stimulated Raman Scattering. *J. Am. Chem. Soc.* **2012**, *134*, 3623–3626. [CrossRef] [PubMed]
63. Kong, L.J.; Ji, M.B.; Holtom, G.R.; Fu, D.; Freudiger, C.W.; Xie, X.S. Multicolor stimulated Raman scattering microscopy with a rapidly tunable optical parametric oscillator. *Opt. Lett.* **2013**, *38*, 145–147. [CrossRef] [PubMed]
64. Fu, D.; Holtom, G.; Freudiger, C.; Zhang, X.; Xie, X.S. Hyperspectral imaging with stimulated Raman scattering by chirped femtosecond lasers. *J. Phys. Chem. B* **2013**, *117*, 4634–4640. [CrossRef]
65. Ozeki, Y.; Umemura, W.; Otsuka, Y.; Satoh, S.; Hashimoto, H.; Sumimura, K.; Nishizawa, N.; Fukui, K.; Itoh, K. High-speed molecular spectral imaging of tissue with stimulated Raman scattering. *Nat. Photonics* **2012**, *6*, 845–851. [CrossRef]
66. Ozeki, Y.; Umemura, W.; Sumimura, K.; Nishizawa, N.; Fukui, K.; Itoh, K. Stimulated Raman hyperspectral imaging based on spectral filtering of broadband fiber laser pulses. *Opt. Lett.* **2012**, *37*, 431–433. [CrossRef]
67. Karpf, S.; Eibl, M.; Wieser, W.; Klein, T.; Huber, R. A time-encoded technique for fibre-based hyperspectral broadband stimulated 711 Raman microscopy. *Nat. Commun.* **2015**, *6*, 6784. [CrossRef] [PubMed]
68. Mohseni, M.; Polzer, C.; Hellerer, T. Resolution of spectral focusing in coherent Raman imaging. *Opt. Express* **2018**, *26*, 10230–10241. [CrossRef] [PubMed]

69. Hellerer, T.; Enejder, A.M.K.; Zumbusch, A. Spectral focusing: High spectral resolution spectroscopy with broad-bandwidth laser pulses. *Appl. Phys. Lett.* **2004**, *85*, 25–27. [CrossRef]
70. Liu, B.; Lee, H.J.; Zhang, D.L.; Liao, C.-S.; Ji, N.; Xia, Y.Q.; Cheng, J.-X. Label-free spectroscopic detection of membrane potential using stimulated Raman scattering. *Appl. Phys. Lett.* **2015**, *106*, 173704. [CrossRef]
71. Liao, C.-S.; Huang, K.-C.; Hong, W.; Chen, A.J.; Karanja, C.; Wang, P.; Eakins, G.; Cheng, J.-X. Stimulated Raman spectroscopic imaging by microsecond delay-line tuning. *Optica* **2016**, *3*, 1377–1380. [CrossRef]
72. Zhang, D.; Wang, P.; Slipchenko, M.N.; Ben-Amotz, D.; Weiner, A.M.; Cheng, J.X. Quantitative vibrational imaging by hyper-spectral stimulated Raman scattering microscopy and multivariate curve resolution analysis. *Anal. Chem.* **2013**, *85*, 98–106. [CrossRef]
73. Bae, K.; Zheng, W.; Huang, Z. Spatial light-modulated stimulated Raman scattering (SLM-SRS) microscopy for rapid multiplexed vibrational imaging. *Theranostics* **2020**, *10*, 312–322. [CrossRef]
74. Camp, C.H., Jr.; Lee, Y.J.; Heddleston, J.M.; Hartshorn, C.M.; Walker, A.R.H.; Rich, J.N.; Lathia, J.D.; Cicerone, M.T. High-speed coherent Raman fingerprint imaging of biological tissues. *Nat. Photonics* **2014**, *8*, 627–634. [CrossRef]
75. Hashimoto, K.; Takahashi, M.; Ideguchi, T.; Goda, K. Broadband coherent Raman spectroscopy running at 24,000 spectra per second. *Sci. Rep.* **2016**, *6*, 21036. [CrossRef]
76. Wang, Z.; Zheng, W.; Huang, Z. Lock-in-detection-free linescan stimulated Raman scattering microscopy for near video-rate Raman imaging. *Opt. Lett.* **2016**, *41*, 3960–3963. [CrossRef]
77. Zhang, C.; Huang, K.C.; Rajwa, B.; Li, J.; Yang, S.; Lin, H.; Liao, C.S.; Eakins, G.; Kuang, S.; Patsekin, V.; et al. Stimulated Raman scattering flow cytometry for labelfree single-particle analysis. *Optica* **2017**, *4*, 103–109. [CrossRef]
78. Lin, H.; Liao, C.S.; Wang, P.; Kong, N.; Cheng, J.X. Spectroscopic stimulated Raman scattering imaging of highly dynamic specimens through matrix completion. *Light Sci. Appl.* **2018**, *7*, 17179. [CrossRef]
79. Wei, L.; Hu, F.; Chen, Z.; Shen, Y.; Zhang, L.; Min, W. Live- Cell Bioorthogonal Chemical Imaging: Stimulated Raman Scattering Microscopy of Vibrational Probes. *Acc. Chem. Res.* **2016**, *49*, 1494–1502. [CrossRef] [PubMed]
80. Wei, L.; Chen, Z.; Shi, L.; Long, R.; Anzalone, A.V.; Zhang, L.; Hu, F.; Yuste, R.; Cornish, V.W.; Min, W. Super-multiplex vibrational imaging. *Nature* **2017**, *544*, 465–470. [CrossRef]
81. Hu, F.; Zeng, C.; Long, R.; Miao, Y.; Wei, L.; Xu, Q.; Min, W. Supermultiplexed optical imaging and barcoding with engineered polyynes. *Nat. Methods* **2018**, *15*, 194–200. [CrossRef] [PubMed]
82. Miao, Y.; Qian, N.; Shi, L.; Hu, F.; Min, W. 9-Cyanopyronin probe palette for super-multiplexed vibrational imaging. *Nat. Commun.* **2021**, *2*, 4518. [CrossRef]
83. Shi, L.; Wei, M.; Miao, Y.; Qian, N.; Shi, L.; Singer, R.A.; Benninger, R.K.; Min, W. Highly-multiplexed volumetric mapping with Raman dye imaging and tissue clearing. *Nat. Biotechnol.* **2022**, *40*, 364–373. [CrossRef] [PubMed]
84. Hell, S.W.; Wichmann, J. Breaking the diffraction resolution limit by stimulated emission: Stimulated-emission-depletion fluorescence microscopy. *Opt. Lett.* **1994**, *19*, 780. [CrossRef]
85. Fujita, K.; Kobayashi, M.; Kawano, S.; Yamanaka, M.; Kawata, S. High-resolution confocal microscopy by saturated excitation of fluorescence. *Phys. Rev. Lett.* **2007**, *99*, 228105. [CrossRef]
86. Gustafsson, M.G.L. Surpassing the lateral resolution limit by a factor of two using structured illumination microscopy. *J. Microsc.* **2000**, *198*, 82. [CrossRef]
87. Huang, F.M.; Zheludev, N.I. Super-resolution without evanescent waves. *Nano Lett.* **2009**, *9*, 1249. [CrossRef] [PubMed]
88. Hajek, K.M.; Littleton, B.; Turk, D.; McIntyre, T.J.; Rubinsztein-Dunlop, H. A method for achieving super-resolved widefield CARS microscopy. *Opt. Express* **2010**, *18*, 19263. [CrossRef] [PubMed]
89. Park, J.H.; Lee, S.-W.; Lee, E.S.; Lee, J.Y. A method for super-resolved CARS microscopy with structured illumination in two dimensions. *Opt. Express* **2014**, *22*, 9854. [CrossRef] [PubMed]
90. Gong, L.; Zheng, W.; Ma, Y.; Huang, Z. Higher-order coherent anti-stokes Raman scattering microscopy realizes label-free super-resolution vibrational imaging. *Nat. Photonics* **2020**, *14*, 115. [CrossRef]
91. Gong, L.; Zheng, W.; Ma, Y.; Huang, Z. Saturated stimulated-Raman-scattering microscopy for far-field superresolution vibrational imaging. *Phys. Rev. Appl.* **2019**, *11*, 034041. [CrossRef]
92. Gong, L.; Wang, H. Breaking the diffraction limit by saturation in stimulated-Raman-scattering microscopy: A theoretical study. *Phys. Rev. A* **2014**, *90*, 013818. [CrossRef]
93. Gong, L.; Wang, H. Suppression of stimulated Raman scattering by an electromagnetically-induced-transparency-like scheme and its application for super-resolution microscopy. *Phys. Rev. A* **2015**, *92*, 023828. [CrossRef]
94. Silva, W.R.; Graefe, C.T.; Frontiera, R.R. Toward label-free super-resolution microscopy. *ACS Photonics* **2016**, *3*, 79. [CrossRef]
95. Yonemaru, Y.; Palonpon, A.F.; Kawano, S.; Smith, N.I.; Kawata, S.; Fujita, K. Super-spatial- and -spectral-resolution in vibrational imaging via saturated coherent anti-stokes Raman scattering. *Phys. Rev. Appl.* **2015**, *4*, 014010. [CrossRef]
96. Beeker, W.P.; Groß, P.; Lee, C.J.; Cleff, C.; Offerhaus, H.L.; Fallnich, C.; Herek, J.L.; Boller, K.-J. A route to sub-diffraction-limited CARS microscopy. *Opt. Express* **2019**, *17*, 22632. [CrossRef]
97. Beeker, W.P.; Lee, C.J.; Boller, K.-J.; Groß, P.; Cleff, C.; Fallnich, C.; Offerhaus, H.L.; Herek, J.L. Spatially dependent Rabi oscillations: An approach to sub-diffraction-limited coherent anti-stokes Raman-scattering microscopy. *Phys. Rev. A* **2010**, *81*, 012507. [CrossRef]

98. Cleff, C.; Groß, P.; Fallnich, C.; Offerhaus, H.L.; Herek, J.L.; Kruse, K.; Beeker, W.P.; Lee, C.J.; Boller, K.-J. Ground-state depletion for subdiffractionlimited spatial resolution in coherent anti-stokes Raman scattering microscopy. *Phys. Rev. A* **2012**, *86*, 023825. [CrossRef]
99. Gong, L.; Lin, J.; Hao, C.; Zheng, W.; Wu, S.Q.Y.; Teng, J.; Qiu, C.-W.; Huang, Z. Supercritical focusing coherent anti-stokes Raman scattering microscopy for high-resolution vibrational imaging. *Opt. Lett.* **2018**, *43*, 5615. [CrossRef] [PubMed]
100. Kim, H.; Bryant, G.W.; Stranick, S.J. Superresolution four-wave mixing microscopy. *Opt. Express* **2012**, *20*, 6042. [CrossRef] [PubMed]
101. Raghunathan, V.; Potma, E.O. Multiplicative and subtractive focal volume engineering in coherent Raman microscopy. *J. Opt. Soc. Am. A* **2010**, *27*, 2365. [CrossRef] [PubMed]
102. Prince, R.C.; Potma, E.O. Going visible: High-resolution coherent Raman imaging of cells and tissues. *Light Sci. Appl.* **2019**, *8*, 10. [CrossRef] [PubMed]
103. Wassie, A.T.; Zhao, Y.; Boyden, E.S. Expansion microscopy: Principles and uses in biological research. *Nat. Methods* **2016**, *16*, 33–41. [CrossRef]
104. Klimas, A.; Gallagher, B.; Wijesekara, P.; Fekir, S.; Stolz, D.; Cambi, F.; Watkins, S.; Barth, A.; Moore, C.; Ren, X.; et al. Magnify is a universal molecular anchoring strategy for expansion microscopy. *Nat. Biotechnol.* **2023**, *41*, 858–869. [CrossRef]
105. Shi, L.; Klimas, A.; Gallagher, B.; Cheng, Z.; Fu, F.; Wijesekara, P.; Miao, Y.; Ren, X.; Zhao, Y.; Min, W. Super-Resolution Vibrational Imaging Using Expansion Stimulated Raman Scattering Microscopy. *Adv. Sci.* **2022**, *9*, 2200315. [CrossRef]
106. Lawrie, B.J.; Lett, P.D.; Marino, A.M.; Pooser, R.C. Quantum sensing with squeezed light. *ACS Photonics* **2019**, *6*, 1307–1318. [CrossRef]
107. Pirandola, S.; Bardhan, B.R.; Gehring, T.; Weedbrook, C.; Lloyd, S. Advances in photonic quantum sensing. *Nat. Photonics* **2018**, *12*, 724–733. [CrossRef]
108. Taylor, M.A.; Bowen, W.P. Quantum metrology and its application in biology. *Phys. Rep.* **2016**, *615*, 1–59. [CrossRef]
109. Slusher, R.E. Quantum optics in the '80s. *Opt. Photonics News* **1990**, *1*, 27–30. [CrossRef]
110. Sewell, R.J.; Napolitano, M.; Behbood, N.; Colangelo, G.; Mitchell, M.W. Certified quantum non-demolition measurement of a macroscopic material system. *Nat. Photonics* **2013**, *7*, 517–520. [CrossRef]
111. Giovannetti, V.; Lloyd, S.; Maccone, L. Advances in quantum metrology. *Nat. Photonics* **2011**, *5*, 222–229. [CrossRef]
112. Andersen, U.L.; Gehring, T.; Marquardt, C.; Leuchs, G. 30 Years of squeezed light generation. *Phys. Scr.* **2016**, *91*, 053001. [CrossRef]
113. Slusher, R.E.; Hollberg, L.W.; Yurke, B.; Mertz, J.C.; Valley, J.F. Observation of squeezed states generated by four-wave mixing in an optical cavity. *Phys. Rev. Lett.* **1985**, *55*, 2409–2412. [CrossRef]
114. Moreau, P.-A.; Toninelli, E.; Gregory, T.; Padgett, M.J. Imaging with quantum states of light. *Nat. Rev. Phys.* **2019**, *1*, 367–380. [CrossRef]
115. Fu, Y.; Wang, H.; Shi, R.; Cheng, J.-X. Characterization of photodamage in coherent anti-Stokes Raman scattering microscopy. *Opt. Express* **2006**, *14*, 3942–3951. [CrossRef]
116. Xu, Z.; Oguchi, K.; Taguchi, Y.; Sano, Y.; Miyawaki, Y.; Cheon, D.; Katoh, K.; Ozeki, Y. Stimulated Raman scattering spectroscopy with quantum-enhanced balanced detection. *Opt. Express* **2022**, *30*, 18589. [CrossRef]
117. Xu, Z.; Oguchi, K.; Taguchi, Y.; Takahashi, S.; Sano, Y.; Mizuguchi, T.; Katoh, K.; Ozeki, Y. Quantum-enhanced stimulated Raman scattering microscopy in a high-power regime. *Opt. Lett.* **2022**, *47*, 5829. [CrossRef] [PubMed]
118. de Andrade, R.B.; Kerdoncuff, H.; Berg-Sørensen, K.; Gehring, T.; Lassen, M.; Andersen, U.L. Quantum-enhanced continuous-wave stimulated Raman scattering spectroscopy. *Optica* **2020**, *7*, 470–475. [CrossRef]
119. Casacio, C.A.; Madsen, L.S.; Terrasson, A.; Waleed, M.; Barnscheidt, K.; Hage, B.; Taylor, M.A.; Bowen, W.P. Quantum-enhanced nonlinear microscopy. *Nature* **2021**, *594*, 201–206. [CrossRef] [PubMed]
120. Gong, L.; Lin, S.; Huang, Z. Super-resolution stimulated Raman scattering microscopy enhanced by quantum light and deconvolution. *Opt. Lett.* **2023**, *48*, 6516–6519. [CrossRef] [PubMed]
121. Beskrovnyy, V.; Kolobov, M. Quantum limits of super-resolution in reconstruction of optical objects. *Phys. Rev. A* **2005**, *71*, 043802. [CrossRef]
122. Yue, S.; Cheng, J.X. Deciphering single cell metabolism by coherent Raman scattering microscopy. *Curr. Opin. Chem. Biol.* **2016**, *33*, 46–57. [CrossRef] [PubMed]
123. Tirinato, L.; Pagliari, F.; Limongi, T.; Marini, M.; Falqui, A.; Seco, J.; Candeloro, P.; Liberale, C.; Di Fabrizio, E. An Overview of Lipid Droplets in Cancer and Cancer Stem Cells. *Stem Cells Int.* **2017**, *2017*, 1656053. [CrossRef] [PubMed]
124. Yue, S.; Li, J.; Lee, S.-Y.; Lee, H.J.; Shao, T.; Song, B.; Cheng, L.; Masterson, T.A.; Liu, X.; Ratliff, T.L.; et al. Cholesteryl ester accumulation induced by PTEN loss and PI3K/AKT activation underlies human prostate cancer aggressiveness. *Cell. Metab.* **2014**, *19*, 393–406. [CrossRef]
125. Dou, W.; Zhang, D.; Jung, Y.; Cheng, J.X.; Umulis, D.M. Label-free imaging of lipid droplet intracellular motion in early Drosophila embryos using femtosecond stimulated Raman loss microscopy. *Biophys. J.* **2012**, *102*, 1666–1675. [CrossRef]
126. AH, H.; Manifold, B.; Fu, D. Tissue imaging depth limit of stimulated Raman scattering microscopy. *Biomed. Opt. Express* **2020**, *11*, 762–774.

127. Ji, M.; Orringer, D.A.; Freudiger, C.W.; Ramkissoon, S.; Liu, X.; Lau, D.; Golby, A.J.; Norton, I.; Hayashi, M.; Agar, N.Y.; et al. Rapid, label-free detection of brain tumors with stimulated Raman scattering microscopy. *Sci. Transl. Med.* **2013**, *5*, 201ra119. [CrossRef]
128. Lu, F.K.; Calligaris, D.; Olubiyi, O.I.; Norton, I.; Yang, W.; Santagata, S.; Xie, X.S.; Golby, A.J.; Agar, N.Y. Label-Free Neurosurgical Pathology with Stimulated Raman Imaging. *Cancer Res.* **2016**, *76*, 3451–3462. [CrossRef] [PubMed]
129. Ji, M.; Arbel, M.; Zhang, L.; Freudiger, C.W.; Hou, S.S.; Lin, D.; Yang, X.; Bacskai, B.J.; Xie, X.S. Label-free imaging of amyloid plaques in Alzheimer's disease with stimulated Raman scattering microscopy. *Sci. Adv.* **2018**, *4*, eaat7715. [CrossRef] [PubMed]
130. Tian, F.; Yang, W.; Mordes, D.A.; Wang, J.Y.; Salameh, J.S.; Mok, J.; Chew, J.; Sharma, A.; Leno-Duran, E.; Suzuki-Uemats Suzuki, N.; et al. Monitoring peripheral nerve degeneration in ALS by label-free stimulated Raman scattering imaging. *Nat. Commun.* **2016**, *7*, 13283. [CrossRef] [PubMed]
131. Fu, D.; Yang, W.; Xie, X.S. Label-free Imaging of Neurotransmitter Acetylcholine at Neuromuscular Junctions with Stimulated Raman Scattering. *J. Am. Chem. Soc.* **2017**, *139*, 583–586. [CrossRef] [PubMed]
132. Liao, C.S.; Wang, P.; Huang, C.Y.; Lin, P.; Eakins, G.; Bentley, R.T.; Liang, R.; Cheng, J.X. In Vivo and in Situ Spectroscopic Imaging by a Handheld Stimulated Raman Scattering Microscope. *ACS Photonics* **2018**, *5*, 947–954. [CrossRef]
133. Tipping, W.J.; Lee, M.; Serrels, A.; Brunton, V.G.; Hulme, A.N. Stimulated Raman scattering microscopy: An emerging tool for drug discovery. *Chem. Soc. Rev.* **2016**, *45*, 2075–2089. [CrossRef]
134. Hong, W.; Zhang, M.; Cheng, J.-X. Rapid determination of antimicrobial susceptibility by SRS single-cell metabolic imaging. In *Chapter 29—Title: Stimulated Raman Scattering Microscopy*; Cheng, J.-X., Min, W., Ozeki, Y., Polli, D., Eds.; Elsevier: Amsterdam, The Netherlands, 2022; pp. 445–461, ISBN 9780323851589. [CrossRef]

photonics

MDPI

Article

Compact Diode-Pumped Solid-State Laser with Intracavity Pump-Enhanced DFG Emitting at ~3.5 Microns

Liam G. K. Flannigan [1,*], Ali Atwi [1], Tyler Kashak [1], Daniel Poitras [2] and Chang-Qing Xu [1]

[1] Department of Engineering Physics, McMaster University, Hamilton, ON L8S 4L8, Canada; atwia@mcmaster.ca (A.A.); cqxu@mcmaster.ca (C.-Q.X.)
[2] Advanced Electronics and Photonics Research Center, National Research Council Canada, Ottawa, ON K1A 0R6, Canada
* Correspondence: flannilg@mcmaster.ca

Abstract: We report a diode-pumped solid-state (DPSS) laser used for intracavity pump-enhanced difference frequency generation (DFG) to create a 3.5-micron laser. Using a 50 mm-long periodically poled lithium niobate (PPLN) crystal inside the cavity of an Nd:YVO_4 solid-state laser at 1064 nm with 4.5 W pump power at 808 nm, and a 310 mW C-band signal at 1529 nm, up to 31 mW of mid-infrared output power at 3499 nm is obtained. The cavity requires no active stabilization and/or locking, and the entire cavity is <8 cm in length. The obtained output power corresponds to a black-box efficiency of 2.20%W^{-1}, which is the highest value reported to date for continuous-wave DFG based on a bulk nonlinear optical crystal with no active stabilization. Potential future applications in free-space optical communication are also discussed.

Keywords: nonlinear optics; difference frequency generation; mid-infrared laser; DPSS laser; linear cavity

1. Introduction

Free-space optical communication has seen a surge in interest, both for terrestrial first and last mile solutions, as well as optical satellite communication [1,2]. Current efforts have focused on utilizing the short-wave infrared (SWIR) atmospheric transmission window, containing the 1530–1565 nm C-band used in telecommunications. Successful demonstrations utilizing these wavelengths include the SOLISS terminal bidirectional link between the International Space Station (ISS) and an optical ground station at the National Institute of Information and Communications Technology (NICT) in 2019, as well as an inter-satellite demonstration between the CubeSOTA and ETS9-HICALI satellites in 2019 [3,4]. However, SWIR communication suffers significantly from scintillation, scattering, and weather effects. The 3–5 μm mid-infrared (MWIR) atmospheric transmission window should offer improved signal-to-noise ratio and resistance to these adverse effects and could lead to higher link availability and reliability [5,6]. A previous review from our research group expanded on the benefits of the MWIR band and identified cascade lasers, as well as difference frequency generation (DFG) using nonlinear optical crystals, as the most promising lasers for use in MWIR transmitters [7–9]. Cascade lasers offer large modulation bandwidths and can cover the entire MWIR spectrum, but typically operate between 77 and 250 K, which is inconvenient for many applications [10,11]. DFG offers room temperature operation and large modulation bandwidths, with the potential to achieve up to watt-level continuous-wave output power [12]. Typically, OPOs are used due to their resonant structure offering high output powers and tunable output wavelengths. However, the use of DFG allows us to inject a C-band signal beam, offering multi-gigabit per second modulation speeds, which can be faster than directly modulating the OPO pump beam [8]. In this paper, we aim to address the lack of high-efficiency intracavity MWIR lasers by presenting a novel DFG laser that uses a simplified linear cavity to reduce the size and sensitivity to alignment and to achieve high efficiency. The laser is stable while free running with no need for external cavity stabilization or locking. In the cavity, only the pump is resonant, as the signal is coupled into the cavity and the idler is coupled out of the cavity in a single pass. We see this as the first step to demonstrating a viable high-power mid-infrared source capable of high-speed modulation.

A common metric used to evaluate DFG frequency converters is DFG efficiency, presented according to the formula in Equation (1):

$$\eta_{DFG} = 100\% \times \frac{P_{DFG}}{P_{pump}P_{signal}}, \quad (1)$$

Efficiency is expressed in units of $\%W^{-1}$. Here, P_{DFG} is the output DFG power, P_{pump} is the input pump power directly before the nonlinear crystal, and P_{signal} is the input signal power directly before the nonlinear crystal. However, this approach does not consider the efficiencies and coupling losses of the pump and signal sources, which can be important in power-limited scenarios where maximizing efficiency is key. So, from a practical application point of view, an interesting metric to compare the efficiency of various DFG structures is the black-box efficiency with units of $\%W^{-1}$, presented here as Equation (2):

$$\eta_{BB} = 100\% \times \frac{\eta_{signal}\eta_{pump}P_{DFG}}{P_{pump}P_{signal}}, \quad (2)$$

where η_{pump} is the diode pumping efficiency for the fiber pump source and η_{signal} is the diode pumping efficiency of the fiber signal source, assuming fiber lasers are used as pump and signal sources. We mention fiber pump and signal sources specifically as these are commonly used for pumping both DFG and OPO structures. Many of the presented structures in the following section make use of 1064 nm Yb-doped fiber lasers (pumped by 9xx nm diode laser) as pump sources. These have optical-to-optical efficiencies of ~60%, so a value of $\eta_{pump} = 0.6$ was assumed for these fiber lasers [13]. For devices using direct 1064 nm diode pumping, there is no additional optical conversion step, and so a value of $\eta_{pump} = 1$ can be used. To summarize, the pump value should be the input optical pump power when treating the system as a black box: For direct diode pumping, no modification is needed, while pumping via an added amplification stage requires one to account for the amplifier efficiency. For the signal fiber, erbium-doped fiber amplifiers (EDFAs) are often used. Unfortunately, it is harder to find a standard operating efficiency for the pump diodes in EDFAs, and most authors only include the output power from the EDFA (P_{signal}). Thus, we have assumed $\eta_{signal} = 1$ for all EDFAs so that they can be easily compared. This does increase the overall efficiency reported in this paper, so we mention this to avoid claims of inflating the efficiency values intentionally. While we do not claim that our device has the highest DFG efficiency reported, we do claim that it has the highest black-box efficiency among the bulk crystal, free-running DFG devices we have reviewed.

We performed a literature review of state-of-the-art single pass and intracavity mid-infrared bulk DFG devices. In 2020, Guha et al. reported a single-pass DFG configuration, emitting 3.55 W at 3400 nm with 43.3 W 1064 nm via fiber laser pump and 31 W 1550 nm signal via EDFA, corresponding to a DFG efficiency of $0.264\%W^{-1}$ and a black-box efficiency of $0.159\%W^{-1}$ [14]. In 2019, Huang et al. demonstrated 60 mW of DFG power at 3393 nm for 1 W of 1064 nm fiber laser pump and 1 W of 1550 nm, corresponding to a DFG efficiency of $5.8\%W^{-1}$ and a black-box efficiency of $2.43\%W^{-1}$, which accounts for the additional loss of the wavelength division multiplexer used [15]. Huang et al.'s cavity is also actively stabilized via dither-locking. These devices represent the highest efficiency single-pass and intracavity efficiencies we found, respectively, but other devices were reviewed with efficiencies in the same order of magnitude [16–20]. These works serve as a representative sample of the state-of-the-art for bulk crystals. Additionally, waveguides allow for tight confinement of the pump and signal beams over the entire length of the nonlinear medium, which boosts efficiency versus bulk crystals. For example, the highest efficiency mid-infrared waveguide nonlinear frequency converter we found has an internal DFG efficiency of 100%W^{-1} and a black-box efficiency of 60%W^{-1} [21]. However, the reported coupling losses are ~−17 dB for both the pump and signal, which would lower the black-box efficiency to less than 1%/W. Hu et al. demonstrated a high-efficiency coupler that could reduce losses for similar thin-film lithium niobate devices to −0.5 dB [22]. Should this low insertion loss be coupled with a high-efficiency waveguide device, it would have much higher efficiency than the bulk crystal devices reviewed above [23,24]. For now, the main niche for bulk devices is relative cost, ease of manufacturing, and ease of use. Nonlinear waveguide properties are critically dependent on their fabrication quality, as even small imperfections can degrade the final performance [25]. So, in the short term, bulk crystals remain easy to source off the shelf with a high level of quality, and linear cavities like the one presented in this manuscript are easy to set up and align. However, as manufacturing continues to improve, waveguide-based mid-infrared sources will be preferable to bulk crystals for many applications. As such, we want to make it clear we limit the novelty claim for our device to bulk crystal devices.

In this work, we implement a hemispherical cavity using Nd:YVO_4 as a gain medium to produce a 1064 nm pump. The resonator contains a periodically poled lithium niobate (PPLN) nonlinear crystal, as well as a dichroic mirror that couples an amplified tunable C-band laser into the cavity. We obtained a black-box efficiency of 2.20%W^{-1}, emitting at 3499 nm with 4.5 W 808 nm pump and 310 mW of 1529 nm signal. This device offers the highest black-box efficiency for bulk nonlinear crystal frequency converters in a linear cavity without the use of active cavity stabilization. Additionally, our modeling work shows good agreement with experimental results.

The manuscript is structured as follows. In Section 2, a schematic of the experimental setup is provided alongside manufacturer part numbers for all equipment used. This section also includes a summary of the theoretical modeling that was used to predict the experimental results, showing good agreement. In Section 3, experimental results for the output 1064 nm and DFG power at 3.5 microns are presented, as well as the output DFG power as a function of temperature. Experimental results are presented with best-fit lines alongside the simulation curves. In Section 4, we discuss the experimental results, as well as potential future work and gaps in the analysis performed. Finally, Section 5 concludes the paper by summarizing the results and important takeaways from the work presented here.

2. Materials and Methods

A schematic diagram of the experimental cavity setup is depicted in Figure 1. It consists of a diode-pumped solid-state (DPSS) laser resonator (Nd:YVO_4, Dientech, Chengdu, China) with a PPLN nonlinear crystal inserted into the cavity. The hemispherical resonator was intentionally selected as it allows for a compact cavity setup and is easy to align. Additionally, this linear cavity requires no active stabilization as we have demonstrated experimentally. This reduces complexity and costs without the need for piezoelectric transducers or lock-in amplifiers to facilitate long-term cavity stability. The following cavity parameters were designed with the help of reZonator, an open-source cavity modeling software capable of modeling arbitrary optical systems [26]. The reZonator model and the beam radius evolution inside of the cavity are summarized in Figure 1:

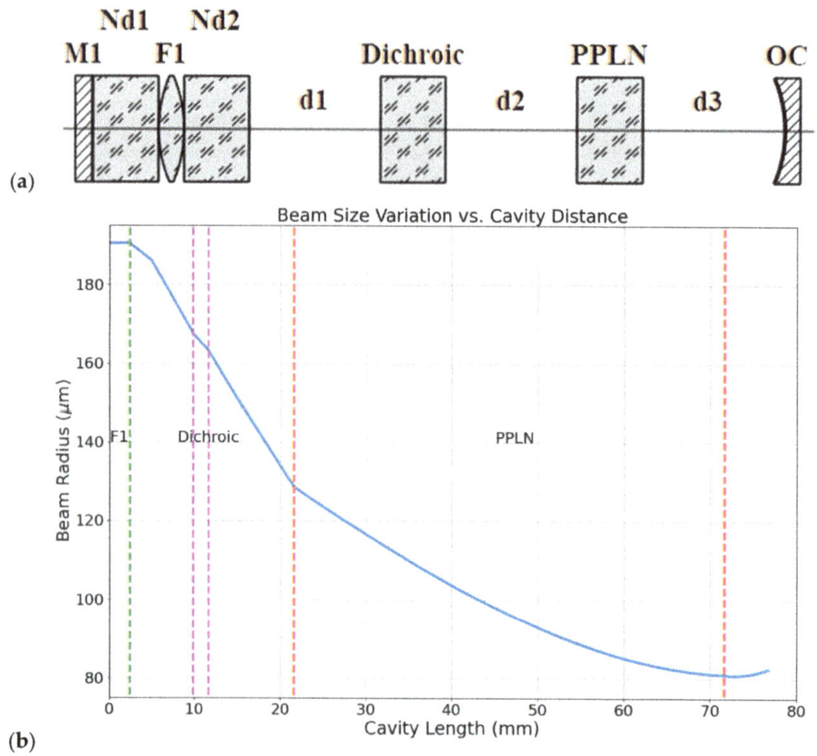

Figure 1. (**a**) Simplified cavity layout as modeled in reZonator. Note that the components are not to scale. (**b**) The beam radius as a function of distance from the plane mirror M1. The solid blue line is the beam radius, the dashed green line denotes the thermal lens F1, the dashed purple lines denote the dichroic mirror facets, and the dashed red lines denote the PPLN facets.

In this model, M1 represents the HR coating at 1064 on the Nd:YVO_4. The 5 mm gain medium has been split into two 2.5 mm slabs with refractive index $n = 2.16$ and a thin lens in the middle, in order to approximate the thermal lens as well. The thermal lens focal length was determined to be ~50 mm, which matched the output beam parameters and output power well, and lines up with similar experimental data [27]. Distances d1, d2, and d3 are 5 mm, 10 mm, and 5 mm, respectively. The dichroic is a 1.8 mm slab with a refractive index of $n = 1.5$, while the PPLN is 50 mm with a refractive index of $n = 2.2$. The OC has a radius of curvature of 100 mm. This results in a 1064 nm pump beam radius in the middle of the PPLN of ~100 μm, which will be used as an input to the nonlinear simulation later. These parameters were selected to achieve a 1064 nm beam waist in the center of the PPLN crystal as close to a confocal parameter of 1 as possible. This value is typically used as a starting point for a good wavelength conversion efficiency [28–30]. For a 50 mm PPLN, this would require a beam waist around 93 μm. Our value of 100 μm was as close as we could obtain while maintaining cavity stability. The 55 μm signal beam waist was selected due to a combination of the space limitations on the optical bench as well as the selection of focusing lenses that were available in the laboratory. As such, the confocal parameter of the signal as well as the beam overlap between the pump and signal is not as high as it could be. In future work, we hope to incorporate a variable beam expander to the signal setup to more easily modify the signal beam waist and, thus, determine the output power and efficiency as a function of beam overlap, as this has been shown to be an important optimization parameter [31].

An 808 nm 125 μm core diameter multimode fiber pigtailed diode laser (AeroDIODE 808LD-2-0-0) emitting up to 4.5 W optical power was imaged onto the Nd:YVO_4 gain medium using a GRIN lens (Edmund Optics #64-544). The Nd:YVO_4 gain medium was HR coated at 1064 nm (R > 99.9%) and high transmission (HT) coated at 808 nm (T > 95%) at the input facet, and anti-reflection (AR) coated at 1064 nm (T > 99.8%) at the output facet. This was followed by a dichroic mirror (Custom unit fabricated at the National Research Council) with a transmission of 99.3% at 1064 nm and a reflectivity of 70% at 1529 nm, both at an angle of incidence of 45°, for s-polarized light. A 50 mm-long 5 mol% MgO doped PPLN crystal (Custom from HCPhotonics, Taiwan, China) with a period of 30.3 μm was mounted to a copper heat sink with a thermoelectric cooler that can be controlled to within 0.05 °C. The input facet of the PPLN was AR-coated for 1064 nm and 1529 nm (T > 99.7%) and the output facet was AR-coated for 1064 nm (T > 99.7%) and 3.5 μm (T > 95%). The output coupler (OC) was a CaF_2 plano-concave mirror (Custom from Fuzhou Witoptics. Fuzhou, China) with a radius of curvature of 100 mm, HR-coated for 1064 nm (R = 99.85%) and AR-coated for ~3.5 μm (T = 92%). A 50 mm focal length AR-coated CaF plano-convex lens (Thorlabs LA5183-E, Newton, NJ, USA) was placed after the OC to collimate the divergent 3499 nm light. A thermal power sensor (Thorlabs S401C) was used to measure the DFG power after filtering out the pump and signal with a Germanium (Ge) filter (Thorlabs WH91050-C9). The dichroic mirror was used to couple in a tunable C-band laser (Agilent 8164B, Santa Clara, CA, USA) at 1529 nm, which was amplified via a polarization-maintained EDFA (CiviLaser EDFA-C-BA-26-PM). To increase the working distance of the C-band laser, the collimated EDFA output was passed through a 5x beam expander (Thorlabs GBE05-C) followed by a 100 mm focal length plano-convex lens (Thorlabs LA1050-C) to focus the laser into the center of the PPLN. This design was simulated in Zemax which predicted a beam waist radius of 55 μm in the center of the PPLN, ensuring that the confocal focusing criterion was met. This was required due to space limitations on the optical bench. A schematic diagram detailing the layout of the above components is presented in Figure 2.

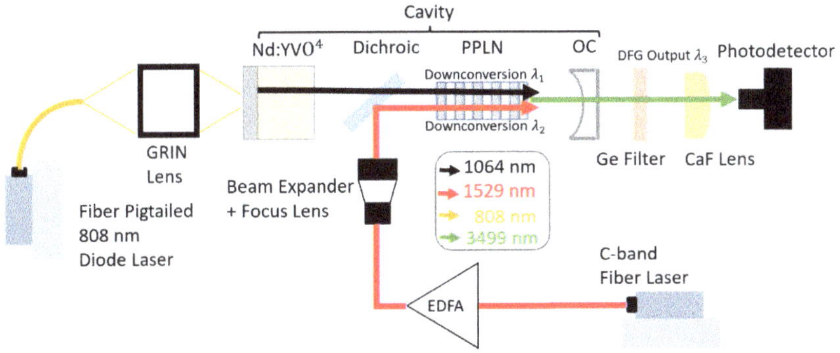

Figure 2. Experimental schematic diagram of the intracavity DFG laser. The fibers for the C-band laser and EDFA are polarization-maintaining and single mode, and the total cavity length is <8 cm.

A scanning slit beam profiler (Thorlabs BP209IR1) was used to verify the beam diameters. For the C-band laser, the beam profiler was placed at the focus after the beam expander assembly. The distance from the beam profiler and the beam waist matched a Zemax simulation well, with a waist diameter of 110 μm at 10.5 cm from the assembly output. The cavity waist was confirmed by taking several measurements of the Gaussian beam diameter outside of the cavity and then deriving the Gaussian waist using the experimental data. The measured waist diameter of 200 μm was close to the predicted value of 190 μm.

We also describe the simulation work presented in the Results section. This is a significantly modified version of work our group has presented previously, although adapted for DFG as opposed to SHG [32]. A flowchart summarizing the flow of the simulation step by step is presented in Figure 3.

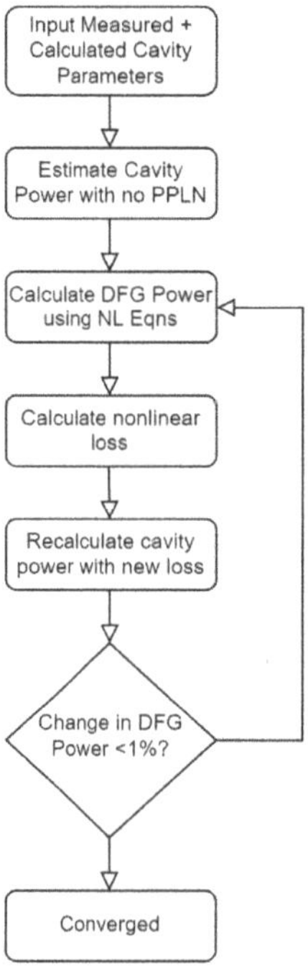

Figure 3. A flowchart summarizing the modeling work. Input parameters are experimentally measured where possible, otherwise, they are simulated or use known values from the literature.

Initially, we must estimate the incident pump and signal power on the nonlinear medium. The signal power is measured experimentally before it is coupled into the cavity, while estimating the intracavity 1064 nm pump power is more involved. A spatially dependent rate equation model was used to derive the intracavity 1064 nm power as detailed by Risk and summarized below in Equation (3) [33]:

$$F = \frac{1 + \frac{B}{fS}\ln(1 + fS)}{f\int_0^\infty \frac{\exp[-(a^2+1)x]}{1+fS\exp(-a^2x)}dx}, \quad (3)$$

where $a = \omega_P/\omega_L$ is the ratio of the 808 nm pump and 1064 nm laser beam waists, f is the total Boltzmann factor of the gain medium, $B = \frac{2N_1^0\sigma l}{L+T}$ is the ratio of reabsorption loss to fixed cavity loss with L being the total cavity loss, T is the transmission loss from the output coupler, σ is the emission cross-section of the gain medium, l is the length of the gain medium, and N_1^0 is the doping concentration of the Nd in the gain medium. F is a normalized variable proportional to pump power, and S is a normalized variable proportional to internal laser power. Then, by solving for dS/dF, we can derive the pump threshold and slope efficiency in terms of S and F and convert them to real pump power and output laser power as follows.

Once dS/dF has been found numerically, the external slope efficiency of the laser can be found using Equation (4):

$$\frac{dP_{out}}{dP_P} = \frac{T}{L+T}\frac{v_L}{v_P}\eta_a\frac{dS}{dF}, \quad (4)$$

where vL/vP is the ratio of frequencies of the lasing and pump light, also known as the quantum defect. η_a is the absorption efficiency of the incident pump photons in the gain medium, and $\frac{T}{L+T}$ is the ratio of laser photons exiting the cavity via the output coupling mirror. We can then solve for the pump threshold of the laser to fully define the expected output power vs. pump power curve, as shown in Equations (5) and (6) below:

$$P_{P,th} = \frac{\pi h v_p(\omega_L^2 + \omega_P^2)(L + T + 2\sigma N_1^0 l)}{4\sigma\tau\eta_a f} \quad (5)$$

$$F_{th} = \frac{(1+a^2)(1+B)}{f}, \quad (6)$$

where τ is the upper manifold lifetime of the gain medium and h is the Planck's constant. Finally, we convert F to P_P by $P_P = \left(\frac{P_{th}}{F_{th}}\right)F$, and we use the derived slope efficiency to determine the expected output power from the laser from the threshold onwards. Then, since the output coupler reflectivity is well known, this allows us to estimate the circulating power within the cavity as well. This determines the input power to the nonlinear medium for the next step in the modeling work.

After determining the input pump and signal power, we use the nonlinear wave equations for DFG to estimate the output idler power as shown in Equations (7) and (8) [34]:

$$\frac{dA_1}{dz} = \frac{2id_{eff}\omega_1^2}{k_1c^2}A_3A_2^*e^{i\Delta kz} \quad (7)$$

$$\frac{dA_2}{dz} = \frac{2id_{eff}\omega_2^2}{k_2c^2}A_3A_1^*e^{i\Delta kz} \quad (8)$$

where A_1 represents the signal wave, A_2 represents the idler wave, and A_3 represents the pump wave. In this form, we have assumed that the cavity pump power is strong and effectively constant, as the conversion efficiency of the DFG process is low enough that nonlinear losses from the conversion are negligible in the device's current form (<1% difference). However, it is important to note that as the signal power is increased, the overall conversion of the pump to signal will increase, and so the pump depletion will be non-negligible. Our model can account for this, which is why we still include the nonlinear loss step in Figure 3. This increased conversion efficiency acts as an additional source of loss within the cavity, so it must be accounted for to accurately model intracavity frequency conversion at high powers. If depletion is significant, we would then take the output DFG power and intracavity 1064 power and estimate a nonlinear conversion loss. This loss is added to the total cavity loss, and then, the 1064 nm cavity power is calculated again. This new lower 1064 nm power will produce a lower estimate of the DFG and, thus, a lower nonlinear conversion loss. This process is repeated until the nonlinear loss converges on the correct value. A similar approach that considers the depletion alongside intracavity DFG is presented by Liu et al. [35].

We swept the operating temperature of the PPLN using the attached TEC until the optimal output power (and thus phase matching) was achieved at ~23 °C, and the temperature dependence of the PPLN is modeled using the Sellmeier equation for MgO-doped PPLN [36]. By using this model, we can accurately predict the output DFG power as a function of input pump power, as well as the temperature dependence of the DFG power via the Sellmeier equation. These results are presented in the Results section alongside our experimental results, and the two are in good agreement. Future simulation expansion is covered in the Discussion section. Finally, we present our simulation inputs used to generate our results in Table 1.

Table 1. Input parameters to the model. Any value measured experimentally or simulated by the authors is designated by "This work". Cited values are from literature.

Parameter	Value	Ref.
Nonlinear coefficient (d_{33})	27 pm/V	[34]
Nd:YVO_4 length (l)	8 mm	This work
Output coupler transmission (T) at 1064 nm	0.15%	This work
Round-trip losses (L)	3%	This work
Absorption efficiency η_a	99%	This work
Boltzmann factor (f)	0.6	[37]
Upper manifold lifetime (τ)	90 µs	[38]
Lasing beam radius (ω_L)	190 µm	This work
Pump beam radius (ω_P)	250 µm	This work
Pump beam diameter in PPLN	100 µm	This work
Signal beam diameter in PPLN	55 µm	This work
Nd doping	0.5 at. %	This work
PPLN refractive index	~2.2	[36]
Stimulated emission cross-section (σ)	25×10^{-19} cm^2	[39]

3. Results

3.1. Output Power Testing

Figure 4 shows the leaking 1064 nm power from the laser cavity as a function of the 808 nm optical pump power. The squares represent the experimental data while the solid line represents the predicted leaking power from the simulation. The leaking 1064 nm power was measured with all components in the resonator and the PPLN temperature tuned to minimize nonlinear losses. The simulation used the measured reflectivity and transmission values for intracavity components to generate a total intracavity loss. The measured and simulated power values are in good agreement, implying that the estimated loss and 1064 nm input power for the DFG simulation are accurate.

After collecting the 1064 nm data, the C-band laser was coupled into the PPLN, and DFG power as high as 31 mW at 3499 nm was obtained at a pump power of 4.5 W 808 nm and a C-band power of 310 mW (measured at the output facet of the C-band collimator). This corresponds to a black-box efficiency of 2.20%W^{-1}. The simulated DFG power is compared to the measured experimental results, as shown in Figure 5. The x-axis is the 808 nm optical pump power, and the y-axis is the DFG output power after the Ge filter. The squares represent experimental data while the solid line is the simulated DFG power. The simulation assumes that both input beams are focused at the center of the PPLN, and the beam diameters and power values are those measured earlier. The trend is in good agreement with some outliers in the 2–3 W pump range. This discrepancy could be attributed to the non-ideal overlap of the 1064 nm and C-band beams, as the hemispherical cavity prevents strong focusing of the 1064 nm beam at the center of the PPLN, and the thermal lens power will change with pump power as well. Nonetheless, the simulated trend can still be used to predict the order of magnitude of experimental DFG performance to a reasonable level of accuracy. Additionally, we measured the power stability over one hour. We observed a stability of ~2.2% or ~0.66 mW with an average power of 30 mW over this period, although we aim to perform a longer analysis in the future.

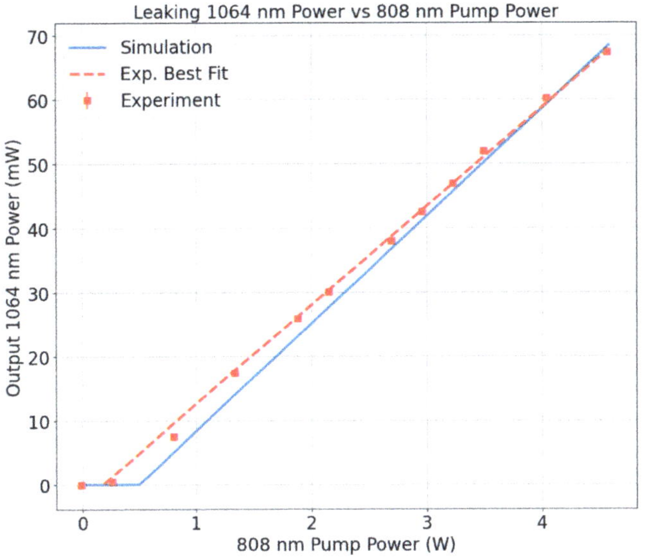

Figure 4. Measured and simulated leaking 1064 nm power vs. 808 nm pump power with all devices in the resonator to account for linear losses. Error bars are present but smaller than the data markers. The solid blue line is the simulation while the dashed red line is the best linear fit for the experimental data.

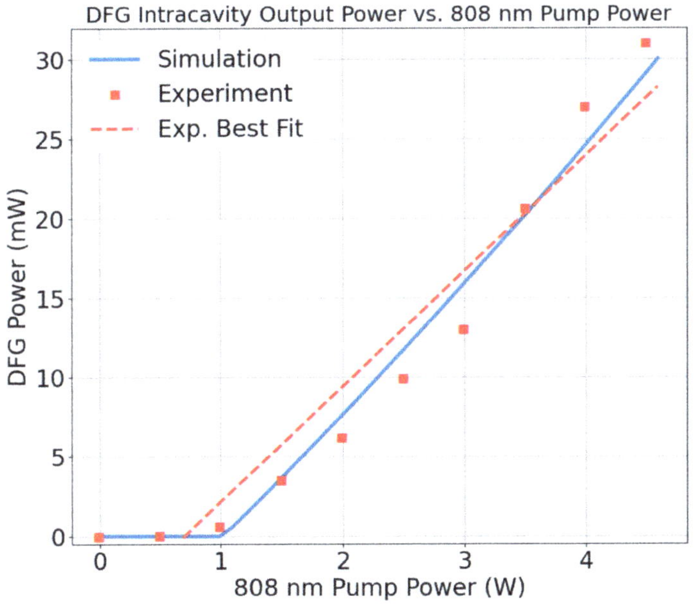

Figure 5. Measured and simulated output DFG power vs. 808 nm pump power with all losses accounted for. The simulation data are represented by the solid blue line, and the dotted red line is the best linear fit for the experimental data.

3.2. Temperature Testing

To study the temperature tolerance of the DFG laser, the temperature tuning curve of the DFG power was measured. We then simulated the temperature dependence of DFG, using the Sellmeier equation for the MgO-doped lithium niobate [36]. The results are presented in Figure 6. The x-axis is the temperature of the PPLN, and the y-axis is the normalized DFG power. The squares are experimental data, while the solid line is the simulated curve. The good agreement of the peak power location and full width at half maximum (FWHM) confirms the presence of DFG output. The tail on the left side can be attributed to a combination of the tightly focused Gaussian signal beam, elliptical focusing, and tilt of the input beams with respect to the optical axis [40]. This was obtained using a standard off-the-shelf thermoelectric cooler and a machined in-house copper heat sink that also served as the housing for the nonlinear crystal. The FWHM of the main peak is ~6 °C, demonstrating that the laser can easily be operated at a stable temperature with standard components. If desired, this temperature bandwidth could be increased via the chirping of the PPLN at the cost of nonlinear conversion efficiency.

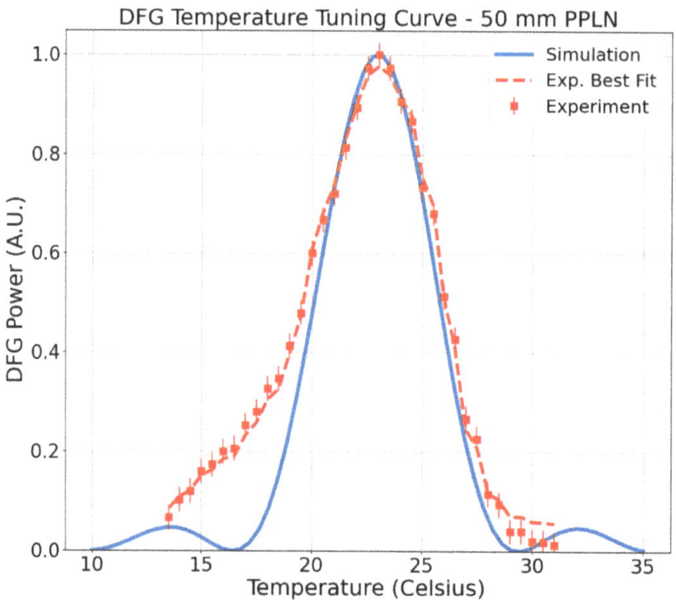

Figure 6. Theoretical temperature tuning curve vs. measured experimental data. The solid blue line is the simulation while the dashed red line is a sinc^2 function fit via a nonlinear solver.

4. Discussion

The results are promising, showing that we can easily achieve state-of-the-art black-box efficiency with a compact, linear cavity without stabilization. According to our modeling, this represents close to the best-case black-box efficiency we can reasonably expect for this design. While it is theoretically possible to improve our efficiency to up to ~2.40–2.50%/W, this was only achieved by further improving the optical coatings, which are already close to the upper limit of performance. A larger efficiency gain may be possible if the pump and signal beam overlap can be improved at the center of the PPLN, as there is a noticeable mismatch with the ~100 um pump derived from the cavity simulation. However, this may require a significant cavity redesign and a change in the focusing setup used for the C-band laser. We will have to investigate in future work whether decreasing the pump beam size is possible, or if increasing the signal beam size is worth the loss in intensity versus the increase in overlap between the pump and signal.

As for why our device efficiency is higher than the reported efficiencies presented in the Introduction section, our preliminary analysis suggests the following. First, all of the presented devices in the introduction make use of a fiber amplification stage before pumping the nonlinear crystal. Since we are not making use of an amplification stage and instead rely on high-efficiency direct-diode pumping, our efficiency is ~60% higher than it would be when using a Yb-based amplifier. Second, the cavity enhancement factor is a good deal higher than the other devices presented, which we attribute to the high-quality optical coatings and high OC reflectivity. For example, the device presented by Witinski et al. has a Finesse of 65, corresponding to a pump enhancement factor of around 10 [19]. Our cavity has a Finesse of ~200, and so our pump enhancement factor is ~30 [41]. The main reason we suspect our device is not 3× more efficient than Witinski et al.'s is that the beam overlap is not as optimized as in their device, and so we have room for improvement. Finally, we are using a large 50 mm PPLN which greatly increases the nonlinear interaction length and thus, the efficiency. However, in future work, we plan to better characterize what parameters affect the efficiency most strongly, including beam overlap, temperature, spectral linewidth, PPLN size, etc.

While our model agrees well with the current experimental results, there are several areas where it could be improved to facilitate future work. First, the model does not consider thermal effects outside of the thermal lensing approximated in the cavity simulation. A future test would involve increasing both the pump's 808 nm power and the signal C-band power to see if higher output power and black-box efficiency can be achieved. However, this will increase the prevalence of thermal lensing in the Nd:YVO4 such that it may affect cavity stability. Additionally, the higher thermal load of a multi-watt level signal coupled with the tens of watts of pump present in the cavity may change the beam properties inside of the nonlinear crystal. Another important factor is the increased nonlinear losses within the cavity at higher efficiencies and pump powers. Currently, the DFG conversion efficiency is low enough that the nonlinear losses are not significant relative to the linear losses within the cavity (scattering, absorption, non-ideal coatings, etc.). Any attempts to increase efficiency will also increase the prevalence of intracavity nonlinear loss. We have previously demonstrated that in a simple nonlinear process like SHG, the nonlinear loss can significantly affect the expected output power and thermal properties of intracavity frequency conversion at higher efficiencies [32]. We have no reason to believe there will not be similar phenomena to observe at higher device powers and higher efficiencies with intracavity DFG as well. These are important points to consider and investigate in future work, as the efficiency we have reported may not be maintained at higher pump and signal powers as a result.

Future work will focus on adapting the device presented here for free-space optical communications. Experiments will include using a 10 W EDFA and 10 W 808 nm pump to push for watt-level DFG power. This is a vital milestone for satellite-to-ground optical communication in low earth orbit, as previous work has identified that 1–2 W CW power is recommended for high link availability and low bit error rates [42]. We will also pursue modulating the C-band source at gigabit speeds, as this opens the door to high-speed weather-resistant free-space optical links for both last mile applications and longer distance links, due to the combination of high-speed and high transmitter power. To the best of our knowledge, a gigabit-speed mid-infrared link has not been demonstrated over distances greater than a few kilometers, so we hope to perform a practical demonstration using drones and high-altitude platforms in the coming months as further proof of principle.

5. Conclusions

In conclusion, a compact intracavity DFG MWIR laser with promise for free-space communication has been demonstrated. A black-box efficiency as high as 2.20%W^{-1} has been achieved with a maximum output power of 31 mW for the input pump and signal powers of 4.5 W and 310 mW, respectively. The cavity is under 8 cm in length and could be made more compact with professional packaging. The linear cavity requires no active stabilization, which is also a notable benefit. In future work, we expect that this laser will be capable of watt-level output power and provide what may be the first opportunity for a long-distance free-space link in the mid-infrared between either a drone or high-altitude platform.

6. Patents

This work is a developed version of the transmitter device laid out in the patent US20210141282A1—Mid-infrared wavelength optical transmitter and receiver.

Author Contributions: Conceptualization, L.G.K.F. and C.-Q.X.; data curation, L.G.K.F., A.A. and T.K.; formal analysis, L.G.K.F. and A.A.; funding acquisition, L.G.K.F. and C.-Q.X.; investigation, L.G.K.F., A.A. and T.K.; methodology, L.G.K.F., A.A., T.K., D.P. and C.-Q.X.; project administration, D.P. and C.-Q.X.; resources, D.P. and C.-Q.X.; software, L.G.K.F.; supervision, C.-Q.X.; validation, L.G.K.F., A.A., T.K. and D.P.; visualization, L.G.K.F.; writing—original draft, L.G.K.F.; writing—review and editing, L.G.K.F., A.A., T.K., D.P. and C.-Q.X. All authors have read and agreed to the published version of the manuscript.

Funding: This research was funded by the Natural Sciences and Engineering Research Council of Canada (grant number RGPIN-2018-05501) and the High Throughput and Secure Networks Challenge Program at the National Research Council of Canada (grant numbers HTSN-621, HTSN-644).

Institutional Review Board Statement: Not applicable.

Informed Consent Statement: Not applicable.

Data Availability Statement: Data underlying the results presented in this paper may be obtained from the authors upon reasonable request.

Acknowledgments: The authors would like to thank Saju Thomas of MDA (Kanata) and Atieh of Optiwave Systems Inc. for their helpful discussions.

Conflicts of Interest: The authors declare no conflict of interest. The funders had no role in the design of the study; in the collection, analyses, or interpretation of data; in the writing of the manuscript; or in the decision to publish the results.

References

1. Trichili, A.; Cox, M.A.; Ooi, B.S.; Alouini, M.-S. Roadmap to Free Space Optics. *J. Opt. Soc. Am. B* **2020**, *37*, A184–A201. [CrossRef]
2. Abadi, M.M.; Ghassemlooy, Z.; Mohan, N.; Zvanovec, S.; Bhatnagar, M.R.; Hudson, R. Implementation and Evaluation of a Gigabit Ethernet FSO Link for "The Last Metre and Last Mile Access Network". In Proceedings of the 2019 IEEE International Conference on Communications Workshops (ICC Workshops), Shanghai, China, 20–24 May 2019; pp. 1–6.
3. Carrasco-Casado, A.; Do, P.X.; Kolev, D.; Hosonuma, T.; Shiratama, K.; Kunimori, H.; Trinh, P.V.; Abe, Y.; Nakasuka, S.; Toyoshima, M. Intersatellite-Link Demonstration Mission between CubeSOTA (LEO CubeSat) and ETS9-HICALI (GEO Satellite). In Proceedings of the 2019 IEEE International Conference on Space Optical Systems and Applications (ICSOS), Portland, OR, USA, 14–16 October 2019; pp. 1–5.
4. Iwamoto, K.; Komatsu, H.; Ohta, S.; Kubo, Y.; Nakao, T.; Yamazoe, H.; Kamata, T.; Munemasa, Y.; Kunimori, H.; Toyoshima, M.; et al. Experimental Results on In-Orbit Technology Demonstration of SOLISS. In Proceedings of the Free-Space Laser Communications XXXIII, Online, 6–12 March 2021; Volume 11678, pp. 51–57.
5. Leitgeb, E.; Plank, T.; Awan, M.S.; Brandl, P.; Popoola, W.; Ghassemlooy, Z.; Ozek, F.; Wittig, M. Analysis and Evaluation of Optimum Wavelengths for Free-Space Optical Transceivers. In Proceedings of the 2010 12th International Conference on Transparent Optical Networks, Munich, Germany, 27 June–1 July 2010; pp. 1–7.
6. Khalighi, M.A.; Uysal, M. Survey on Free Space Optical Communication: A Communication Theory Perspective. *IEEE Commun. Surv. Tutor.* **2014**, *16*, 2231–2258. [CrossRef]
7. Flannigan, L.; Yoell, L.; Xu, C. Mid-Wave and Long-Wave Infrared Transmitters and Detectors for Optical Satellite Communications—A Review. *J. Opt.* **2022**, *24*, 043002. [CrossRef]
8. Prasad, N.S. Optical Communications in the Mid-Wave IR Spectral Band. In *Free-Space Laser Communications: Principles and Advances*; Majumdar, A.K., Ricklin, J.C., Eds.; Optical and Fiber Communications Reports; Springer: New York, NY, USA, 2008; pp. 347–391. ISBN 978-0-387-28677-8.
9. Grillot, F.; Spitz, O.; Didier, P.; Dely, H.; Bonazzi, T.; Awwad, E.; Vasanelli, A.; Sirtori, C. Recent Advances in High-Speed Data Communications Using Mid-Infrared Quantum Cascade Lasers. In Proceedings of the Novel In-Plane Semiconductor Lasers XXII, San Francisco, CA, USA, 28 January–3 February 2023; Volume 12440, pp. 97–102.

10. Hinkov, B.; Hugi, A.; Beck, M.; Faist, J. Rf-Modulation of Mid-Infrared Distributed Feedback Quantum Cascade Lasers. *Opt. Express* **2016**, *24*, 3294–3312. [CrossRef]
11. Calvar, A.; Amanti, M.I.; Renaudat St-Jean, M.; Barbieri, S.; Bismuto, A.; Gini, E.; Beck, M.; Faist, J.; Sirtori, C. High Frequency Modulation of Mid-Infrared Quantum Cascade Lasers Embedded into Microstrip Line. *Appl. Phys. Lett.* **2013**, *102*, 181114. [CrossRef]
12. Xi, C.; Wang, P.; Li, X.; Liu, Z. Highly Efficient Continuous-Wave Mid-Infrared Generation Based on Intracavity Difference Frequency Mixing. *High Power Laser Sci. Eng.* **2019**, *7*, e67. [CrossRef]
13. Unger, S.; Schwuchow, A.; Jetschke, S.; Reichel, V.; Scheffel, A.; Kirchhof, J. Optical Properties of Yb-Doped Laser Fibers in Dependence on Codopants and Preparation Conditions. In Proceedings of the Optical Components and Materials V, San Jose, CA, USA, 19–24 January 2008; Volume 6890, pp. 270–280.
14. Guha, S.; Barnes, J.O.; Gonzalez, L.P. Multiwatt-Level Continuous-Wave Midwave Infrared Generation Using Difference Frequency Mixing in Periodically Poled MgO-Doped Lithium Niobate. *Opt. Lett.* **2014**, *39*, 5018–5021. [CrossRef]
15. Huang, K.; Gan, J.; Zeng, J.; Hao, Q.; Yang, K.; Yan, M.; Zeng, H. Observation of Spectral Mode Splitting in a Pump-Enhanced Ring Cavity for Mid-Infrared Generation. *Opt. Express* **2019**, *27*, 11766–11775. [CrossRef]
16. Hao, Q.; Zhu, G.; Yang, S.; Yang, K.; Duan, T.; Xie, X.; Huang, K.; Zeng, H. Mid-Infrared Transmitter and Receiver Modules for Free-Space Optical Communication. *Appl. Opt.* **2017**, *56*, 2260–2264. [CrossRef]
17. Su, Y.; Wang, W.; Hu, X.; Hu, H.; Huang, X.; Wang, Y.; Si, J.; Xie, X.; Han, B.; Feng, H.; et al. 10 Gbps DPSK Transmission over Free-Space Link in the Mid-Infrared. *Opt. Express* **2018**, *26*, 34515–34528. [CrossRef]
18. Gray, A.C.; Berry, S.A.; Carpenter, L.G.; Gates, J.C.; Gawith, C.B.E.; Smith, P.G.R. Upconversion Detection of 1.25 Gb/s Mid-Infrared Telecommunications Using a Silicon Avalanche Photodiode. *Opt. Express* **2020**, *28*, 34279–34289. [CrossRef] [PubMed]
19. Witinski, M.F.; Paul, J.B.; Anderson, J.G. Pump-Enhanced Difference-Frequency Generation at 3.3 μm. *Appl. Opt.* **2009**, *48*, 2600–2606. [CrossRef] [PubMed]
20. Wang, K.; Li, X.; Wang, P.; Hua, W.; Wang, Z.; Han, K. Broadband, Continuous-Wave, Mid-Infrared Generation Based on ASE Fiber Source. *Photonics* **2022**, *9*, 724. [CrossRef]
21. Mishra, J.; McKenna, T.P.; Ng, E.; Stokowski, H.S.; Jankowski, M.; Langrock, C.; Heydari, D.; Mabuchi, H.; Fejer, M.M.; Safavi-Naeini, A.H. Mid-Infrared Nonlinear Optics in Thin-Film Lithium Niobate on Sapphire. *Optica* **2021**, *8*, 921–924. [CrossRef]
22. Hu, C.; Pan, A.; Li, T.; Wang, X.; Liu, Y.; Tao, S.; Zeng, C.; Xia, J. High-Efficient Coupler for Thin-Film Lithium Niobate Waveguide Devices. *Opt. Express* **2021**, *29*, 5397–5406. [CrossRef] [PubMed]
23. Ledezma, L.; Roy, A.; Costa, L.; Sekine, R.; Gray, R.; Guo, Q.; Nehra, R.; Briggs, R.M.; Marandi, A. Octave-Spanning Tunable Infrared Parametric Oscillators in Nanophotonics. *Sci. Adv.* **2023**, *9*, eadf9711. [CrossRef] [PubMed]
24. Hwang, A.Y.; Stokowski, H.S.; Park, T.; Jankowski, M.; McKenna, T.P.; Langrock, C.; Mishra, J.; Ansari, V.; Fejer, M.M.; Safavi-Naeini, A.H. Mid-Infrared Spectroscopy with a Broadly Tunable Thin-Film Lithium Niobate Optical Parametric Oscillator. *arXiv* **2023**, arXiv:2307.04199.
25. Santandrea, M.; Stefszky, M.; Ansari, V.; Silberhorn, C. Fabrication Limits of Waveguides in Nonlinear Crystals and Their Impact on Quantum Optics Applications. *New J. Phys.* **2019**, *21*, 033038. [CrossRef]
26. Chunosov, N.I. reZonator. Available online: http://rezonator.orion-project.org/ (accessed on 6 June 2023).
27. Detailed Study of Thermal Lensing in Nd:YVO/Sub 4/under Intense Diode End-Pumping. Available online: https://ieeexplore.ieee.org/document/947669/ (accessed on 25 September 2023).
28. Boyd, G.D.; Kleinman, D.A. Parametric Interaction of Focused Gaussian Light Beams. *J. Appl. Phys.* **2003**, *39*, 3597–3639. [CrossRef]
29. Guha, S.; Falk, J. The Effects of Focusing in the Three-frequency Parametric Upconverter. *J. Appl. Phys.* **2008**, *51*, 50–60. [CrossRef]
30. Guha, S.; Wu, F.-J.; Falk, J. The Effects of Focusing on Parametric Oscillation. *IEEE J. Quantum Electron.* **1982**, *18*, 907–912. [CrossRef]
31. Zhang, G.-J.; Gong, M.-L.; Zhang, W.-Q. Numerical Analysis of a Dual-Pass Pumping Laser with Weak Absorption. *Chin. Phys. B* **2017**, *26*, 050203. [CrossRef]
32. Flannigan, L.; Kashak, T.; Xu, C.-Q. Study of Fundamental Wave Depletion in Intracavity Second Harmonic Generation. *Opt. Express* **2021**, *29*, 6810–6823. [CrossRef]
33. Risk, W.P. Modeling of Longitudinally Pumped Solid-State Lasers Exhibiting Reabsorption Losses. *J. Opt. Soc. Am. B* **1988**, *5*, 1412–1423. [CrossRef]
34. Boyd, R. *Nonlinear Optics*, 4th ed.; Academic Press: Cambridge, MA, USA, 2020.
35. Self-Raman Nd-Doped Vanadate Laser: A Pump Source of Organic Crystal Based Difference Frequency Generation. *J. Mod. Opt.* **2020**, *67*, 914–919. [CrossRef]
36. Gayer, O.; Sacks, Z.; Galun, E.; Arie, A. Temperature and Wavelength Dependent Refractive Index Equations for MgO-Doped Congruent and Stoichiometric $LiNbO_3$. *Appl. Phys. B* **2008**, *91*, 343–348. [CrossRef]
37. Bai, Y. Pumping Wavelength Related Population Inversion in Nd:Doped Laser. *AIP Adv.* **2020**, *10*, 105309. [CrossRef]
38. Fornasiero, L.; Kück, S.; Jensen, T.; Huber, G.; Chai, B.H.T. Excited State Absorption and Stimulated Emission of Nd_{3+} in Crystals. Part 2: YVO_4, $GdVO_4$, and $Sr_5(PO_4)_3F$. *Appl. Phys. B* **1998**, *67*, 549–553. [CrossRef]
39. Chen, Y.-F.; Lan, Y.P. Comparison between C-Cut and a-Cut Nd:YVO4 Lasers Passively Q-Switched with a Cr^{4+}:YAG Saturable Absorber. *Appl. Phys. B* **2002**, *74*, 415–418. [CrossRef]

40. Tukker, T.W.; Otto, C.; Greve, J. Elliptical-Focusing Effect on Parametric Oscillation and Downconversion. *J. Opt. Soc. Am. B* **1998**, *15*, 2455–2461. [CrossRef]
41. Suter, M.; Dietiker, P. Calculation of the Finesse of an Ideal Fabry–Perot Resonator. *Appl. Opt.* **2014**, *53*, 7004–7010. [CrossRef] [PubMed]
42. Gagnon, S.; Sylvestre, B.; Gagnon, L.; Koujelev, A.; Gratton, D.; Hranilovic, S. Recent Developments in Satellite Laser Communications: Canadian Context. In Proceedings of the International Conference on Space Optical Systems and Applications (ICSOS), Corsica, France, 9–10 October 2012; pp. 1–8.

photonics

MDPI

Article

Processing Accuracy of Microcomb-Based Microwave Photonic Signal Processors for Different Input Signal Waveforms

Yang Li [1], Yang Sun [1], Jiayang Wu [1,*], Guanghui Ren [2], Bill Corcoran [3], Xingyuan Xu [4], Sai T. Chu [5], Brent. E. Little [6], Roberto Morandotti [7], Arnan Mitchell [2] and David J. Moss [1,*]

1 Optical Sciences Center, Swinburne University of Technology, Hawthorn, VIC 3122, Australia; yangli@swin.edu.au (Y.L.); yangsun@swin.edu.au (Y.S.)
2 Integrated Photonics and Applications Centre, School of Engineering, RMIT University, Melbourne, VIC 3000, Australia; guanghui.ren@rmit.edu.au (G.R.); arnan.mitchell@rmit.edu.au (A.M.)
3 Photonic Communications Laboratory, Department Electrical and Computer Systems Engineering, Monash University, Clayton, VIC 3800, Australia; bill.corcoran@monash.edu
4 State Key Laboratory of Information Photonics and Optical Communications, Beijing University of Posts and Telecommunications, Beijing 100876, China; xingyuanxu@bupt.edu.cn
5 Department of Physics, City University of Hong Kong, Hong Kong; saitchu@cityu.edu.hk
6 The State Key Laboratory of Transient Optics and Photonics, Xi'an Institute of Optics and Precision Mechanics, Chinese Academy of Sciences, Xi'an 710119, China; brent.little@qxptech.com
7 Institut National de la Recherche Scientifique—Énergie, Matériaux et Télécommunications, Varennes, QC J3X 1S2, Canada; roberto.morandotti@inrs.ca
* Correspondence: jiayangwu@swin.edu.au (J.W.); dmoss@swin.edu.au (D.J.M.)

Abstract: Microwave photonic (MWP) signal processors, which process microwave signals based on photonic technologies, bring advantages intrinsic to photonics such as low loss, large processing bandwidth, and strong immunity to electromagnetic interference. Optical microcombs can offer a large number of wavelength channels and compact device footprints, which make them powerful multi-wavelength sources for MWP signal processors to realize a variety of processing functions. In this paper, we experimentally demonstrate the capability of microcomb-based MWP signal processors to handle diverse input signal waveforms. In addition, we quantify the processing accuracy for different input signal waveforms, including Gaussian, triangle, parabolic, super Gaussian, and nearly square waveforms. Finally, we analyse the factors contributing to the difference in the processing accuracy among the different input waveforms, and our theoretical analysis well elucidates the experimental results. These results provide guidance for microcomb-based MWP signal processors when processing microwave signals of various waveforms.

Keywords: optical microcombs; microwave photonic; signal processing

1. Introduction

Microwave signal processors have found wide applications in telecommunication and radar systems [1–4]. Traditional microwave signal processors relying on electronic devices exhibit significant loss and strong crosstalk when handling high-frequency microwave signals, which make them suffer from limited operation bandwidths. To overcome this restriction, microwave photonic (MWP) signal processors that perform signal processing functions based on MWP technologies have attracted great interest [3–6].

A variety of MWP signal processors have been demonstrated by exploiting different optical filtering modules to process microwave signals that are modulated onto a single optical carrier [7–17]. Although these approaches perform well regarding achieving specific processing functions, they face limitations in their reconfigurability to realize diverse processing functions based on a single system. On the contrary, in MWP signal processors implemented based on the transversal filter structure [18], input microwave signals are modulated onto multiple optical carriers with adjustable time delays and tap weights before summing via photodetection. This enables high reconfigurability to achieve various processing functions without changing any hardware [2,18].

For MWP signal processors implemented by the transversal filter systems, a large number of taps, or the wavelength channels provided by multi-wavelength optical sources, are required to improve their performance. Compared to other multi-wavelength optical sources, such as discrete laser arrays [19–21], fibre Bragg grating arrays [22–24], laser frequency combs generated by electro-optic (EO) modulation [25–27], and mode-locked fibre lasers [28,29], optical microcombs can provide significantly increased numbers of wavelength channels by using compact micro-scale resonators [30,31], which is critical for improving the processing accuracy of MWP transversal signal processors. They also have the ability to offer broad Nyquist zones, which allow for large processing bandwidths [4,32,33]. With these advantages, a variety of signal processing functions have been successfully demonstrated using microcomb-based MWP signal processors, such as differentiation [34], integration [35], Hilbert transform [36], arbitrary waveform generation [37], and convolutional processing [38,39]. In addition to applications in microwave photonics, optical microcombs have also found wide applications in communications [40–42], logic operation [43], and sensing [44,45].

Although a range of signal processing functions have been realized, they only used Gaussian input waveforms for demonstrations, while the ability to handle various input signal waveforms is essential for practical applications. In this paper, we experimentally demonstrate the capability of microcomb-based MWP signal processors for dealing with various input signal waveforms. We investigate the processing accuracy of different input waveforms, including Gaussian, triangle, parabolic, super Gaussian, and nearly square waveforms. We also perform theoretical analysis and discuss the reasons for the difference in the processing accuracy among the different input waveforms. These results offer a valuable guide for microcomb-based MWP signal processors to handle microwave signals with different waveforms.

2. Microcomb-Based MWP Signal Processors

MWP signal processors based on the transversal filter are implemented based on MWP technologies, which can overcome the electrical bandwidth bottleneck by providing a substantially increased processing bandwidth. An MWP transversal signal processor has high reconfigurability in terms of its spectral transfer function, which can be expressed as [4]

$$H(\omega) = \sum_{n=0}^{M-1} a_n e^{-j\omega n \Delta T}, \tag{1}$$

where ω is the angular frequency of the input microwave signal to be processed, M is the tap number, a_n ($n = 0, 1, 2, \ldots, M-1$) is the tap coefficient of the nth tap, and ΔT is the time delay between adjacent taps. By properly designing the various tap coefficients a_n ($n = 0, 1, 2, \ldots, M-1$), different signal processing functions can be realized by using a single system without changing the hardware.

Optical microcombs enable MWP transversal systems to have small size, weight, and power consumption (SWaP) [30,31,46,47], and excellent compatibility with monolithic integration, making them powerful alternatives to conventional multi-wavelength optical sources. Figure 1 shows the schematic of a microcomb-based MWP signal processor. An optical microcomb is used to generate multiple wavelength channels that act as discrete taps for the transversal signal processor. The generated optical microcomb is spectrally shaped according to the designed tap coefficients a_n ($n = 0, 1, 2, \ldots, M-1$). Next, all of the wavelength channels of the shaped optical microcomb are imprinted with the input microwave signal via an electro-optic modulator (EOM), leading to the generation of multiple microwave replicas. Following this, the modulated optical signals transmit through a dispersive medium to introduce time delays ΔT, which progressively separate the microwave replicas. Finally, the delayed replicas are summed upon photodetection via a photodetector.

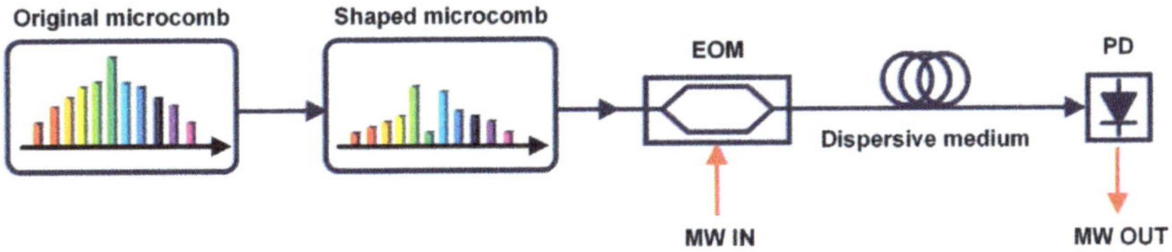

Figure 1. Schematic diagram of a microcomb-based microwave photonic (MWP) signal processor. EOM: electro-optic modulator. MW: microwave. PD: photodetector.

In principle, the microcomb-based MWP signal processor in Figure 1 can work with different types of optical microcombs, such as single soliton [48–50], dark soliton [51,52], and soliton crystal [53,54]. Before spectral shaping is used to achieve the desired tap coefficients, the initial comb with non-uniform power distributions in the comb lines can be amplified and pre-shaped to obtain a uniform power distribution. Therefore, the initial spectral shape of optical microcomb is not an essential factor for the realization of MWP transversal signal processors. It is also worth mentioning that as the optical microcomb primarily functions as a multi-wavelength source for the transversal signal processor, it can operate under relatively incoherent conditions. However, in many practical applications, high-coherence optical combs are still preferred. Optical microcombs operating under different states, including different types of solitons, as well as incoherent and especially chaotic states [49], can introduce additional noise and instability, thus leading to a decrease in the processing accuracy [30,46,47].

The operation bandwidth of the microcomb-based MWP transversal signal processor in Figure 1 is determined by both the comb spacing and time delay between adjacent wavelength channels. Generally, it remains constant unless there are changes to the hardware within the system, such as using an MRR with a different free spectral range (FSR) or employing a dispersion medium with a different length or dispersion parameter. According to the Nyquist sampling theorem, the sampling rate of a continuous-time bandwidth-limited signal needs to exceed twice its maximum frequency component to avoid aliasing. This constraint sets an upper threshold for the bandwidth of the input microwave signal to be processed, which should not surpass half of the microcomb's comb spacing. On the other hand, the bandwidth of the input microwave signal should not exceed half of the FSR of the RF spectral response (FSR_{RF}), i.e., $1/2 \cdot FSR_{RF} = 1/(2\Delta T)$. This sets another limitation for the maximum bandwidth of the input microwave signal. In our work, we focus on the influence of different temporal waveforms of input microwave signals on the processing accuracy of the signal processors. Detailed discussion on the influence of the input signal bandwidth on the processing accuracy can be found in Ref. [30].

3. Experimental Results

In our experimental demonstration, we implemented the microcomb-based MWP signal processors based on the setup shown in Figure 2, which consisted of a microcomb generation module and a transversal signal processing module. In the microcomb generation module, the optical microcomb was generated by a microring resonator (MRR) made from high-index doped silica glass [55]. The high-index doped silica glass offers attractive material properties for microcomb generation, including ultra-low linear loss (~0.06 dB/cm), a moderate nonlinear parameter (~233 $W^{-1} \cdot km^{-1}$), and a negligible nonlinear loss even at extremely high intensities (~25 $GW \cdot cm^{-2}$) [55]. The MRR had a quality factor of ~1.5×10^6. A continuous-wave (CW) light was amplified to ~32.1 dBm by an erbium-doped fibre amplifier (EDFA) and used to pump the MRR. The polarization of the CW pump was adjusted to TE polarization, which aligned with the TE-polarized resonance of the MRR at ~1551.23 nm. When the pump power of the CW laser was sufficiently high and its wavelength was swept across the MRR's resonance at ~1551.23 nm, optical parametric oscillation occurred, resulting in the generation of a palm-like soliton crystal microcomb [56], as shown in Figure 3a. The MRR was designed to have a radius of ~592 μm, which corresponded to a comb spacing of ~0.4 nm or ~49 GHz. In our experimental demonstration, 20 comb lines were employed as discrete taps. The initially generated microcomb exhibited non-uniform power distributions among the comb lines, and so it was shaped by the first waveshaper (WS1, Finisar) to flatten the comb lines. This was carried out to achieve a higher signal-to-noise ratio and reduce the required loss control range for the second waveshaper in the transversal signal processing module, which further shaped the comb lines according to the designed tap coefficients.

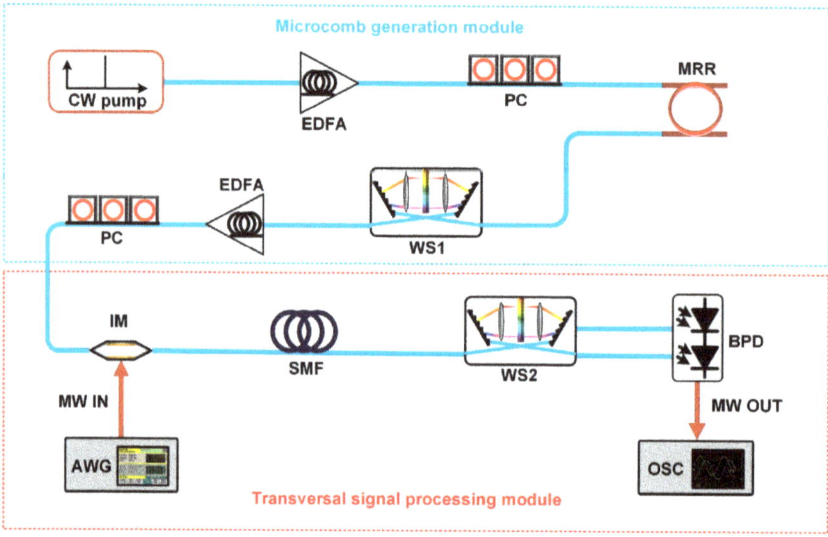

Figure 2. Experimental schematic of a microcomb-based MWP signal processor. CW pump: continuous-wave pump. EDFA: erbium-doped fibre amplifier. PC: polarization controller. MRR: microring resonator. WS: wave shaper. IM: intensity modulator. MW: microwave. BPD: balanced photodetector. SMF: single-mode fibre. AWG: arbitrary waveform generator. OSC: oscilloscope.

In the transversal signal processing module, the shaped microcomb was modulated by the input microwave signal via an intensity modulator (IM) (iXblue). The input microwave signal was multicast onto different wavelength channels, resulting in the generation of multiple microwave replicas. Next, the microwave replicas were transmitted through a spool of single mode fibre (SMF), which served as the dispersive medium that introduced a time delay between adjacent wavelength channels, i.e., ΔT in Equation (1). The time delay ΔT can be further expressed as [4]

$$\Delta T = L \times D_2 \times \Delta\lambda \tag{2}$$

where L is the fibre length, D_2 is the second-order dispersion parameter, and $\Delta\lambda$ is the comb spacing. In our experiments, these parameters were L = ~5.124 km, D_2 = ~17.4 ps/nm/km, and $\Delta\lambda$ = ~0.4 nm, which resulted in a time delay ΔT = ~35.7 ps.

After passing the dispersive medium, the comb lines were spectrally shaped by the second waveshaper (WS2, Finisar) according to the designed tap coefficients a_n (n = 0, 1, 2, ..., $M - 1$). Finally, the delayed microwave replicas were summed upon photodetection via a balanced photodetector (BPD, Finisar). The BPD separated the wavelength channels into two categories according to the sign of tap coefficients, achieving both positive and negative tap coefficients.

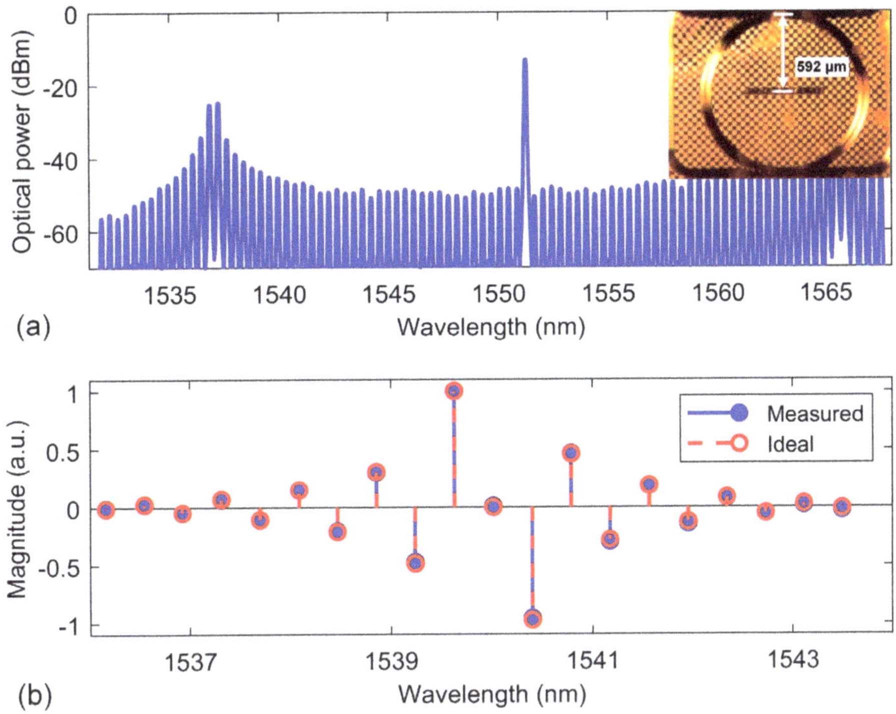

Figure 3. (**a**) Optical spectrum of soliton crystal microcomb generated by an MRR made from high-index doped silica glass. Inset shows a microscope image of the MRR. (**b**) Ideal and measured tap coefficients after optical spectral shaping.

We took the first-order differentiation as an example to investigate the influence of different input signal waveforms on the processing accuracy of microcomb-based MWP signal processors. The spectral transfer function of the first-order differentiation can be described by [3]

$$H(\omega) = j\omega, \tag{3}$$

where $j = \sqrt{-1}$, and ω is the angular frequency. The ideal tap coefficients were calculated by performing an inverse Fourier transform of Equation (3), and the results are shown in Figure 3b. For comparison, the measured tap coefficients after spectral shaping of the comb lines are also shown. As can be seen, the measured tap coefficients closely matched with the ideal tap coefficients, indicating the achievement of effective spectral shaping.

We selected five different temporal waveforms for the input microwave signal, including Gaussian, triangle, parabolic, super Gaussian, and nearly square waveforms. The input microwave signals were generated by an arbitrary waveform generator (AWG, Keysight). As mentioned in Section 2, to prevent aliasing, the sampling rate of the input microwave signal needs to exceed twice its maximum frequency component. Thus, the bandwidth of the input microwave signal to be processed should not surpass half of the microcomb's comb spacing, i.e., ~24.5 GHz. In addition, the FSR of the RF spectral response (FSR_{RF}) of the differentiator was $1/\Delta T$ = ~28 GHz. Therefore, the bandwidth of the input microwave signal should not exceed 28/2 =~14 GHz. Considering these factors, in our experiments, we employed input microwave signals with a full width at half maximum (FWHM) of ~0.2 ns (Figure 4a) and the primary frequency components resided within 14 GHz.

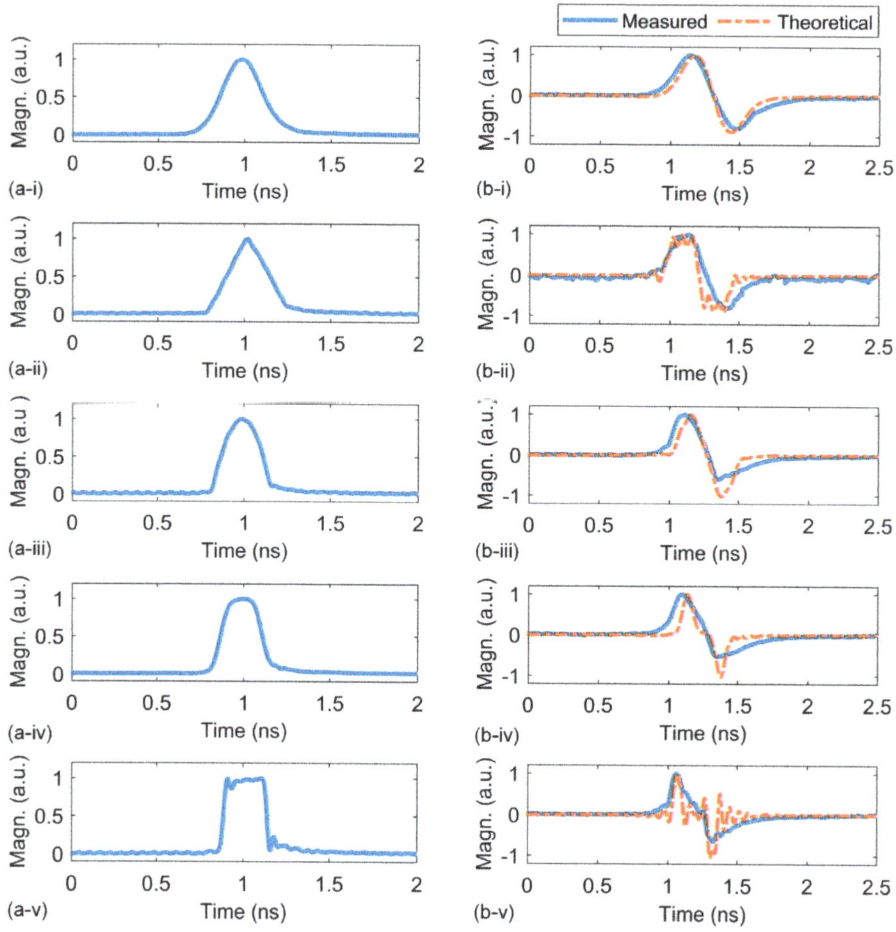

Figure 4. (**a**) Measured input microwave signal waveforms of (**i**) Gaussian, (**ii**) triangle, (**iii**) parabolic, (**iv**) super Gaussian, and (**v**) nearly square waveforms with full width at half maximum (FWHM) of ~0.2 ns. (**b**) Measured output waveforms from the MWP signal processor that performs first-order differentiation. The theoretical output results are also shown for comparison.

The signal processing results are shown in Figure 4b, which were measured by a high-speed real-time oscilloscope (OSC, Keysight). The theoretical outputs are also shown for comparison, which were calculated based on Equations (1)–(3). To facilitate a fair comparison, we used the recorded waveforms generated by the AWG as the input signal waveforms to calculate the theoretical outputs. As can be seen, all the measured outputs match with their corresponding theoretical outputs. Nevertheless, different input waveforms exhibit differences in the discrepancies between them. The Gaussian input waveform shows the lowest discrepancies, whereas the nearly square waveform displays the highest.

To quantify the processing accuracy of the processing results, the concept of root mean square error (RMSE) is introduced, which is defined as [30]

$$\text{RMSE} = \sqrt{\sum_{i=1}^{k} \frac{(Y_i - y_i)^2}{k}} \quad (4)$$

where $Y_1, Y_2, \ldots, Y_k$ are the values of theoretical processing results; $y_1, y_2, \ldots, y_k$ are the values of measured output waveforms.

Figure 5a shows the RMSEs between the measured output waveforms and the theoretical processing results for different input signal waveforms. The Gaussian and nearly square waveforms have the lowest and highest RMSE values, showing agreement with the results in Figure 4b.

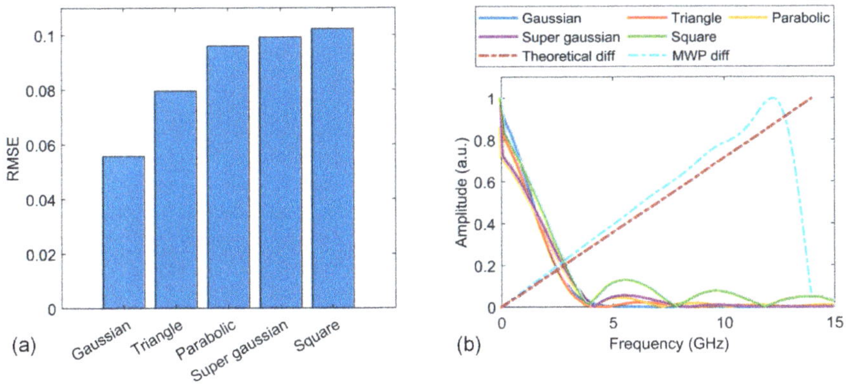

Figure 5. (**a**) RMSEs between theoretical differentiation results and the processor's output waveforms for different input microwave signal waveforms in Figure 4. (**b**) Amplitude frequency response of theoretical differentiation and the processor, together with the amplitude spectra of different input microwave signals including with Gaussian, triangle, parabolic, super Gaussian, and nearly square waveforms.

To analyse the reason for the differences in the processing accuracy for different waveforms, we further plot the amplitude frequency response of the processor and a theoretical differentiator in Figure 5b, together with the spectra of input signals with different waveforms. It can be seen that the deviations between the response of the transversal signal processor and the theoretical differentiator become more significant in the high-frequency range. On the other hand, the nearly square waveform contains greater high-frequency components than other waveforms, resulting in a reduction in its processing accuracy. In contrast, the Gaussian waveform has the least high-frequency components, enabling the highest level of processing accuracy.

4. Discussion

Based on the experimental results in Section 3, it can be seen that the processing accuracy varies for different input signal waveforms, even when performing the same processing function. The processing accuracy improves when there is better overlap between the high-intensity frequency components of the input signal and the low-error region of the MWP processor's response spectrum.

In our system, uncertainties and noise are introduced by several sources due to the imperfect response of practical devices [57]. These sources can be classified into two categories, relying on whether amplitude or phase errors are introduced to the taps. The sources for amplitude errors include the amplitude noise of microcombs, chirp of EOM, high-order dispersion of SMF, shaping errors of waveshapers, intensity noise of PD, and limited response bandwidth and uneven transmission response of EOM and PD. The sources of phase errors include the phase noise of microcombs, high-order dispersion of SMF, and shot noise of PD.

To reduce the amplitude and phase noises of microcombs, mode-locking approaches, such as power kicking [58], self-injection locking [59], forward and backward tuning [60], filter-driven four-wave mixing (FWM) [61], two-colour pumping [62], integrated heaters [51],

EO modulation [63], and self-referencing [64], can be employed to achieve microcombs with high coherence and stability [3,4]. Recently, some mode-locking approaches such as filter-driven FWM and self-injection locking have experienced significant improvement [61,65]. Some advanced mode-locking approaches have also been proposed, such as nonlinear dynamics engineering [66], cryogenic cooling [67], and auxiliary laser heating [68]. There are two main directions that are crucial for future improvement in practical applications. One involves achieving easily operated mode-locking without complex procedures [66] or self-emergence microcombs [69], the other is realizing mode-locking systems on a single chip, where, for example, significant advances have been achieved through the integration of actuators for the piezoelectric control of microcombs [70].

Introducing feedback control can effectively alleviate processing errors induced by slowly varying noises, such as the uneven gain or transmission response and shaping errors [3,30]. A feedback control loop can be adopted to calibrate the tap coefficients set for the second waveshaper in Figure 2, where the powers of comb lines are detected and then compared with the desired tap coefficients, yielding error signals for calibration. To further reduce the processing errors, two-stage feedback control can be introduced. One feedback loop can be employed in the microcomb generation module to achieve a uniform power distribution of the comb lines. This allows for uniform wavelength channel link gain and can also reduce the loss control range for spectral shaping in the transversal signal processing module. To minimize the intensity noise, multiple iterations along the entire feedback loop can be conducted. To reduce the phase noise, calibration can be achieved by using the phase-modulating capabilities of the waveshaper to compensate for the discrepancies between the measured and ideal phase response.

The photodetector in Figure 2 induces both intensity and phase noise. The intensity noise can be largely cancelled out by employing a BPD based on a balanced detection scheme. The phase noise arises from the shot noise, which causes random power fluctuations in the output microwave signal. The shot noise restricts the lowest achievable phase noise floor, and can be mitigated by using highly sensitive PDs [71].

Comb multiplexing technologies [72] hold promise in supplying additional discrete wavelength channels for the MWP transversal signal processors, which is beneficial for improving their processing accuracy. However, the augmentation of wavelength channels would also lead to a reduction in the channel spacing or processing bandwidth for the MWP transversal signal processors. Consequently, there is a trade-off that needs to be taken into account in practical applications [30,46,47].

For on-chip microcomb-based MWP signal processors, to improve the processing accuracy, feedback control is also required. Self-calibrating photonic integrated circuits have been proposed recently [73,74], where the calibration of impulse response was realized by introducing an optical reference path to establish a Kramers–Kronig relationship, followed by the computation of amplitude and phase errors. Apart from feedback control, there are some other ways to further alleviate the uncertainties and noise in the system. The chirp of silicon EOM can be minimized through the adoption of push–pull configurations and a p-n depletion mode structure [75], together with proper bias point calibration methods [76]. Gradient-descent control and calibration procedures can be employed to mitigate the shaping errors of integrated optical spectral shapers [76]. Integrated optical delay elements can introduce additional loss, especially when waveguides with high propagation loss are used. The incorporation of adiabatic Euler bends can be beneficial for achieving low-loss and low-crosstalk waveguide bends [77]. In addition, to enable large-scale integration, a wavelength-addressable serial integration scheme can be employed [78].

5. Conclusions

In summary, we experimentally demonstrate that microcomb-based MWP signal processors are capable of processing microwave signals with different temporal waveforms. We characterize the processing accuracy for different input signal waveforms, including Gaussian, triangle, parabolic, super Gaussian, and nearly square waveforms. We find that the difference in the processing accuracy for various input waveforms mainly results from the difference in their frequency components, as well as their overlap with the processor's frequency response that exhibits different degrees of deviation from the ideal response. These results provide a useful guidance for microcomb-based MWP signal processors to process microwave signals with various waveforms.

Author Contributions: Conceptualization, methodology, J.W. and Y.L.; Experiments, Y.L. and Y.S.; Integrated devices, S.T.C., B.E.L. and R.M.; Experimental instruments, G.R., B.C., X.X. and A.M.; writing—original draft preparation, Y.L. and Y.S.; writing—review and editing, J.W. and D.J.M.; supervision, J.W., A.M. and D.J.M. All authors have read and agreed to the published version of the manuscript.

Funding: This work was supported in part by the funded ARC Centre of Excellence in Optical Microcombs for Breakthrough Science (grant number CE230100006), in part by the Australian Research Council Discovery Projects Programs (grant numbers DP150104327, DP190102773, and DP190101576), and in part by the Early Career Researcher-Swinburne University Postgraduate Research Award program.

Institutional Review Board Statement: Not applicable.

Informed Consent Statement: Not applicable.

Data Availability Statement: The authors declare that the data supporting the findings of this study are available on request.

Conflicts of Interest: The authors declare no conflict of interest.

References

1. Capmany, J.; Novak, D. Microwave photonics combines two worlds. *Nat. Photonics* **2007**, *1*, 319–330. [CrossRef]
2. Yao, J. Microwave Photonics. *J. Light. Technol.* **2009**, *27*, 314–335. [CrossRef]
3. Wu, J.Y.; Xu, X.Y.; Nguyen, T.G.; Chu, S.T.; Little, B.E.; Morandotti, R.; Mitchell, A.; Moss, D.J. RF Photonics: An Optical Microcombs' Perspective. *IEEE J. Sel. Top. Quantum Electron.* **2018**, *24*, 20. [CrossRef]
4. Sun, Y.; Wu, J.; Tan, M.; Xu, X.; Li, Y.; Morandotti, R.; Mitchell, A.; Moss, D.J. Applications of optical microcombs. *Adv. Opt. Photonics* **2023**, *15*, 86. [CrossRef]
5. Marpaung, D.; Yao, J.; Capmany, J. Integrated microwave photonics. *Nat. Photonics* **2019**, *13*, 80–90. [CrossRef]
6. Capmany, J.; Mora, J.; Gasulla, I.; Sancho, J.; Lloret, J.; Sales, S. Microwave Photonic Signal Processing. *J. Light. Technol.* **2013**, *31*, 571–586. [CrossRef]
7. Liu, W.; Li, M.; Guzzon, R.S.; Norberg, E.J.; Parker, J.S.; Lu, M.; Coldren, L.A.; Yao, J. A fully reconfigurable photonic integrated signal processor. *Nat. Photonics* **2016**, *10*, 190–195. [CrossRef]
8. Ferrera, M.; Park, Y.; Razzari, L.; Little, B.E.; Chu, S.T.; Morandotti, R.; Moss, D.J.; Azaña, J. On-chip CMOS-compatible all-optical integrator. *Nat. Commun.* **2010**, *1*, 29. [CrossRef]
9. Yao, J.; Zhang, W. Fully reconfigurable waveguide Bragg gratings for programmable photonic signal processing. *J. Light. Technol.* **2019**, *38*, 202–214. [CrossRef]
10. Zhang, W.; Yao, J. Photonic integrated field-programmable disk array signal processor. *Nat. Commun.* **2020**, *11*, 406. [CrossRef]
11. Berger, N.K.; Levit, B.; Fischer, B.; Kulishov, M.; Plant, D.V.; Azaña, J. Temporal differentiation of optical signals using a phase-shifted fiber Bragg grating. *Opt. Express* **2007**, *15*, 371–381. [CrossRef]
12. Rutkowska, K.A.; Duchesne, D.; Strain, M.J.; Morandotti, R.; Sorel, M.; Azaña, J. Ultrafast all-optical temporal differentiators based on CMOS-compatible integrated-waveguide Bragg gratings. *Opt. Express* **2011**, *19*, 19514–19522. [CrossRef]
13. Liu, F.; Wang, T.; Qiang, L.; Ye, T.; Zhang, Z.; Qiu, M.; Su, Y. Compact optical temporal differentiator based on silicon microring resonator. *Opt. Express* **2008**, *16*, 15880–15886. [CrossRef]
14. Wu, J.; Cao, P.; Hu, X.; Jiang, X.; Pan, T.; Yang, Y.; Qiu, C.; Tremblay, C.; Su, Y. Compact tunable silicon photonic differential-equation solver for general linear time-invariant systems. *Opt. Express* **2014**, *22*, 26254–26264. [CrossRef]
15. Wu, J.; Liu, B.; Peng, J.; Mao, J.; Jiang, X.; Qiu, C.; Tremblay, C.; Su, Y. On-Chip Tunable Second-Order Differential-Equation Solver Based on a Silicon Photonic Mode-Split Microresonator. *J. Light. Technol.* **2015**, *33*, 3542–3549. [CrossRef]
16. Zheng, A.; Dong, J.; Zhou, L.; Xiao, X.; Yang, Q.; Zhang, X.; Chen, J. Fractional-order photonic differentiator using an on-chip microring resonator. *Opt. Lett.* **2014**, *39*, 6355. [CrossRef]
17. Zheng, A.; Yang, T.; Xiao, X.; Yang, Q.; Zhang, X.; Dong, J. Tunable fractional-order differentiator using an electrically tuned silicon-on-isolator Mach-Zehnder interferometer. *Opt. Express* **2014**, *22*, 18232–18237. [CrossRef]
18. Capmany, J.; Ortega, B.; Pastor, D.; Sales, S. Discrete-time optical processing of microwave signals. *J. Light. Technol.* **2005**, *23*, 702–723. [CrossRef]
19. Mansoori, S.; Mitchell, A. RF transversal filter using an AOTF. *IEEE Photonics Technol. Lett.* **2004**, *16*, 879–881. [CrossRef]

20. Zhang, J.; Yao, J. Photonic true-time delay beamforming using a switch-controlled wavelength-dependent recirculating loop. *J. Light. Technol.* **2016**, *34*, 3923–3929. [CrossRef]
21. Zhang, L.; Li, M.; Shi, N.; Zhu, X.; Sun, S.; Tang, J.; Li, W.; Zhu, N. Photonic true time delay beamforming technique with ultra-fast beam scanning. *Opt. Express* **2017**, *25*, 14524–14532. [CrossRef] [PubMed]
22. Yu, G.; Zhang, W.; Williams, J. High-performance microwave transversal filter using fiber Bragg grating arrays. *IEEE Photonics Technol. Lett.* **2000**, *12*, 1183–1185. [CrossRef]
23. Hunter, D.B.; Minasian, R.A.; Krug, P.A. Tunable optical transversal filter based on chirped gratings. *Electron. Lett.* **1995**, *31*, 2205–2207. [CrossRef]
24. Liu, Y.; Yao, J.; Yang, J. Wideband true-time-delay unit for phased array beamforming using discrete-chirped fiber grating prism. *Opt. Commun.* **2002**, *207*, 177–187. [CrossRef]
25. Supradeepa, V.R.; Long, C.M.; Wu, R.; Ferdous, F.; Hamidi, E.; Leaird, D.E.; Weiner, A.M. Comb-based radiofrequency photonic filters with rapid tunability and high selectivity. *Nat. Photonics* **2012**, *6*, 186–194. [CrossRef]
26. Hamidi, E.; Leaird, D.E.; Weiner, A.M. Tunable Programmable Microwave Photonic Filters Based on an Optical Frequency Comb. *IEEE Trans. Microw. Theory Tech.* **2010**, *58*, 3269–3278. [CrossRef]
27. Metcalf, A.J.; Kim, H.J.; Leaird, D.E.; Jaramillo Villegas, J.A.; McKinzie, K.A.; Lal, V.; Hosseini, A.; Hoefler, G.E.; Kish, F.; Weiner, A.M. Integrated line-by-line optical pulse shaper for high-fidelity and rapidly reconfigurable RF-filtering. *Opt. Express* **2016**, *24*, 23925–23940. [CrossRef]
28. Ortigosa-Blanch, A.; Mora, J.; Capmany, J.; Ortega, B.; Pastor, D. Tunable radio-frequency photonic filter based on an actively mode-locked fiber laser. *Opt. Lett.* **2006**, *31*, 709–711. [CrossRef]
29. Maram, R.; Onori, D.; Azaña, J.; Chen, L.R. Discretely programmable microwave photonic filter based on temporal Talbot effects. *Opt. Express* **2019**, *27*, 14381–14391. [CrossRef]
30. Sun, Y.; Wu, J.; Li, Y.; Tan, M.; Xu, X.; Chu, S.T.; Little, B.E.; Morandotti, R.; Mitchell, A.; Moss, D.J. Quantifying the Accuracy of Microcomb-Based Photonic RF Transversal Signal Processors. *IEEE J. Sel. Top. Quantum Electron.* **2023**, *29*, 1–17. [CrossRef]
31. Lin, G.; Liu, F.; Coillet, A.; Gomila, D.; Menyuk, C.R.; Chembo, Y.K. Subharmonic instabilities in Kerr microcombs. *Opt. Lett.* **2023**, *48*, 578–581. [CrossRef] [PubMed]
32. Pasquazi, A.; Peccianti, M.; Razzari, L.; Moss, D.J.; Coen, S.; Erkintalo, M.; Chembo, Y.K.; Hansson, T.; Wabnitz, S.; Del'Haye, P.; et al. Micro-combs: A novel generation of optical sources. *Phys. Rep.* **2018**, *729*, 1–81. [CrossRef]
33. Xu, X.; Tan, M.; Wu, J.; Morandotti, R.; Mitchell, A.; Moss, D.J. Microcomb-Based Photonic RF Signal Processing. *IEEE Photonics Technol. Lett.* **2019**, *31*, 1854–1857. [CrossRef]
34. Xu, X.; Wu, J.; Shoeiby, M.; Nguyen, T.G.; Chu, S.T.; Little, B.E.; Morandotti, R.; Mitchell, A.; Moss, D.J. Reconfigurable broadband microwave photonic intensity differentiator based on an integrated optical frequency comb source. *APL Phontonics* **2017**, *2*, 096104. [CrossRef]
35. Xu, X.; Tan, M.; Wu, J.; Boes, A.; Corcoran, B.; Nguyen, T.G.; Chu, S.T.; Little, B.E.; Morandotti, R.; Mitchell, A.; et al. Photonic RF and Microwave Integrator Based on a Transversal Filter With Soliton Crystal Microcombs. *IEEE Trans. Circuits Syst. II-Express Briefs* **2020**, *67*, 3582–3586. [CrossRef]
36. Tan, M.; Xu, X.; Corcoran, B.; Wu, J.; Boes, A.; Nguyen, T.G.; Chu, S.T.; Little, B.E.; Morandotti, R.; Mitchell, A.; et al. Microwave and RF Photonic Fractional Hilbert Transformer Based on a 50 GHz Kerr Micro-Comb. *J. Light. Technol.* **2019**, *37*, 6097–6104. [CrossRef]
37. Wang, B.; Yang, Z.; Sun, S.; Yi, X. Radio-frequency line-by-line Fourier synthesis based on optical soliton microcombs. *Photonics Res.* **2022**, *10*, 932. [CrossRef]
38. Xu, X.; Tan, M.; Corcoran, B.; Wu, J.; Boes, A.; Nguyen, T.G.; Chu, S.T.; Little, B.E.; Hicks, D.G.; Morandotti, R.; et al. 11 TOPS photonic convolutional accelerator for optical neural networks. *Nature* **2021**, *589*, 44–51. [CrossRef]
39. Feldmann, J.; Youngblood, N.; Karpov, M.; Gehring, H.; Li, X.; Stappers, M.; Le Gallo, M.; Fu, X.; Lukashchuk, A.; Raja, A.S.; et al. Parallel convolutional processing using an integrated photonic tensor core. *Nature* **2021**, *589*, 52–58. [CrossRef]
40. Pfeifle, J.; Brasch, V.; Lauermann, M.; Yu, Y.; Wegner, D.; Herr, T.; Hartinger, K.; Schindler, P.; Li, J.; Hillerkuss, D. Coherent terabit communications with microresonator Kerr frequency combs. *Nat. Photonics* **2014**, *8*, 375–380. [CrossRef]
41. Zhou, Z.; Ou, X.; Fang, Y.; Alkhazraji, E.; Xu, R.; Wan, Y.; Bowers, J.E. Prospects and applications of on-chip lasers. *eLight* **2023**, *3*, 1. [CrossRef]
42. Corcoran, B.; Tan, M.; Xu, X.; Boes, A.; Wu, J.; Nguyen, T.G.; Chu, S.T.; Little, B.E.; Morandotti, R.; Mitchell, A.; et al. Ultra-dense optical data transmission over standard fibre with a single chip source. *Nat. Commun.* **2020**, *11*, 7. [CrossRef] [PubMed]
43. Li, Y.; An, N.; Lu, Z.; Wang, Y.; Chang, B.; Tan, T.; Guo, X.; Xu, X.; He, J.; Xia, H.; et al. Nonlinear co-generation of graphene plasmons for optoelectronic logic operations. *Nat. Commun.* **2022**, *13*, 3138. [CrossRef] [PubMed]
44. Tan, T.; Yuan, Z.; Zhang, H.; Yan, G.; Zhou, S.; An, N.; Peng, B.; Soavi, G.; Rao, Y.; Yao, B. Multispecies and individual gas molecule detection using Stokes solitons in a graphene over-modal microresonator. *Nat. Commun.* **2021**, *12*, 6716. [CrossRef] [PubMed]
45. Zhang, C.; Wang, J.; Kang, G.; Gao, J.; Qu, Z.; Wan, S.; Dong, C.; Pan, Y.; Qu, J. Soliton microcomb-assisted microring photonic thermometer with ultra-high resolution and broad range. *Photonics Res.* **2023**, *11*, A44–A53. [CrossRef]
46. Sun, Y.; Wu, J.; Li, Y.; Xu, X.; Ren, G.; Tan, M.; Chu, S.K.; Little, B.E.; Morandotti, R.; Mitchell, A.; et al. Optimizing the performance of microcomb based microwave photonic transversal signal processors. *J. Light. Technol.* **2023**, *41*, 1–15.

47. Li, Y.; Sun, Y.; Wu, J.; Ren, G.; Morandotti, R.; Tan, M.; Xu, X.; Mitchell, A.; Moss, D.J. Performance analysis of microwave photonic spectral filters based on optical microcombs. *Opt. Lasers Eng.* 2023, in press.
48. Kippenberg, T.J.; Gaeta, A.L.; Lipson, M.; Gorodetsky, M.L. Dissipative Kerr solitons in optical microresonators. *Science* **2018**, *361*, eaan8083. [CrossRef]
49. Wang, W.Q.; Wang, L.R.; Zhang, W.F. Advances in soliton microcomb generation. *Adv. Photonics* **2020**, *2*, 034001. [CrossRef]
50. Zhang, H.; Chang, B.; Li, Z.; Liang, Y.-P.; Qin, C.-Y.; Wang, C.; Xia, H.-D.; Tan, T.; Yao, B.-C. Coherent optical frequency combs: From principles to applications. *J. Electron. Sci. Technol.* **2022**, *20*, 100157. [CrossRef]
51. Xue, X.; Xuan, Y.; Liu, Y.; Wang, P.H.; Chen, S.; Wang, J.; Leaird, D.E.; Qi, M.; Weiner, A.M. Mode-locked dark pulse Kerr combs in normal-dispersion microresonators. *Nat. Photonics* **2015**, *9*, 594–600. [CrossRef]
52. Liu, M.; Huang, H.; Lu, Z.; Wang, Y.; Cai, Y.; Zhao, W. Dynamics of dark breathers and Raman-Kerr frequency combs influenced by high-order dispersion. *Opt. Express* **2021**, *29*, 18095–18107. [CrossRef] [PubMed]
53. Cole, D.C.; Lamb, E.S.; Del'Haye, P.; Diddams, S.A.; Papp, S.B. Soliton crystals in Kerr resonators. *Nat. Photonics* **2017**, *11*, 671–676. [CrossRef]
54. Lu, Z.; Chen, H.; Wang, W.; Yao, L.; Wang, Y.; Yu, Y.; Little, B.E.; Chu, S.T.; Gong, Q.; Zhao, W.; et al. Synthesized soliton crystals. *Nat. Commun.* **2021**, *12*, 3179. [CrossRef]
55. Razzari, L.; Duchesne, D.; Ferrera, M.; Morandotti, R.; Chu, S.; Little, B.E.; Moss, D.J. CMOS-compatible integrated optical hyper-parametric oscillator. *Nat. Photonics* **2010**, *4*, 41–45. [CrossRef]
56. Xu, X.; Tan, M.; Wu, J.; Nguyen, T.G.; Chu, S.T.; Little, B.E.; Morandotti, R.; Mitchell, A.; Moss, D.J. Advanced Adaptive Photonic RF Filters with 80 Taps Based on an Integrated Optical Micro-Comb Source. *J. Light. Technol.* **2019**, *37*, 1288–1295. [CrossRef]
57. Salzenstein, P.; Pavlyuchenko, E. Uncertainty evaluation on a 10.52 GHz (5 dBm) optoelectronic oscillator phase noise performance. *Micromachines* **2021**, *12*, 474. [CrossRef]
58. Yi, X.; Yang, Q.F.; Yang, K.Y.; Suh, M.G.; Vahala, K. Soliton frequency comb at microwave rates in a high-Q silica microresonator. *Optica* **2015**, *2*, 1078–1085. [CrossRef]
59. Del'Haye, P.; Beha, K.; Papp, S.B.; Diddams, S.A. Self-Injection Locking and Phase-Locked States in Microresonator-Based Optical Frequency Combs. *Phys. Rev. Lett.* **2014**, *112*, 043905. [CrossRef]
60. Guo, H.; Karpov, M.; Lucas, E.; Kordts, A.; Pfeiffer, M.H.; Brasch, V.; Lihachev, G.; Lobanov, V.E.; Gorodetsky, M.L.; Kippenberg, T.J. Universal dynamics and deterministic switching of dissipative Kerr solitons in optical microresonators. *Nat. Phys.* **2017**, *13*, 94–102. [CrossRef]
61. Bao, H.; Cooper, A.; Rowley, M.; Di Lauro, L.; Totero Gongora, J.S.; Chu, S.T.; Little, B.E.; Oppo, G.-L.; Morandotti, R.; Moss, D.J.; et al. Laser cavity-soliton microcombs. *Nat. Photonics* **2019**, *13*, 384–389. [CrossRef]
62. Strekalov, D.V.; Yu, N. Generation of optical combs in a whispering gallery mode resonator from a bichromatic pump. *Phys. Rev. A* **2009**, *79*, 041805. [CrossRef]
63. Del'Haye, P.; Papp, S.B.; Diddams, S.A. Hybrid Electro-Optically Modulated Microcombs. *Phys. Rev. Lett.* **2012**, *109*, 263901. [CrossRef] [PubMed]
64. Del'Haye, P.; Coillet, A.; Fortier, T.; Beha, K.; Cole, D.C.; Yang, K.Y.; Lee, H.; Vahala, K.J.; Papp, S.B.; Diddams, S.A. Phase-coherent microwave-to-optical link with a self-referenced microcomb. *Nat. Photonics* **2016**, *10*, 516–520. [CrossRef]
65. Voloshin, A.S.; Kondratiev, N.M.; Lihachev, G.V.; Liu, J.; Lobanov, V.E.; Dmitriev, N.Y.; Weng, W.; Kippenberg, T.J.; Bilenko, I.A. Dynamics of soliton self-injection locking in optical microresonators. *Nat. Commun.* **2021**, *12*, 235. [CrossRef]
66. Shen, B.; Chang, L.; Liu, J.; Wang, H.; Yang, Q.; Xiang, C.; Wang, R.; He, J.; Liu, T.; Xie, W.; et al. Integrated turnkey soliton microcombs. *Nature* **2020**, *582*, 365–369. [CrossRef]
67. Moille, G.; Chang, L.; Xie, W.; Rao, A.; Lu, X.; Davanço, M.; Bowers, J.E.; Srinivasan, K. Dissipative Kerr Solitons in a III-V Microresonator. *Laser Photon. Rev.* **2020**, *14*, 2000022. [CrossRef]
68. Zhou, H.; Geng, Y.; Cui, W.; Huang, S.W.; Zhou, Q.; Qiu, K.; Wei Wong, C. Soliton bursts and deterministic dissipative Kerr soliton generation in auxiliary-assisted microcavities. *Light Sci. Appl.* **2019**, *8*, 50. [CrossRef]
69. Rowley, M.; Hanzard, P.-H.; Cutrona, A.; Bao, H.; Chu, S.T.; Little, B.E.; Morandotti, R.; Moss, D.J.; Oppo, G.L.; Totero Gongora, J.S.; et al. Self-emergence of robust solitons in a microcavity. *Nature* **2022**, *608*, 303–309. [CrossRef]
70. Liu, J.; Tian, H.; Lucas, E.; Raja, A.S.; Lihachev, G.; Wang, R.N.; He, J.; Liu, T.; Anderson, M.H.; Weng, W.; et al. Monolithic piezoelectric control of soliton microcombs. *Nature* **2020**, *583*, 385–390. [CrossRef]
71. Miao, J.; Zhang, F. Recent progress on highly sensitive perovskite photodetectors. *J. Mater. Chem. C* **2019**, *7*, 1741–1791. [CrossRef]
72. Zhang, H.; Tan, T.; Chen, H.-J.; Yu, Y.; Wang, W.; Chang, B.; Liang, Y.; Guo, Y.; Zhou, H.; Xia, H.; et al. Soliton Microcombs Multiplexing Using Intracavity-Stimulated Brillouin Lasers. *Phys. Rev. Lett.* **2023**, *130*, 153802. [CrossRef] [PubMed]
73. Xu, X.; Ren, G.; Feleppa, T.; Liu, X.; Boes, A.; Mitchell, A.; Lowery, A.J. Self-calibrating programmable photonic integrated circuits. *Nat. Photonics* **2022**, *16*, 595–602. [CrossRef]
74. Xu, X.; Ren, G.; Dubey, A.; Feleppa, T.; Liu, X.; Boes, A.; Mitchell, A.; Lowery, A.J. Phase retrieval of programmable photonic integrated circuits based on an on-chip fractional-delay reference path. *Optica* **2022**, *9*, 1401. [CrossRef]
75. Sinatkas, G.; Christopoulos, T.; Tsilipakos, O.; Kriezis, E.E. Electro-optic modulation in integrated photonics. *J. Appl. Phys.* **2021**, *130*, 010901. [CrossRef]
76. Bai, B.; Yang, Q.; Shu, H.; Chang, L.; Yang, F.; Shen, B.; Tao, Z.; Wang, J.; Xu, S.; Xie, W.; et al. Microcomb-based integrated photonic processing unit. *Nat. Commun.* **2023**, *14*, 66. [CrossRef]

77. Ji, X.; Liu, J.; He, J.; Wang, R.N.; Qiu, Z.; Riemensberger, J.; Kippenberg, T.J. Compact, spatial-mode-interaction-free, ultralow-loss, nonlinear photonic integrated circuits. *Commun. Phys.* **2022**, *5*, 84. [CrossRef]
78. Kuramochi, E.; Nozaki, K.; Shinya, A.; Takeda, K.; Sato, T.; Matsuo, S.; Taniyama, H.; Sumikura, H.; Notomi, M. Large-scale integration of wavelength-addressable all-optical memories on a photonic crystal chip. *Nat. Photonics* **2014**, *8*, 474–481. [CrossRef]

MDPI AG
Grosspeteranlage 5
4052 Basel
Switzerland
Tel.: +41 61 683 77 34

Photonics Editorial Office
E-mail: photonics@mdpi.com
www.mdpi.com/journal/photonics

www.ingramcontent.com/pod-product-compliance
Lightning Source LLC
Chambersburg PA
CBHW061606210926
51850CB00009B/2234
* 9 7 8 3 7 2 5 8 5 5 5 1 3 *